Integrated Science Applied for Nurses

INTEGRATED SCIENCE APPLIED FOR NURSES

Barry G. Hinwood BSc, MSc, DipEd

Lecturer, School of Biological Sciences
Sydney Technical College

Cassell Australia

Author's note

I would like to thank all my students, both past and present, my colleagues from Sydney Technical College and the N.S.W. College of Nursing, family and the staff of Collier Macmillan for their encouragement, criticism and suggestions. In particular, I would like to thank Robyn Williams, Perry McIntyre and Suzanne Baker for typing the manuscript; K. Banks-Hughes, Margaret Manny, Dr John Pearce and Philippe Vandervaere for their suggestions and criticism; Helen Wortham for editing; Anthony Dwyer, Lyndy Macready, Jackie Gaff, Steven Dunbar and Meryl Potter for encouragement, guidance and production of the book. I especially would like to thank my wife Danielle for her understanding, encouragement, criticisms and suggestions and Jenni Hardy for her reviewing, criticisms and suggestions.

Cassell Australia Limited
44 Waterloo Road, North Ryde, New South Wales, 2113
30 Curzon Street, North Melbourne, Victoria, 3051

First published 1981
Designed by Steven Dunbar
Illustrations by Jan Düttmer and Brett Cullen
Set, printed and bound by
Hedges and Bell, Maryborough, Victoria
F.1081

National Library of Australia
Cataloguing-in-Publication Data
Hinwood, B. G.
Integrated science applied for nurses.
Bibliography.
Includes index.
ISBN 0 7269 3739 8.
1. Science. 2. Nursing. I. Title.
502.4613

Contents in brief

Preface

Recent years have seen the rapid development of technology and a greater understanding of the mechanisms of normal and abnormal body function. Paralleling these changes there has been an increased need for nurses and others in the health care team to develop a wider range of skills. An increased educational requirement has also been associated with these changes which has resulted in nurses needing a greater knowledge and understanding of science. The changes in educational levels have affected students entering the profession as well as those beginning refresher courses. This book has been written with the aim of assisting nurses and others in the health care team to understand science at a level that satisfies current needs.

The integration of the strands of science is designed to assist the student in understanding the total effect of science on nursing practice and the body function. Separation of biology, chemistry, and physics results in students often having difficulty in combining these strands of science in practical nursing. Throughout this book, the theme of homeostasis is used to integrate these strands of science and their relationship to body function.

The target audience of this book includes students entering the nursing profession and those nurses seeking continuing education for which a background in science is necessary. The book assumes no previous knowledge in all the topics mentioned by the author. This text also aims to be a supplementary book for students studying physiology. Since physiology requires a sound understanding of scientific principles, the use of physiological examples will enable an introductory physiology student studying in a health related area to use this book as a reference aide.

The comprehensive scope of this book and its layout have been selected to enable health care students to use it as a reference book throughout their education and work. The following aspects of *Integrated Science Applied for Nurses* are designed to minimize the difficulties that nurses and other health care workers often have in applying science to their work.

1. The book is arranged into units containing a common theme derived from biology, chemistry, and physics.
2. The independent nature of each unit enables the reader to select the general order of contents to suit individual needs.
3. Each chapter contains a list of objectives and a summary.
4. The inclusion of applications is intended to assist the student in relating science to nursing and body function. The applications are intended to be supplementary and as such they can be omitted without influencing the general context of a particular aspect of science.
5. To overcome the difficulty often encountered in understanding chemical reactions, a system of diagrams is used to show how atoms are combined in chemical reactions.
6. To assist students who have difficulty with S.I. units, the non-S.I. unit is placed in brackets beside any corresponding unit.
7. A glossary of items is included that covers many of the important terms mentioned in the book.

Contents in detail

Unit nine
Digestion and metabolism of chemicals 167

DEDICATION

To my mother, late father and wife without whose encouragement this book would not have been possible.

Unit One
Homeostasis

This unit discusses homeostasis in relation to the health of a patient. The relationship between homeostasis and science is introduced as a principle theme of this book.

Chapter 1

Homeostasis

Objectives

At the completion of this chapter the student should be able to:
1. define homeostasis,
2. relate homeostasis to the health of a patient,
3. discuss the differences in homeostasis that exist between single-cellular and multicellular organisms,
4. relate cell homeostasis to cell membrane function,
5. describe the components of a homeostatic mechanism and relate these components to homeostatic mechanisms in the body,
6. explain how stress can effect an imbalance in the homeostasis of an individual.

1.1 Homeostasis

HOMEOSTASIS AND THE HEALTH OF A PATIENT

Homeostasis is a relative state of equilibrium between the internal environment of an organism and a changing external environment. This state of equilibrium or optimum conditions usually ranges between an upper and lower limit. Beyond these limits harmful effects can occur. Homeostatic mechanisms assist in returning abnormal conditions back to the normal optimum range. A continuation of abnormal conditions usually results in ill health, and in some instances can lead to death. Treatment of a body imbalance generally involves either eliminating the causative factors and/or replenishing deficiencies of substances.

An understanding of homeostatic mechanisms is important in determining the impact of a particular treatment on the entire body. Throughout this book an emphasis is therefore placed on homeostasis and its relationship to scientific principles. Examples of the following aspects of homeostasis are found in the cited chapters.

1. Factors that indicate the health of a person, that is, normal versus abnormal homeostatic state. EXAMPLES: Measurement and units—Chapter 3; Concept of concentration—Chapter 4; The composition and properties of body fluids—Chapter 12.
2. The properties of matter that are involved in homeostatic mechanisms. EXAMPLES: Atoms and matter—Chapter 2; Physical states of matter—Chapter 6; How atoms combine—Chapter 8; Properties of pressure—Chapter 13; Diffusion—Chapter 14.
3. The properties of matter that influence the care and treatment of patients. EXAMPLES: Mechanical properties of matter—Chapter 7; Biologically important organic compounds—Chapter 9; Properties of fluid mixtures—Chapter 10; Properties of pressure—Chapter 13.
4. The properties of energy that assist diagnosis and therapy of abnormal homeostatic conditions. EXAMPLE: Matter and energy—Chapter 5.
5. The homeostatic mechanisms that maintain a healthy person. EXAMPLES: Acid-base, balance in Chapter 11; Osmosis and body fluid balance—Chapter 15; Plasma membrane function—Chapter 17.
6. The effect of abnormal homeostatic mechanisms. EXAMPLES: Genetic abnormalities in Chapter 18; Endocrine malfunction—Chapter 28.
7. Factors that can cause homeostatic imbalances. EXAMPLES: Excess energy in Chapter 5; Infections in Chapter 21; Stress in Chapter 26.
8. Inter-relationship of homeostatic mechanisms. EXAMPLES: Blood gases in Chapter 20; Interconversion of biochemicals in Chapter 24; General organization and function of the endocrine system—Chapter 28.
9. Integration and control of homeostatic mechanisms. EXAMPLES: General organization and function of the nervous system—Chapter 26; Neuroendocrine connections—Chapter 27; General organization and function of the endocrine system—Chapter 28.

The examples mentioned above are only intended to indicate a sample of the relationships between homeostasis and health. Many chapters contain information that relate to more than one of the aspects of homeostasis that are listed.

HOMEOSTASIS IN SINGLE-CELLED ORGANISMS

Single-celled organisms such as bacteria and

protozoa are capable of withstanding relatively large alterations in their external environment. This is due to their ability to perform all the functions that are necessary for an organism to exist. The ability to obtain all the substances that are essential for their existence eliminates the need for a dependence on other cells. These single-celled organisms are thus able to maintain a relative state of equilibrium within their internal environment.

HOMEOSTASIS IN MULTICELLULAR ORGANISMS

In complex multicellular organisms such as humans, homeostasis can be divided into that occurring within the cell and that occurring outside the cell but within the body. The latter contains fluid referred to as extracellular fluid.

Cell Homeostasis

Cell homeostasis involves the maintenance of an optimum environment for a particular cell to function. This cellular environment consists of optimum concentrations of gases, nutrients, chemical regulators (hormones and enzymes), electrically charged particles (ions) and water. The homeostasis of cells is maintained by the properties of the barrier that separates the cell's contents or intracellular fluid from the different concentrations of substances in the extracellular fluid. The general differences between intracellular and extracellular fluid are discussed in Chapter 12.

Cell membranes act as a selective barrier for the exchange of substances between the intracellular and extracellular fluids. This mechanism of cell membrane function is described in Chapter 17. The cell membrane is thus an essential homeostatic mechanism for the maintenance of cell homeostasis.

All cell membranes have similar homeostatic properties in that they regulate the internal environment according to the needs of a particular cell. The specialized nature and the diversity of function of cell types in the body are described in Chapter 19. This diversity of cell function requires cell membrances that have properties related to the specialized function of a particular cell. EXAMPLE: In nerve cells, a momentary flow of sodium ions into the cell is required for a nerve impulse to occur. The membranes of these cells have the specialized property of allowing sodium to pass through them for brief periods of time. Without this mechanism the function of both the cell and the body functions that it influences would be altered.

Body Homeostasis

The integrated function of the body is essential for a complex multicellular organism such as humans. The actions of highly specialized cells need to be coordinated for efficient body function.

The maintenance of cell homeostasis in a particular cell usually depends upon the integrated action of other cell types throughout the body. The different needs of specific cells also require coordinated homeostatic mechanisms for the individual cell's needs to be satisfied. EXAMPLE: Nerve cells constantly depend upon a high concentration of glucose in the extracellular fluid, whereas muscle cells only require high concentrations during exercise. An increase in the flow of blood to regions of muscle contraction results in the necessary elevated supply of glucose. A homeostatic mechanism is used to selectively increase the flow of blood to the muscle cells. This mechanism includes:

1. an increase in heart rate and its pumping ability to force blood and thus glucose at a rapid rate to the relevant region,
2. a localized dilation in the blood vessels resulting in an increased flow of blood in the vicinity of the contracting muscles.

Homeostasic mechanisms have therefore increased the supply of glucose to the relevant muscle cells by increasing the volume of blood to the region. Note that the concentration of glucose, that is the quantity of glucose molecules in a certain volume, has not been significantly altered.

Body homeostasis involves the maintenance of a relatively constant concentration of substances in the extracellular fluid. The supply of nutrients to different regions of the body is altered by homeostatic mechanisms according to the needs of particular cells. The composition of extracellular fluid depends upon the integrated action of several different homeostatic mechanisms. EXAMPLE: The concentration of oxygen in the extracellular fluid relies upon the coordinated and efficient function of the respiratory tract, lungs, heart and blood vessels, along with a sufficient quantity of red blood cells for transporting the oxygen. An imbalance between these factors can alter the supply of oxygen to the cells. The integration of these factors is thus necessary for homeostasis to occur.

Body homeostasis differs from cell homeostasis in that it relies upon an intercellular communication network. Without this network an imbalance in body homeostasis would occur. In addition, the coordinated action of various homeostatic mechanisms requires an integrating centre to assess the effects of the actions of any homeostatic mechanism on the entire body.

The brain acts as the integrating centre for most homeostatic mechanisms. It receives information, assesses it and then sends out instructions that regulate homeostatic imbalances. The communication networks that it utilizes are the nervous system and the endocrine system. The role that they play in body homeostasis is described in Chapters 26, 27 and 28.

1.2 Factors involved in homeostatic mechanisms

GENERALIZED HOMEOSTATIC MECHANISMS

In any homeostatic mechanism several essential components are necessary:
1. Receptor—sensor or detector used to inform the body of any imbalances.
2. Integrating centre—used to coordinate sensor information and determine the required response to any imbalance. If the response is insufficient then a further response will occur.
3. Effector—these effect the response. EXAMPLES: Hormones, muscle.
4. Communication pathways—a pathway is necessary to link the information of the receptor to the integrating centre. This is called the afferent pathway. An efferent pathway connects the integrating centre with the effector.

The relationship of these components is shown in Figure 1.1. This generalized diagram can be superimposed on any homeostatic mechanism. Note that

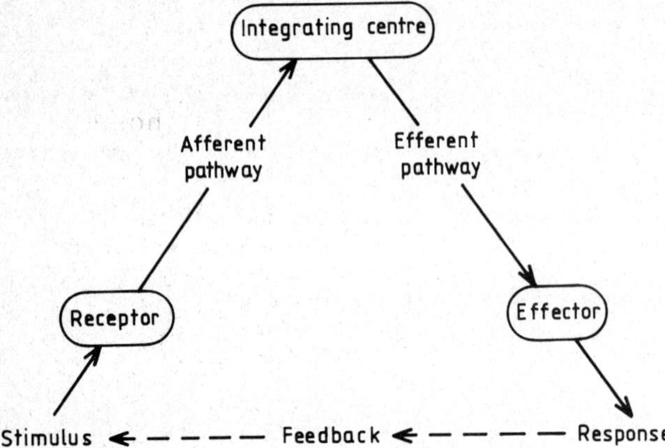

Figure 1.1 Generalized diagram of a homeostatic mechanism.

all of these components are essential for a particular homeostatic mechanism to work. A loss of any component will result in an inability of the homeostatic mechanism to function. EXAMPLE: An inability of the effector to function in the blood sugar homeostatic mechanisms results in this mechanism being unable to function. The result is diabetes mellitus.

An understanding of the principle of homeostasis and the mechanisms involved will assist in the understanding of the sites of malfunction and the possible effects of a particular imbalance on body function.

An individual homeostatic mechanism usually involves several parts of the body, and in many cases this results in an overlap with other homeostatic mechanisms. EXAMPLE: The homeostasis of blood pressure involves the heart, arterioles and veins, large arteries, sympathetic and vagus nerves, brain and any factors that either require or cause an altered blood pressure. All of these components can be combined to form a homeostatic mechanism that is based on Figure 1.1.

APPLICATION: HOMEOSTASIS OF BLOOD CALCIUM

The homeostasis of blood calcium consists of a relatively simple mechanism. The general mechanism is shown in Figure 1.2. This diagram indicates that a decrease in blood calcium concentration stimulates the receptor in the parathyroid gland. This gland probably acts as the integrating centre which compares this low level of calcium with the normal concentration in order to determine the magnitude of response. An increased quantity of parathormone is secreted from the gland and into the blood. This hormone acts on bone to release the appropriate quantity of calcium into the blood. The elevated concentration of calcium is then detected by receptors and compared with normal values to determine the next response. Both the afferent and efferent pathways involve the vascular system and not the nervous system.

Homeostasis of calcium is important, as calcium plays an important role in nerve and muscle function and in the coagulation of blood.

STRESS

Stress is any stimulus that causes an imbalance in the homeostasis of an individual. The stress may

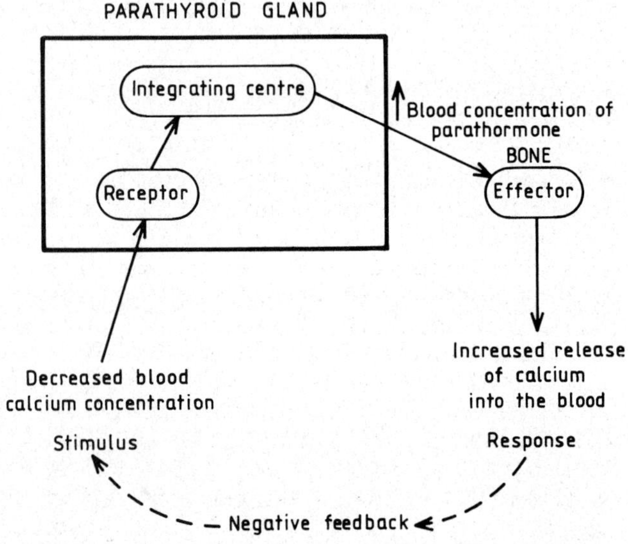

Figure 1.2 Homeostatic mechanism for the regulation of blood calcium concentration.

result from physical and chemical factors, psychological factors or a combination of factors. The effect of a particular stress on different people will vary according to their individual homeostatic abilities to cope with the stress. Stress can originate from the external environment. EXAMPLES: Temperature variation, loud noise, pollution and exposure to harmful microorganisms. Stress can also originate internally. EXAMPLES: High blood pressure, pain, tumours, emotional upset and endocrine disturbances. Often a combination of external and internal stresses occur, with the internal stress being directly or indirectly related to the external stress.

The greater the stress, the greater the demands placed on the body's homeostatic mechanisms to counteract the effects of the stress. The ability of the body to withstand and correct imbalances will depend upon the overall state of health of the patient. EXAMPLE: An ill person may have a lower capacity to cope with an emotional problem or an infection than a healthy person. Both symptoms may therefore ensue from stresses that a healthy person has little difficulty in coping with.

FEEDBACK CONTROLS

Imbalances in the body's homeostasis require carefully monitored adjustments by the integrating centres. The relationship between an imbalance and the response of the integrating centre to this imbalance is called feedback.

There are two forms of feedback, negative feedback and positive feedback. Negative feedback results in the response of the body being opposite to the imbalance. EXAMPLE: Blood sugar levels—a decreased blood sugar level results in negative feedback causing homeostatic mechanisms to raise the blood sugar level, whereas if an increased level exists the mechanisms will cause a decrease in blood sugar levels. Nearly all homeostatic mechanisms involve negative feedback, that is a system that attempts to return imbalances to the normal ranges.

Positive feedback results in the response of the body being the same as the imbalance. Positive feedback therefore acts as an amplifier, that is, it enhances the imbalance which results in the imbalance being further away from the normal range. Clearly, positive feedback is an undesirable mechanism for homeostasis! Some homeostatic mechanisms involve positive feedback but generally its use in the body is limited. EXAMPLE: In the menstrual cycle, an increase in the blood level of the hormone oestrogen stimulates the release of another hormone called luteinizing hormone. Blood oestrogen levels cause a positive feedback on the pituitary gland to increase the levels of luteinizing hormone.

Summary

Homeostasis is a relative state of equilibrium between the internal environment of an organism and a changing external environment. Stress can cause imbalances to the normal homeostatic state. These imbalances can result from factors in either the external or internal environments.

Homeostatic mechanisms return any imbalance to the normal optimum range of conditions. The greater the stress and the resulting imbalance, the greater is the extent to which a homeostatic mechanism is required to function. When the homeostatic mechanism is unable to cope with the stress, ill health can result. The inability to cope with a particular stress may be due to the following factors:

1. the level of stress is greater than the capabilities of the homeostatic mechanisms of an individual, and these capabilities vary between individuals,
2. continued stress may result in a decrease in the efficiency of a homeostatic mechanism to correct an imbalance,
3. the state of health of the patient may decrease the ability to cope with any additional stresses,
4. a malfunction in the homeostatic mechanism prevents the mechanism from functioning.

Treatment of a particular body imbalance will usually involve eliminating the causative factors and/or correcting abnormal levels of substances. Both these approaches assist homeostatic mechanisms in returning the body to normal function.

In single-celled organisms such as bacteria the cell performs all the vital functions for the organism, whereas in multicellular organisms such as humans a coordinated action of many cells is required for the organism to live. Single-called organisms depend upon the cell membrane for their homeostasis whereas humans rely upon both cell homeostasis and the coordinated action of cells or body homeostasis.

Homeostatic mechanisms require the following components for the mechanism to function: receptor, integrating centre, effector and communication pathways linking these. The integrating centre relies upon these pathways for feedback on the effect of its instructions on the body. In most homeostatic mechanisms negative feedback is used, that is, a system that attempts to return imbalances to the normal ranges.

Unit Two
Matter: Its structure and measurement

This unit introduces the structural units of matter and the methods of measuring matter. Emphasis is placed on the use of S.I. units as a system of measurement.

Chapter 2

Atoms and matter

Objectives

At the completion of this chapter the student should be able to:

1. define the terms matter, mass, element and atomic number,
2. describe the structure of an atom and relate this structure to the properties that identify a particular element,
3. explain how cations and anions are formed,
4. describe the properties of isotopes and relate these properties to their benefits and dangers for humans.

2.1 Matter and mass

The term mass is simply the quantity of matter that a substance contains when compared with a fixed piece of platinum kept by the International Bureau of Weights and Measures in France. This standard mass contains a mass defined as 1 kilogram (kg) in the international units known as S.I. units. A substance with half the amount of matter as the standard mass has a mass of 0.5 kg, whereas a substance of 2 kg mass has twice the amount of matter as the standard mass. Matter is anything that occupies space and has some mass.

The mass of an object or the quantity of matter in an object is the same wherever that object is taken in the universe. This book has the same mass in water as it does in a classroom or on Mars. The amount of matter or mass does not change.

MEASUREMENT OF MASS

A fundamental property of matter is that relatively small pieces of matter are attracted to relatively large objects in the universe. This property is used to determine the mass of an object. The attraction of matter to the earth is approximately five times that of the moon's attraction for matter. The term gravity is often used to refer to the earth's attraction of matter.

Gravity is nearly constant anywhere on the earth, and when it is multiplied with mass a measure of matter is formed which is known as weight:

weight = Mass × acceleration due to gravity (Eqn 2.1)

The weight of an object is five times greater on the earth than on the moon, that is, weight varies according to the gravity. Since gravity is nearly constant, the terms mass and weight are interchanged when referring to the amount of matter being measured on the planet Earth.

By exactly opposing the attraction of an object to the earth the mass or weight of an object is determined. EXAMPLE: An instrument known as a balance is designed to oppose exactly the object's attraction to the earth and produce a reading in grams or kilograms.

2.2 Atoms

All matter consists of small particles. A knowledge of the fundamental structure of nature is important for the understanding of the physical properties of matter, e.g. how matter moves.

Matter is mainly composed of three subatomic particles known as protons, neutrons and electrons. These particles are arranged to form a structure called an atom which acts as the building block or structural unit of matter. All atoms are composed of a central region of relatively large mass known as a

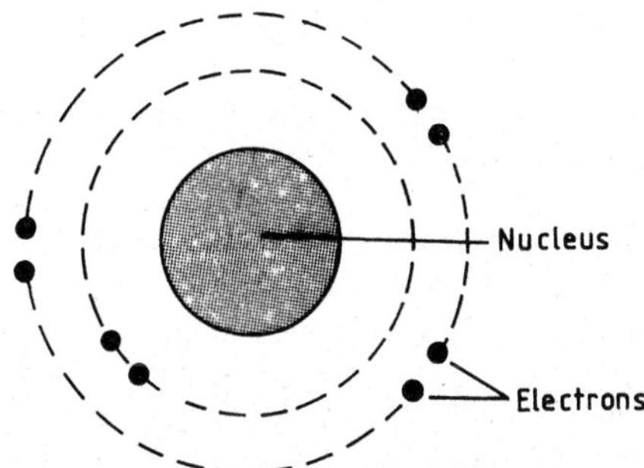

Figure 2.1 Two-dimensional representation of an atom in which the central positively charged nucleus contains protons and neutrons, and the negatively charged electrons orbit the nucleus.

nucleus which contains protons and neutrons, and an outer cloud of orbiting particles with little mass called electrons (Figure 2.1).

PROTONS

The nucleus consists of positively charged subatomic particles called protons. Each proton has a charge of +1, and thus five protons would have a charge of +5. To date, the maximum number of protons found in a nucleus is 105. The properties of an atom are determined by the number of positive charges within the nucleus, that is, the number of protons.

ELECTRONS

All atoms contain negatively charged subatomic particles called electrons. Each electron contains approximately 1/2000th the mass of a proton. These minute amounts of mass each have a charge of -1, and thus five electrons would have a charge of -5.

Electrons spin around the nucleus in oval-shaped orbits, which are also called electron clouds or electron shells. When there are many electrons, more than one electron cloud exists. Each orbit of electrons represents an amount of energy, since energy is required to hold the electrons in orbit. Orbits or electron clouds are thus often referred to as energy levels.

The maximum number of electrons that can be held in an orbit is limited by the amount of energy present. The quantity of energy in each orbit varies with the distance from the nucleus. An orbit close to the nucleus has less energy to support electrons than an orbit further away from the nucleus. Since the electrons carried by an orbit are limited by the amount of energy available to hold electrons, it follows that orbits further away from the nucleus contain more electrons than inner orbits.

Another property of energy levels is that the maximum number of electrons that can be carried by the outermost orbit is independent of the inner orbits. The first orbit will have a maximum of two electrons even when it is the only orbit. If more than one orbit exists the outermost orbit will contain a maximum of eight electrons.

A general property of electrons is that they occupy the lowest available energy level first, in order to obtain the most stable electron distribution. The first energy level or electron shell is thus filled before electrons begin to fill the second shell. Whenever possible, electrons orbit the nucleus in pairs.

To determine the distribution of electrons in atoms that contain more than ten electrons (two in the first shell, eight in the second) it is necessary to combine the properties of holding a maximum of eight electrons in the outermost shell and the property of filling successive energy levels. When an atom has eleven electrons (sodium) the orbits will contain two in the first, eight in the second and one in the third (Figure 2.2).

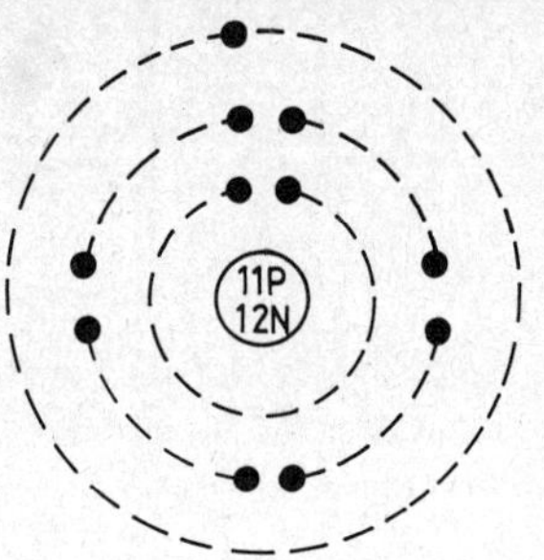

Figure 2.2 Sodium contains eleven electrons which requires the formation of a third electron shell.

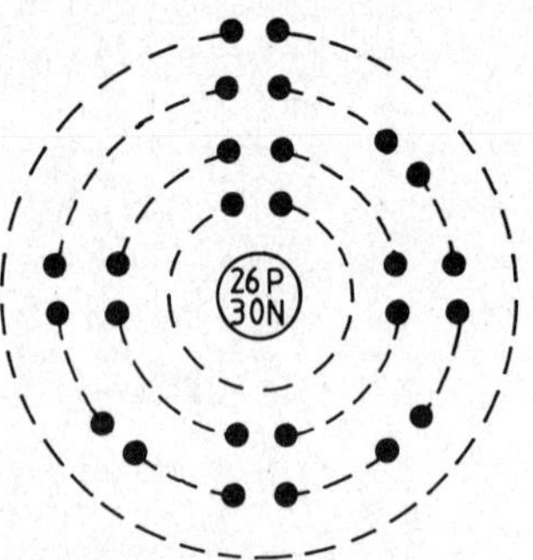

Figure 2.3 Iron contains twenty-six electrons which results in the third orbit containing fourteen electrons.

If the third shell is the outermost shell a maximum of eight electrons can be held. However, if an atom contains more than eighteen electrons (first shell—two; second shell—eight; outermost—eight) a fourth shell is formed which automatically becomes the outermost shell. Now the third shell is not restricted to the rule governing outermost orbits and thus it can hold a greater number of electrons. EXAMPLE: Iron (Figure 2.3). When four or more orbits exist, the third orbit contains eighteen electrons. Likewise, when a fifth orbit is present the fourth level contains enough energy to hold thirty-two electrons.

2.3 Ions

A fundamental property of nature is that all atoms attempt to have a full outer orbit of electrons. This property confers important physical and chemical effects on matter and is thus of great importance in body function. These chemical effects on matter are described in Unit 4.

Atoms lose or gain electrons in an attempt to have a complete outer electron shell. Generally atoms with more than four electrons in their outer orbit gain electrons in order to fill the outer orbit (Figure 2.4a). Less energy is involved in gaining one, two or three electrons than in losing seven, six or five electrons respectively.

An atom with seven electrons in the outer orbit will gain one electron so that the outer shell is complete with eight electrons. This extra electron results in the atom now having an excess of electrons with respect to protons, that is, the atom has a charge of -1. Likewise, an atom with six electrons in the outer orbit has a charge of -2 when the outer shell is completed. An anion is any atom or group of atoms that has a negative charge of at least -1.

Generally, atoms with less than four electrons in their outer shell find it easier to lose these outer shell electrons than gain five or more electrons. When these atoms lose the outer shell electrons the second outermost orbit (which is usually full) becomes the outer orbit. EXAMPLE: Sodium (Figure 2.4b). The loss of electrons results in the atom containing a relative excess of protons which thus confers a positive charge on the atom. The loss of one electron results in a charge of + 1 whereas the loss of two electrons results in a charge of + 2. A cation is an atom or a group of atoms that has a positive charge of at least + 1.

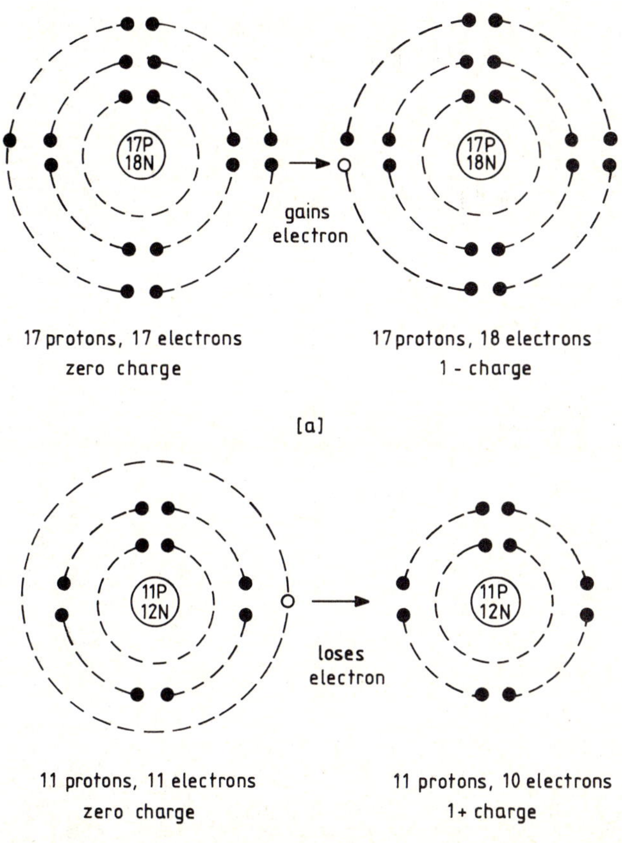

Figure 2.4 (a) Chlorine gains one electron to complete the outer shell, (b) sodium loses one electron to complete the outer shell.

Atoms with four or eight electrons in their outer shell do not form ions and these atoms will be discussed in detail in Unit 4.

In summary, atoms with five, six or seven electrons in their outer shell will form anions as they are electron acceptors. Atoms with one, two or three electrons in their outer shell will form cations as they are electron donors.

2.4 Molecules

Molecules consist of a combination of either atoms or ions. For any particular molecule the combination exists in a specific proportion. EXAMPLE: Water always consists of two hydrogen atoms combined to one oxygen atom, that is H_2O. The properties of a molecule are different from those of its constituents. EXAMPLE: Water is used to extinguish fires whereas its constituents, hydrogen and oxygen, are highly inflammable. A more detailed discussion of molecules and their properties is presented in Chapter 8.

2.5 Attraction of charged particles

A fundamental property of nature is that all matter attempts to have a neutral or zero charge. This results in opposite charges attracting and like charges repelling in an attempt to neutralize the charges. Positively-charged particles are thus attracted to negatively-charged particles in an attempt to become neutral, whereas the positively-charged particles are repelled by other positively-charged particles. EXAMPLE: a charge of + 1 combines with a charge of -1 to produce 0 or no charge. This force of attraction is important in determining properties of matter, e.g. how substances combine with other substances, electricity and body electrolyte composition.

In an atom this force results in the atom containing the same number of protons (charge + 1) as electrons (charge -1). The number of electrons orbiting a nucleus, and thus the number of orbits, is determined by the quantity of protons fixed in the nucleus.

2.6 Atomic number

The properties of an atom are determined by the number of protons located in the nucleus. An atom with one proton (hydrogen) differs in its properties to an atom with eight protons (oxygen). The importance of the number of protons present in an atom has resulted in the need for a term to describe the quantity of protons that are contained in an atom. This is referred to as atomic number, that is, all atoms with an atomic number of one contain one proton, whereas atoms with an atomic number of eight, contain eight protons.

2.7 Elements and the periodic table

WHAT IS AN ELEMENT?

The term element is used to describe all atoms with the same atomic number, that is, number of protons. If the number of protons is altered a new element is formed. An element is normally not capable of being broken down into a different element by ordinary chemical means, but can be changed by physical means such as radioactivity. All matter is derived from one of the 105 known elements.

THE PERIODIC TABLE

The Periodic Table lists all the 105 known elements (see pages 230 to 231). The symbols of the elements used in this text are listed in Table 2.1. The Periodic Table is arranged so that the elements with similar properties are grouped together in a vertical column (see Periodic Table) with the elements being referred to as a group or family of elements. The atoms in a group contain the same number of outer shell electrons. Elements in the same horizontal row are all members of the same period and contain the same number of electron shells.

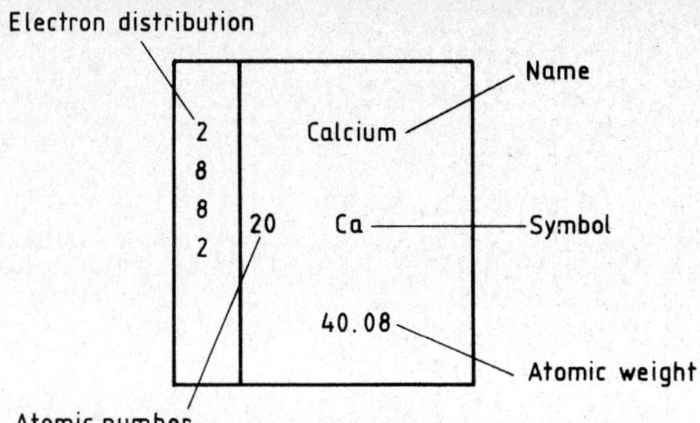

Figure 2.5 Periodic Table key.

The Periodic Table is also used as an information sheet for each element. The characteristics of an element can be determined by the information that is contained within the element's box (Figure 2.5). The position of an element in the table enables a person to predict various properties of that element. EXAMPLE: All elements in the eighth vertical column have completed outer shells which results in an inability of these elements to react with most substances. These elements are called inert gases or noble gases. All elements in the seventh vertical column contain seven electrons in their outer orbit and are referred to as halogens.

The transition elements are substances such as iron. The second outermost orbit in these elements is incomplete. The explanation for this occurrence is described in 2.2

Table 2.1 Symbols of elements cited in this text

Element	Symbol	Atomic number	Element	Symbol	Atomic number
Aluminium	Al	13	Nickel	Ni	28
Arsenic	As	33	Nitrogen	N	7
Barium	Ba	56	Oxygen	O	8
Bromine	Br	35	Phosphorus	P	15
Calcium	Ca	20	Platinum	Pt	78
Caesium	Cs	55	Plutonium	Pu	94
Carbon	C	6	Potassium	K	19
Chlorine	Cl	17	Radium	Ra	88
Cobalt	Co	27	Radon	Rn	86
Copper	Cu	29	Silicon	Si	14
Fluorine	F	9	Silver	Ag	47
Gold	Au	79	Sodium	Na	11
Helium	He	2	Strontium	Sr	38
Hydrogen	H	1	Sulfur	S	16
Iodine	I	53	Technetium	Tc	43
Iron	Fe	26	Tin	Sn	50
Lead	Pb	82	Uranium	U	92
Lithium	Li	3	Yttrium	Y	39
Magnesium	Mg	12	Zinc	Zn	30
Mercury	Hg	80			

Table 2.2 Characteristics of metals and non-metals

Metals	Non-metals
Conduct heat and electricity	Poor conductors of heat and electricity
Donate electrons	Accept electrons
Usually solid	Solid, liquid or gas
Shiny surface when cut	Dull surface when cut

Table 2.3 Common metals and non-metals found in the body

Metals	Non-metals
Calcium	Carbon
Potassium	Chlorine
Sodium	Hydrogen
	Nitrogen
	Oxygen
	Phosphorus
	Sulfur

2.8 Metals and non-metals

Metals and non-metals differ as a result of the number of electrons in the outer shell. Most metals contain one, two or three electrons in the outer shell and are thus electron donors. Most elements that contain four, five, six or seven electrons in their outer orbit are non-metals and are thus electron acceptors (see Periodic Table). Hydrogen exhibits properties of both metals and non-metals. The characteristic physical properties of metals and non-metals are listed in Table 2.2. The common metals and non-metals found in the body are listed in Table 2.3.

2.9 Neutrons and isotopes

NEUTRONS

The nucleus also contains subatomic particles known as neutrons. A neutron has no charge associated with it since it is composed of a proton (+1) and an electron (-1). Electrons within a nucleus always form part of a neutron and thus they do not exert the properties of an orbiting electron. Since neutrons have a neutral charge, they are unable to alter the number of free protons in the nucleus and the quantity of electrons in the orbit. Neutrons therefore do not influence the chemical properties of an atom and thus matter. Neutrons can influence the physical properties of atoms by emitting radiation.

ISOTOPES

Isotopes are elements that contain the same number of protons but a varying number of neutrons. An element may therefore have several isotopes without changing the atomic number. Since isotopes of the same element only vary according to the number of neutrons present, they are both chemically and physiologically indistinguishable. EXAMPLE: Iodine atoms can contain either seventy-four, seventy-eight or eighty-two neutrons. These three isotopes all have similar chemical characteristics and they each exist in a fixed proportion in nature.

MASS NUMBER

As all isotopes of a given element have the same atomic number, they cannot be identified according to their atomic number. The term mass number is used to identify isotopes. Mass number represents the number of protons plus neutrons present in an atom. EXAMPLE: Iodine contains fifty-four protons and either seventy-four, seventy-eight or eighty-two neutrons. The mass numbers of iodine are therefore 127, 131 or 135.

The mass number and atomic number of any element can be represented as follows:

mass number
(protons and neutrons)
ELEMENT
atomic number
(protons)

EXAMPLE: Iodine is represented by the symbol I so its three isotopes are: ${}^{127}_{53}\mathrm{I}$ ${}^{131}_{53}\mathrm{I}$ ${}^{135}_{53}\mathrm{I}$

2.10 Radioisotopes

Most elements contain isotopes that have a ratio of neutrons to protons which results in a stable nucleus, however some isotopes contain a disproportionate number of neutrons which results in an unstable nucleus. An unstable nucleus attempts to achieve stability by losing neutrons and in the process emits energy in the form of radiation. Radioisotopes or radioactive isotopes are isotopes that emit radiation. An element may contain both stable isotopes and radioisotopes according to the mass numbers of each isotope for that element. All isotopes of an element have essentially the same chemical and physiological function regardless of their nuclear stability or instability. However the energy emitted from radioactive isotopes may be harmful to living tissue.

RADIOACTIVE EMISSIONS

Radioactivity is the process whereby radioisotopes breakdown or decay in an attempt to obtain a more stable nucleus. During the decay of the nucleus, energy is emitted in the form of either moving particles or energy waves.

Beta Particles

Nuclear instability can lead to the breakdown of neutrons resulting in the formation of protons and electrons. Nuclear forces eject the electron, with its charge of -1, from the nucleus. The affected electron is called a beta particle. Since an extra proton now exists in the nucleus, a new element has been formed with an atomic number that is one greater than the original element.

Alpha Particles

Elements that contain more than eighty-three protons spontaneously eject particles that consist of two protons and two neutrons. These large positively charged particles are referred to as alpha particles. The ejection of an alpha particle results in the atom changing into a new element that has an atomic number two less than before the emission.

Gamma Rays

The production of alpha and beta particles results in the release of energy which is emitted in the form of waves of energy known as gamma rays. These rays have no mass or charge and are similar to X-rays (see Section 5.4). Gamma rays are emitted with either alpha or beta rays. Gamma emission does not alter the number of protons or neutrons in a nucleus and thus does not alter the atomic number.

HALF-LIFE

Every radioisotope has a unique rate of nucleus decay. Some radioisotopes decay rapidly, whereas others decay at a very slow rate. The rate of decay for a particular radioisotope is constant. Half-life is the time taken for half of a given number of atoms of an isotope to decay. Since the rate of decay varies for different radioisotopes, the half-life must also vary.

All isotopes have a unique half-life and thus a substance can be determined by its half-life (Table 2.4). EXAMPLE: An unknown substance has decayed from 100 mg to 50 mg in 8.1 days. Looking up the half-life table it is found that the only radioisotope with this property is the radioisotope of iodine known as ^{131}I. Furthermore, every 8.1 days half of the remaining atoms will have decayed or disintegrated into a new element (Table 2.5).

The range of half-life between different radioisotopes can vary from billions of years to fractions of a second. The shorter the half-life, the greater is the rate at which atoms decay and thus the greater is the quantity of nucleus emissions.

Table 2.4 The half-life of common radioisotopes

Radioisotope	Half-life	Radioisotope	Half-life
Calcium-47	4.5 day	Molybdenum-99	2.78 day
Carbon-14	5730 year	Phosphorus-32	14.3 day
Chromium-51	27.8 day	Sodium-24	15.0 hour
Cobalt-57	270 day	Strontium-87	2.8 hour
Gold-198	2.7 day	Technetium-99	6.0 hour
Iodine-125	60 day	Yttrium-90	64.2 hour
Iodine-131	8.1 day		
Iron-59	45.1 day		
Krypton-79	1.5 day		
Mercury-197	2.7 day		

Table 2.5 Half-life of ^{131}I showing the rate of decay or disintegration of ^{131}I

Quantity ^{131}I (mg)	Time (day)
100.00	0
50.00	8
25.00	16
12.50	24
6.25	32
3.13	40

MEASUREMENT OF RADIOACTIVITY

The detection of radiation is very important in medical work, particularly with respect to protecting personnel and patients from excessive exposure to radiation and for therapeutic and diagnostic purposes.

Different detecting systems are required for determining the type of nuclear emission (alpha particles, beta particles and gamma rays). The type and number of nuclear emissions can be determined for any radioisotope and thus radioisotopes can be identified.

The physical unit of radioactivity is expressed in terms of the number of nuclear disintegrations per second from a radioactive source. The products of these disintegrations are emitted in the form of alpha particles, beta particles or gamma rays. The S.I. unit (see 3.2) is the becquerel (Bq), which represents one disintegration per second. The becquerel has recently replaced the non-S.I. unit known as the curie. One curie equals 3.7×10^{10} Bq. For an explanation of this mathematical notation see 3.2.

The becquerel is not a useful unit in medical work since it only indicates the number of disintegrations per second and not the effect of the radiation upon tissue. The S.I. unit that relates the amount of radiation absorbed by tissue is the gray (Gy). One gray equals 1 joule of energy that is absorbed by 1 kg of tissue. The gray has recently replaced the non-S.I. unit called the rad. One rad equals 0.01Gy. The gray and rad describe irradiation of tissue regardless of whether the tissue is plant or animal. The damage to tissue by absorbing a given amount of energy varies according to the type of radiation.

A correction factor known as the relative biological effectiveness (rbe) is used to relate radiation damage to human tissue. The unit that relates specifically to the biological effect of radiation absorbed by humans is the sievert. The sievert replaces the non-S.I. unit known as the rem. One rem equals 0.01 sievert, whereas 1 sievert equals 100 rem. The relationship between the sievert-gray and the rem-rad are shown in equations 2.2 and 2.3 respectively.

$$1 \text{ sievert} = 1 \text{ gray} \times \text{rbe} \qquad \text{(Eqn. 2.2)}$$

$$1 \text{ rem} = 1 \text{ rad} \times \text{rbe} \qquad \text{(Eqn 2.3)}$$

EFFECTS OF RADIATION ON HUMANS

The effect of radiation on humans varies according to the type of radiation emitted and the tissue that has been irradiated.

The relatively large size and mass of alpha particles (two protons and two neutrons) restricts the ability of these particles to penetrate matter. Alpha particles are relatively harmless when they strike our skin as they cannot penetrate the dead outer layer of skin and therefore are unable to irradiate vital organs. However, when alpha particles enter the

Table 2.6 Recommended radiation dose limits for different body structures of personnel working in radiation-related occupations

Structure	Maximum dosage (millisievert/year)
Blood-forming organs	50*
Bone marrow	50*
Eye lens	50*
Gonads	50*
Lymph	50*
Skin and thyroid	300
Arms, hands and feet	750
Other organs	150

* After age eighteen, no more than 25mSv/3-month period

body by either inhaling or ingesting radioactive substances, serious tissue damage can result at the site of radioactive irradiation.

The smaller size of a beta particle (one electron) enables these particles to penetrate a few millimetres of skin which will result in an area that appears similar to a burn. They can only cause tissue damage inside the body when they have been taken into the body by breathing or ingestion.

Gamma rays have no mass and travel at the speed of light. These properties enable gamma rays to have a penetrating power about 10,000 times that of alpha particles. Exposure to gamma rays is far more harmful to the body than either alpha or beta particles, since they can penetrate the entire body.

The effect of radiation varies according to the type of tissue. The most sensitive tissues are bone marrow, the gonads, lymph tissue and the lens of the eye. Rapidly dividing cells, such as in a foetus, are very susceptible to radiation, and thus exposure to radiation during pregnancy should be avoided. The International Commission of Radiological Protection (ICRP) recognizes the differences in tissue sensitivity and recommends separate dose limits for different tissues (Table 2.6).

Low doses of radiation alter the structure of the chemicals involved in cell division. The alteration of hereditary information in the sex cells can result in future generations with mutations. Damage to other cells in the body can result in regions of abnormal cells which may lead to cancer. With increasing doses of radiation cell death occurs. Further increases in the amount of radiation absorbed will eventually result in rapid death.

The damage to body tissue as a result of radiation exposure is similar whether the tissue has been exposed to one large dose of radiation or a series of smaller doses which equals the large dose. The effects of radiation exposure are thus cumulative. Personnel who work in areas where radiation is used are required to wear badges that record the accumulated amount of radiation to which they have

Table 2.7 Some common diagnostic radioisotope

Radioisotape	Symbol	Use
Chromium-51	^{51}Cr	Red blood cell analysis
Iodine-131	^{131}I	Thyroid function
Technetium-99m	^{99m}Tc	Scans of the brain, lung perfusion, bone, renal function
Iron-59	^{59}Fe	Evaluation of body-iron concentration
Phosphorus-32	^{32}P	Detection of cancer cells

been exposed. These badges are checked at regular intervals to determine if the quantity of accumulated radiation is above the recommended safety levels.

DIAGNOSTIC USE OF ISOTOPES

More than a hundred different radioisotopes have been used to assist in a wide variety of diagnostic purposes. These radioisotopes are mainly selected according to the following criteria:

1. ability to emit gamma radiation,
2. short half-life,
3. ability of the radioisotope to be eliminated from the body shortly after completion of the diagnostic test,
4. smallest possible detectable dose.

Firstly, gamma radiation is used since these highly penetrating rays are easily detected outside the body. The use of long half-life radioisotopes causes more serious tissue damage because the low level of radiation produced is not easily detectable and thus higher doses are required. Short half-life substances minimize tissue damage since most of the disintegrations occur over a short period of time. Tissue damage is also minimized by eliminating the radioisotope soon after completing the test and by using a minimal quantity of radioisotope.

Elements play varying roles in body function. To evaluate a particular body function a specific radioisotope may be used since the body cannot distinguish between different radioisotopes of an element. A particular isotope will provide specific information for a more precise diagnosis. Some of the important diagnostic radioisotopes are listed in Table 2.7.

APPLICATION: THYROID FUNCTION TESTS

To assist in the diagnosis of overactive and underactive thyroid conditions the radioisotope ^{131}I may be administered to a patient. A normal functioning thyroid gland will take up about twelve per cent of a specific iodine dose within a few hours.

A typical thyroid scan involves a dose of 20 mega Bq (500 microcuries) which produces 0.05 Gy (5 rads—a mega Bq represents one million Bq whereas a microcurie is one millionth of a curie) to the thyroid. The neck is then scanned in the region of the thyroid gland with an appropriate detector. If the level of radioactivity over one hour is greater than fifteen per cent of the administered dose an overactive condition may exist. When the monitored thyroid level is less than four per cent an underactive condition probably exists. Note that this range of I uptake can be altered by a variety of conditions and is thus intended only as a relative guide to thyroid activity.

THERAPEUTIC USE OF ISOTOPES

The main objective of therapeutic radioisotope use is the selective destruction of abnormal cells such as uncontrolled rapidly dividing cancer cells. Destruction of these cells requires a highly localized intense dose of radiation. Failure to localize the radiation will result in damage of nearby healthy cells. Radioisotopes are administered to patients by one of three methods.

Firstly, specific radioisotopes that are selectively taken up by particular tumours can be administered orally. The administered dose of the radioisotopes is such that, when it is diluted throughout the body fluids, damage to general body fluids is minimal. Accumulation of the radioisotopes by the very localized tumour cells results in a radioisotope concentration that is sufficient to destroy these cells. EXAMPLE: Iodine-131 is selectively concentrated by the thyroid gland and any secondary tumours (metastases) by up to 200 times that of the rest of the body. The administration of ^{131}I orally at a higher level than that used for diagnosis, results in the tumour being irradiated with a dose of radiation that causes the death of tumour cells.

Secondly, radioisotopes can be implanted in a sealed metal container beside the tumour. Radioisotopes such as cobalt-60 (^{60}Co) and radium-226 (^{226}Ra) are encased in metal cases made of aluminium, gold, nickel or platinum. The metal cases can be in the shape of needles, beads or wires. Gamma radiation passes through the tissue and irradiates the nearby tumour cells. The long-life of ^{60}Co and ^{226}Ra requires these substances to be removed when the tumour has been destroyed to prevent unnecessary tissue damage. The radioisotope yttrium-90 (^{90}Y) has a shorter half-life (64.2 hour) than ^{226}Ra and ^{60}Co, is a beta emitter, and the low penetration of beta rays results in only the immediate area of the implant being irradiated (^{90}Y can be implanted directly into a tumour).

Thirdly, radioisotopes can be applied from a fine beam that emanates from a radioactive source This source is housed in a protective shield of lead and is usually either of the gamma emitters, cobalt-60 (^{60}Co) or caesium-137 (^{137}Cs). These radiation sources emit gamma rays that are directed as a fine beam to the cancer site, destroying the cancer cells.

Radioisotopes can also be used as a power source, by converting the emitted energy from nuclear decay into electrical energy. EXAMPLE: The radioisotope plutonium-238 (^{238}Pu) acts as a tiny nuclear battery in heart pacemakers. The pacemaker is implanted in the chest wall or into the subcutaneous tissue of the upper abdomen and is replaced about every ten years.

Summary

Matter is anything that occupies space and has some mass. It is composed of three subatomic particles known as protons, neutrons and electrons. These particles are arranged to form a structure called an atom which acts as the building block of matter.

An atom consists of a positively charged nucleus and an outer cloud of orbiting electrons. The nucleus contains protons with a positive charge and neutrons with no charge. Electrons have a negative charge, and the number orbiting the nucleus equals the number of protons in the nucleus. The equal number of protons and electrons results in atoms having no net charge.

The properties of atoms that determine their ability to combine with other atoms are:

1. The number of electrons in the outer orbit. All atoms with more than one orbit attempt to have a completed outer orbit of eight electrons. When less than four electrons are present an atom will attempt to lose these electrons and form a positively charged ion called a cation. When atoms contain four or more electrons in the outer orbit they attempt to gain electrons and form anions.
2. All atoms attempt to have a neutral charge. When ions are formed the oppositely charged ions are attracted to each other in an attempt to become neutral.

Atoms can contain up to 105 protons. An element consists of atoms with the same proton number or atomic number since these atoms all have identical chemical properties.

The number of neutrons in a particular element can vary between atoms. Atoms of an element that have different numbers of neutrons are called isotopes. When isotopes contain a disproportionate number of neutrons, the nucleus is unstable. In attempting to become stable by losing or emitting neutrons, these isotopes emit energy in the form of alpha particles, beta particles and gamma rays. These isotopes are called radioisotopes and the energy that they emit can be used for diagnostic and therapeutic purposes. Excessive exposure to this energy can result in damage to tissue and may ultimately result in death.

Chapter 3

Measurement and units

Objectives

At the completion of this chapter the student should be able to:

1. differentiate between qualitative, semi-qualitative and quantitative measurements,
2. name the fundamental S.I. units,
3. express numbers in scientific notation,
4. express measurements of weight, volume, size, temperature, pressure and energy in S.I. units.

3.1 Measurement

Measurement plays an important role in patient care. The recording of observations often relies on the need to measure what you are observing. EXAMPLES: Blood pressure, urine output, fluid intake and temperature. The recorded measurements need to be written in a manner that other hospital staff can interpret. The measure of an observation is separated into one of three categories.

QUALITATIVE MEASUREMENTS

Qualitative measurements are based on the presence or absence of signs. Signs are the evidence of a disease to an observer who records the signs in acceptable terms such as pale, very pale. A combination of descriptive signs can assist in the understanding of a patient's problems.

SEMI-QUALITATIVE MEASUREMENTS

A semi-qualitative measurement records information on a relative scale. The recorded information is usually in the form of + or relative words. The relative number of + indicates the extent of the sign. EXAMPLE: + + +, + +, +, 0 is used to indicate the sensitivity of microorganisms to antibiotics, and also the presence of glucose with clinistix tests. Descriptive words such as absent/moderate/severe are used to indicate the extent of shock. Semi-qualitative measurements are estimates of an observation. The recorded measurement relies on the experience and training of the person who is observing a sign and interpreting it.

QUANTITATIVE MEASUREMENTS

Quantitative measurements use numbers to express the quantity of a variable. These measurements require the use of instruments which enable a precise numerical measurement to be recorded. The observed number that is recorded by an instrument is a precise measurement as long as the instrument is operating and functioning correctly. A number is often useless by itself unless an indication is given as to what the number is referring! EXAMPLE: A pulse rate of fifty fails to indicate whether there are fifty pulses per second, per fifteen seconds, per sixty seconds etc. Most quantitative measurements require an additional piece of information which is referred to as a unit.

A unit is a standardized, descriptive word which specifies the dimension of a number. A fundamental unit is a defined physical characteristic. The seven fundamental physical properties of matter that can be measured independently of each other to provide information about the relative properties of a piece of matter are:

1. duration of measurement or time,
2. the physical size of an object or length,
3. the mass of an object or mass,
4. the amount of electric current passing through an object or current,
5. the temperature of an object or temperature,
6. the amount of substance present,
7. the brightness of an object or luminous intensity.

Each of these fundamental units is standardized by a term which is defined by a precise physical phenomena that can be measured. Any measurement of a fundamental unit is based on a reference to the standardized physical phenomena. EXAMPLE: The universal term for time is the second. A second is defined as the duration of 9,192,631,770 cycles of radiation associated with a specified transition of the caesium-133 atom. The time of any watch or clock in the world is based upon this definition. Clearly an individual is unable to determine a second directly by measuring the cycles of radiation associated with a caesium-133 atom. An individual can measure seconds indirectly by checking their watch with the standard time which is announced over telephones, radios and televisions.

All other physical properties of matter can be distinguished from each other by derivation from a combination of fundamental units. A derived unit can always be written as a combination of several fundamental units. EXAMPLE: The S.I. unit of force is the newton. It is derived by combining the units mass, length and time. Unfortunately over the past few centuries the development of scientific knowledge in different parts of the world has led to the formation of several distinct measurement systems. The two principle systems that were formed are commonly referred to as the English system and the metric system. Problems were encountered when information was reported in one system but was required to be converted into the other, and often a lot of unnecessary time was spent converting measurements from one system to another.

3.2 S.I. Units

WHAT IS AN S.I. UNIT?

Several years ago an international system of units was agreed upon by most major countries. This Système International d'unites or S.I. units is a modern system of units that relates present scientific knowledge to a uniform simplified system of units. The introduction of the S.I. units will bring uniformity of terminology, simplicity and a trend towards more uniform practice.

The fundamental S.I. units are listed in Table 3.1. In addition to the seven fundamental units there are a number of derived units such as litre, joule and pascal that are used in hospitals (Table 3.2).

Table 3.1 Fundamental S.I. units

Quantity	Name	Symbol
Length	metre	m
Mass	kilogram	kg
Time	second	sec or s
Current	ampere	A
Temperature	Kelvin	K
Amount of substance	mole	mol
Luminous intensity	candela	cd

Table 3.2 Common derived S.I. units

Physical quantity	Name	Symbol
Force	newton	N
Energy	joule	J
Pressure	pascal	Pa
Potential difference	volt	V
Frequency	hertz	Hz
Volume	litre	L

MULTIPLES AND SUBMULTIPLES

The magnitude or size of a measurement may result in an extremely large number for a specified unit. EXAMPLE: 3600 seconds is difficult to comprehend in terms of time. Other submultiples and multiples of a unit are used to eliminate cumbersome aggregates of numbers. EXAMPLE: The multiples of time are the minute, hour, day, week and year. These are useful to use, as one hour represents 3600 seconds. Imagine the unwieldy size of writing 86,400 seconds when referring to one day! Subdivisions of a second include milli- (one thousandth) and micro- (one millionth).

The S.I. system is based on multiples of ten, that is, multiples and submultiples of each unit are used when describing the magnitude of a measurement. The names of these multiples and submultiples are listed in Table 3.3.

The expression of a number such as one millionth, which is one divided by a million, can be written as either 1/1,000,000 or 0.000001. Both of these forms of expressing numbers appear unnecessarily large and complex. Since sophisticated pathology equipment can measure quantities as low as one millionth of a millionth, a simplified form of numerical expression is used to record all medical results. This simplified expression is known as scientific notation.

Table 3.3 Names and abbreviations of S.I. unit multiples and submultiples

Prefix	Symbol	Meaning	Scientific notation
tera	T	one million million	10^{12}
giga	G	one thousand million	10^{9}
mega	M	one million	10^{6}
kilo	k	one thousand	10^{3}
hecto	h	one hundred	10^{2}
deca	da	ten	10^{1}
deci	d	one tenth	10^{-1}
centi	c	one hundredth	10^{-2}
milli	m	one thousandth	10^{-3}
micro	μ	one millionth	10^{-6}
nano	n	one thousandth of a million	10^{-9}
pico	p	one millionth of a million	10^{-12}
femto	f	one thousandth of a pico	10^{-15}
atto	a	one millionth of a pico	10^{-18}

SCIENTIFIC NOTATION

Very large and very small numbers are often encountered when describing chemical and biological measurements. When these numbers are written in their ordinary way a visually large set of numbers occurs. EXAMPLES: There are 8,500,000,000 white blood cells per litre of blood in a normal person. In a fasting patient there exists a quantity of insulin in the range 0.0000000001 to 0.00000000001 mole per litre of serum. These numbers are difficult to write without making mistakes, hard to understand and almost impossible to say. The use of submultiple and multiple prefixes can assist in describing a number, but a simplified system of numbers is necessary when mathematical calculations are required.

Scientific notation is the writing of numbers in terms of a power of ten. EXAMPLES:

$$100 = 10 \times 10 = 10^2$$
$$1000 = 10 \times 10 \times 10 = 10^3$$
$$1{,}000{,}000 = 10 \times 10 \times 10 \times 10 \times 10 \times 10 = 10^6$$

Note that with large numbers, the figure placed above the ten is equal to the quantity of zeros.

When a number contains other numerals the number is written as follows:

$$25 = 2.5 \times 10^1$$
$$2500 = 2.5 \times 1000 = 2.5 \times 10^3$$
$$2{,}500{,}000 = 2.5 \times 1{,}000{,}000 = 2.5 \times 10^6$$

The accepted form of scientific notation is to place only one numeral to the left of the decimal point. The quantity of numerals that are on the right of the decimal point is written above the ten, e.g. 5400 is 5.4×10^3 since three numbers are to the right of five.

Numbers that are less than one are written the following way. Place a decimal point to the right of the first non-zero digit. Multiply this number by ten and place above the ten the number of digits that have been moved past the original position of the decimal point. A negative sign is placed in front of the raised numeral to indicate that the powers of ten are being divided into the digit.

EXAMPLES:

$$0.1 = 1.0 \times 10^{-1} \text{ or } 1/10$$
$$0.01 = 1.0 \times 10^{-2} \text{ or } 1/100$$
$$0.001 = 1.0 \times 10^{-3} \text{ or } 1/1000$$
$$0.0065 = 6.5 \times 10^{-3} \text{ or } 6.5/1000$$
$$0.000084 = 8.4 \times 10^{-5} \text{ or } 8.4/100{,}000$$

IMPORTANT S.I. UNITS

Kilogram

The kilogram is the unit of weight and is represented by the symbol kg. Table 3.4 shows the relationship between other units and the kilogram. There are 1000 grams to a kilogram. The gram is often used as a base unit in hospitals. One milligram is thus 10^{-3} of a gram and not 10^{-3} of a kilogram. The symbol g is used when referring to grams.

Litre

The S.I. unit for volume is the litre which is represented as L (note L is the preferred symbol although l is often used). The relationship between the litre and other units is shown in Table 3.5.

Table 3.4 Comparative measures of weight

1 kilogram	=	1000 gram
1 gram	=	1000 milligram
1 milligram	=	10^{-3} gram
1 microgram	=	10^{-6} gram
1 pound	=	0.454 kilogram
1 pound	=	454 gram
1 ounce	=	28.35 gram
25 gram	=	0.9 ounce

Table 3.5 Comparative measures of volume

1 litre	=	1000 millilitre
100 millilitre	=	1 decilitre
1 millilitre	=	1000 microlitre
1 microlitre	=	1 millilitre
1 UK gallon	=	4.5 litre
1 pint	=	568 millilitre
1 fluid ounce	=	28.42 millilitre

Table 3.6 Comparative measures of length

1 metre	=	10^{-3} kilometre
1 centimetre	=	10^{-2} metre
1 millimetre	=	10^{-3} metre
1 square metre	=	10^2 metre
1 metre	=	39.37 inch
1 mile	=	1.6 kilometre
1 yard	=	0.9 metre
1 foot	=	0.3 metre
1 inch	=	25.4 millimetre

Metre

The metre or m is the S.I. unit of length. There are 39.37 inches in one metre. A comparison of the common measures of length is shown in Table 3.6.

Kelvin

The Kelvin (K) is the S.I. unit of temperature. Absolute zero or 0 K is equal to −273°C (see 5.5). A change of 1 K is identical to a change of 1°C. Note that Kelvin is not preceded by a ° degrees symbol. To determine K temperature from a Celsius temperature, simply add 273 to the Celsius temperature. EXAMPLE: A body temperature of 38°C is equal to 311 K as 38°C + 273 = 311 K. For practical purposes degees Celsius will continue to be used for the present time.

The Fahrenheit scale of 32 to 212°F encompasses the same temperature range as 0 to 100°C. A change of 1°F is not equal to 1°C. The conversion of a reading in Fahrenheit into Celsius is performed by subtracting thirty-two, multiplying by five and dividing by nine. EXAMPLE: 97°F - 32 = 65 × 5 = 325 ÷ 9 = 36.1°C. The conversion of a reading in Celsius into Fahrenheit is performed by reversing the above procedure, that is multiply by nine, divide by five and add thirty-two. EXAMPLE: 36.1°C × 9 = 325 ÷ 5 = 65 + 32 = 97°F.

Pascal

The pascal or Pa is the S.I. unit of pressure. The kilopascal or kPa is being introduced as the unit in blood gas estimations. A kilopascal is determined by multiplying the old units of millimeters of mercury (mm Hg) by 0.133. The old method of recording blood pressure in mm Hg will continue to be used until practical considerations permit the use of kPa sphygmomanometers.

The British unit of pounds per square inch (lb/in^2) or psi is converted into kPa by multiplying by 6.88.

Joule

A joule or J is described as the potential energy which is released when 1 kg weight falls through 1 m by the force of gravity. The joule is to replace the calorie as the unit of energy. Since the joule is smaller than the Calorie, the larger kilojoule is preferred. A Calorie is converted into a kilojoule by multiplying by 4.184. The common comparative units are listed in Table 3.7.

Table 3.7 Comparative measures of energy

1 calorie	=	4.184 joules
1000 calories	=	1 Calorie or kilocalorie
1 Calorie	=	4184 joules or 4.184 kilojoule
1000 Calorie	=	4184 kilojoule
1 kilojoule	=	0.238 Calories

Mole

The mole is the S.I. unit for the amount of substance present. This unit is discussed in Chapter 4.

Summary

Measurement of a patients condition can be classified into three forms:

1. Qualitative measurement is used to describe the presence or absence of signs.
2. Semi-qualitative measurement records information on a relative scale using + or relative words. This measurement relies on experience and training for the estimation of an observation.
3. Quantitative measurement records information in terms of numbers. This measurement usually requires the use of an instrument.

The seven fundamental physical properties of matter and their S.I. units are: time—second, length—metre, mass—kilogram, current—ampere, temperature—Kelvin, amount of substance present—mole, luminous intensity—candela.

Important S.I. units that can be derived from the seven fundamental units are: force—newton, energy—joule, pressure—pascal, potential difference—volt, frequency—hertz and volume—litre.

A simplified system of expressing numbers is used in the S.I. system. Prefixes such as micro- and nano- can be used to describe a number. Alternatively, numbers are expressed as multiples of ten. EXAMPLE: One million is written as 10^6 and one millionth as 10^{-6}.

Chapter 4

Concept of concentration

Objectives

At the completion of this chapter the student should be able to:
1. explain the term concentration,
2. express concentration in S.I. units,
3. define the terms mole, molar and atomic weight,
4. convert other concentration expressions into moles per litre,
5. explain the difference between S.I. units and international units of concentration.

4.1 Concentration and the body

Concentration refers to the quantity of substance present in a given volume. In S.I. units this volume is one litre. The concentration of substances within different healthy individuals is approximately the same, even though the amount of substances may be different. A large person generally contains more substances than a small person. However the large person also contains a proportionately greater volume which results in both people containing approximately the same concentration.

Body function relies upon a balanced concentration of many different body substances. These substances form the basis of the biochemical tests that are performed by pathology laboratories. Within a population, a normal concentration range has been determined for many body chemicals. Variations in the concentration of a particular substance from the normal range usually indicates abnormal body function. Biochemical tests are used to assist diagnosis and in addition they can be used for monitoring the response of a patient to treatment.

4.2 The S.I. unit for concentration

The S.I. unit termed moles per litre has resulted in a uniform expression of concentration. Prior to the introduction of moles per litre and millimoles per litre many forms of expressing the concentration of chemicals were found in hospitals. EXAMPLES: milliequivalents per litre, percentage solutions, gram per litre, gram per decilitre, gram per 100 millilitre, volume per cent, Bessey units etc. In most cases these units have been abolished. The confusion in understanding results was enhanced by the use of several units for the same substance. EXAMPLE: Plasma calcium was measured in either milligram per 100 millilitre or milliequivalents per litre, with a result in the latter being twice the former.

DETERMINING CONCENTRATION

Concentration refers to the number of particles present in a given volume. The methods used to determine the quantity of particles in a container are:

1. Determine the total weight of the particles and divide this weight by the weight of one particle. ANALOGY: A box of eggs has a weight of 600 g and the weight of each egg is approximately 50 g. The number of eggs present is calculated by dividing 50 into 600, that is, the box contains 12 eggs.
2. Count the total number of particles present. ANALOGY: If the eggs could be counted in the container the quantity of eggs is easily determined.

With these methods it is necessary to either weigh or count the particles in order to determine the concentration of a substance.

Often in the past, the person dealing with chemicals had to calculate the particle concentration, since the units of concentration were not uniform and did not represent the total particle concentration. With the introduction of S.I. units a uniform and precise system of representing particle concentration has been introduced. This system has the advantage of eliminating the need for calculating the true concentration from units that express a relative concentration such as grams per litre (see 4.3). With the S.I. system the particle concentration in a container is stated and thus the number of particles is known. ANALOGY: At a supermarket, the number of eggs in a carton is printed on the container which eliminates the need to count the number of eggs or to weigh the total contents and divide by the weight of one egg. The major advantage of this system is that the true particle concentration of different substances can be compared without the need for calculations.

In chemistry, a very large number of molecules exists in any small sample of a substance. Instead of labelling chemical containers with very large confusing numbers a simplified system is now used. The enormous number 6.023×10^{23} is replaced in chemistry by the term one mole which is abbreviated as mol. This large, convenient number is referred to as Avogadro's number. In a container there is one mole of molecules present for every 6.023×10^{23} molecules. ANALOGY: When referring to large quantities of paper it is convenient to use the term ream. A ream represents 480 sheets of paper whereas in chemistry a mole represents 6.023×10^{23} molecules. Three reams of paper indicates a quantity of 3×480 whereas three moles of substance represents $3 \times 6.023 \times 10^{23}$ molecules.

A container with three moles of a particular substance contains the same number of molecules as another container with three moles of another substance. ANALOGY: A package with three reams of foolscap paper contains the same number of pages as package with three reams of a different size paper. The term mole is also used to indicate the number of ions present in a solution. The concentration of molecules refers to the number of molecules of a substance that is present in a volume of one litre. The relationship between moles and concentration is as follows: *the concentration of a solution is determined by the number of moles of substance present in one litre of that solution*. This concentration is molar. EXAMPLE: A one molar solution contains one mole of substance in one litre of the solution, whereas a two molar solution contains two moles of substance in one litre of the solution. Note that a solution consists of substances dissolved in a liquid. A more detailed description is given in 10.2.

The concentration of substances within the body and in solutions added to the body is generally lower than one mole. This results in other units such as millimole per litre, micromole per litre and nanomole per litre being used. The most frequently used unit is the millimole (one thousandth of a mole).

So far we have determined that the term mole represents a specific number of molecules or ions. Since the mole refers to the number of particles in any substance, then the total number of particles in a mixture of substances can be determined by adding the moles of each substance. EXAMPLE: The intravenous amino acid solution "synthamin 17" contains 100 mmol/L of amino acids, 70 mmol/L of acetate and 40 mmol/L chloride. The total quantity of substances present is $100 + 70 + 40 = 210$ mmol/L. Every litre of "synthamin 17" contains 210 mmol of molecules and ions. A solution of 500 mL contains half the volume and thus half the quantity of these substances, i.e. 105 mmol. The term molarity refers to the number of moles of a substance or substances present in one litre of a solution. EXAMPLE: "synthamin 17" has a molarity of 210 mmol.

The number of millimoles in a bottle may not be the required quantity that has to be administered to a patient. The volume of solution that represents the required amount of millimoles can be determined by using the following equation:
volume of solution needed = required moles substance/molarity of solution.
EXAMPLE: A patient requires a quantity of 200 mmol of sodium chloride from a one litre bottle that contains 308 mmol/L (1.8 per cent) sodium chloride. Volume of solution required = 200 mmol/308 mmol = two-thirds of a litre. The volume required is therefore approximately 670 mL.

Sometimes it is necessary to calculate the total quantity of moles of a substance that has been administered for a given volume. The total number of moles can be calculated as follows:
millimoles of substance = solution molarity $\times$ volume in litres.
EXAMPLE: How many millimoles of sodium chloride has been administered in 500 mL of a 150 mmol sodium chloride solution? The number of millimoles of substance = $150 \times 0.5 = 75$ mmol.

PREPARING A MOLAR SOLUTION

Most preparations are labelled with the quantity of millimoles present. To assist in the understanding of molarity it is necessary to explain how a solution has been prepared. Clearly a person is unable to count a given number of molecules and make up a one litre solution.

Initially it is necessary to weigh an amount of a substance. The required weight will vary according to the weight of one molecule. There is an enormous variation in the weight of the molecules of different substances. For a given quantity of molecules, a substance that consists of a relatively heavy molecule will require a greater amount than a substance that is composed of a lighter molecule. ANALOGY: The weight of ten oranges is greater than ten cherries, and therefore a greater weight is required to obtain ten oranges than ten cherries.

A molecule is extremely small and may not be weighed on a balance. The weight of an individual molecule is determined by adding the weights of the constituent elements. The weight of each element varies according to the number of protons and neutrons present since the mass of an atom results from its protons and neutrons. The weight of each proton is defined as one, as is the weight of each neutron. Nearly all carbon atoms contain six protons and six neutrons, although a very small percentage contain six protons and eight neutrons. The carbon atom is defined as having an atomic weight of twelve.

The atomic weight of any atom is the relative weight of that atom compared to an atom of carbon. This means that the atomic weight is not an actual weight, but rather a weight that is relative to carbon. As atomic weight is a comparative weight it therefore lacks a unit. EXAMPLES: Hydrogen mainly consists of one proton and therefore has an atomic weight of one. Oxygen contains a total of sixteen protons and neutrons which results in an atomic weight of sixteen. The atomic weight of each element is listed in the Periodic Table (see 2.7). An element such as chlorine exists mainly as two isotopes, one of which contains two additional neutrons. The atomic weight of chlorine is 35.5 which is derived by averaging the proportions of the weight of each isotope.

The atomic weight is a useful chemical term as it indicates the relative weight of an atom compared to another element. When twelve grams of carbon are weighed a total of 6.032×10^{23} atoms are present, that is, one mole of atoms is present. Note that a sample of a pure element only contains one type of atom. A mole refers to the number of atoms and not molecules in these situations. *Whenever the atomic weight of an element has the same number as the number of grams, one mole of atoms is present.* EXAMPLES: One mole of hydrogen atoms is present in one gram of hydrogen, whereas sixteen grams of oxygen forms one mole of oxygen atoms since the atomic weight of hydrogen is one and oxygen is sixteen.

The weight of a molecule is termed the molecular weight. It is the total weight of all the constituent atoms. EXAMPLE: A glucose molecule is composed of six carbon, twelve hydrogen and six oxygen atoms. The molecular weight of glucose is calculated as follows:

carbon	6×12	= 72
hydrogen	12×1	= 12
oxygen	6×16	= 96
total		= 180

One mole of glucose molecules is present in 180 g. A one molar solution of glucose contains 180 g of glucose in one litre of solution. A one millimolar solution of glucose contains 0.180 g or 180 mg of glucose in one litre of solution. A 0.5 millimolar solution is prepared from 90 mg of glucose.

The S.I. unit of concentration is moles per litre. This unit can only be expressed if the molecular weight of a particular molecule is known, since the molar solution can only be prepared by using the atomic and molecular weights. There are numerous substances for which the molecular weight has not been obtained, EXAMPLE: many proteins. These substances require other forms of expression to indicate their concentration.

4.3 Other expressions for concentration

Prior to the introduction of S.I. units several methods existed for expressing concentrations. These methods were replaced by millimoles per litre, except where the molecular weight of a substance was unknown.

MILLIEQUIVALENTS

These units have been replaced by the millimole. A milliequivalent is determined by relating the ability of an element to combine with, or displace, hydrogen. Unlike millimoles, one milliequivalent (meq) of one substance does not necessarily contain the same number of particles as one milliequivalent of another substance. EXAMPLE: One milliequivalent of calcium ions contains half the number of particles as one milliequivalent of sodium ions.

Care must be taken when converting plasma calcium (Ca^{2+}) and magnesium (Mg^{2+}) into mmol/L since 1 meq/L of these substances represents 0.5 mmol/L. With most of the other chemicals that were expressed in meq/L, the concentration in mmol/L is represented by the same numbers. EXAMPLE: 1 meq/L of sodium ions equals 1 mmol/L.

WEIGHT — VOLUME

Prior to the introduction of S.I. units, the concentration of a substance was sometimes expressed as the number of grams per 100 mL of water. A concentration expressed in this form is a percentage solution. EXAMPLE: A five per cent dextrose (glucose) solution indicates that there is five grams of dextrose in a solution of 100 mL. This form of concentration does not indicate how many glucose molecules are present. A ten per cent dextrose solution indicates that there are twice as many glucose molecules present than in the five per cent solution.

A five per cent solution of sodium chloride does not contain the same quantity of molecules as a five per cent dextrose solution as sodium chloride has a molecular weight of 58.5 and dextrose one of 180. The sodium chloride molecule is approximately one-third the weight of a dextrose molecule, and thus 5 g of sodium chloride will contain approximately three times the number of molecules of 5 g of glucose. ANALOGY: In a purchase of 5000 g of oranges and 5000 g of cherries a far greater quantity of cherries would be expected since the weight of a cherry is less than an orange.

No comparison can be made between percentage solutions of different substances without taking into account the molecular weight of each substance. Percentage solutions fail to indicate the concentration of molecules in a solution, and for this reason they have been replaced by molar solutions wherever the molecular weight of a substance is known.

The use of percentage solutions is now restricted to substances where the weight of a molecule is unknown. EXAMPLES: Pathology reports — plasma proteins, enzymes, hormones; intravenous solutions such as intralipid (ten and twenty per cent) and various dextran solutions. In addition, the introduction of S.I. units has resulted in an alteration in expression of percentage solutions from grams per 100 millilitre to grams per litre. EXAMPLE: A total protein concentration of 7.0g/100mL is now recorded as 70g/L.

RATIO SOLUTIONS

Ratio solutions express concentration in the terms of a ratio. EXAMPLE: A ratio of 1:1000 represents the number of grams per volume of solution. The one is in grams and the 1000 in millilitres. Ratio solutions lack a unit. These solutions can also be expressed in terms of the proportion of one substance to another. EXAMPLE: Dextrose—nitrogen ratio represents the ratio between dextrose and nitrogen in the urine. Different ratio solutions cannot be compared.

INTERNATIONAL UNITS

With some substances where the molecular weight is unknown the concentration is reported in terms of international units. An international unit is a measure of the effect of a concentration of a particular substance upon a standard solution with which that substance reacts. These units have been standardized by an International Conference for the Unification of Formulae. The substances where the biological effect is the unit of concentration, are usually either biological substances such as enzymes, hormones, vitamins, or drugs such as penicillin. EXAMPLE: The international unit of insulin is 1/22 of a milligram of the pure crystalline product. International units differ from substance to substance, and thus the international units from two substances cannot be compared or added.

Summary

Concentration refers to the quantity of substance present in a given volume. In the body, a balanced concentration of substances is necessary for homeostasis.

The S.I. unit of concentration is moles per litre. This unit represents the number of molecules or ions present in one litre of solution, with one mole being 6.023×10^{23} particles (Avogadro's number).

The total number of particles in a mixture of substances can be determined by adding the moles of each substance.

The term molarity refers to the number of moles of a substance or substances present in one litre of solution.

The atomic weight of an element is the relative weight of an atom compared to another element. The atomic weight is approximately equal to the number of protons and neutrons present in the most abundant isotope of an element. When the atomic weight of an element is the same number as the number of grams, one mole of atoms is present.

The molecular weight is the total atomic weight of the constituent atoms of a molecule.

The molarity of a solution can only be determined if the atomic and molecular weights are known. When these weights are not known the concentration can be expressed as the number of grams per litre.

Some substances with an unknown molecular weight have their concentration expressed in international units. An international unit is a measure of the effect of a particular substance on a standard solution. International units will differ from substance to substance, and thus the international units of two substances cannot be compared or added. This method of expressing concentration can be used for hormones, enzymes, vitamin and drugs.

Unit Three
Physical properties of matter

This unit integrates the physical properties of matter and applies these properties to nursing, normal and abnormal body function, the uses and the dangers of diagnostic and therapeutic equipment. You will learn how the properties of matter are influenced by various forms of energy and how matter moves.

Chapter 5

Matter and energy

Objectives

At the completion of this chapter the student should be able to:
1. describe the different forms of energy and their properties,
2. relate the properties of energy to nursing, normal and abnormal body function and the uses and dangers of diagnostic and therapeutic equipment,
3. describe how an action potential is formed,
4. describe the requirements for a flow of current.

5.1 Potential energy and kinetic energy

Matter is anything that occupies space and has some mass. To move matter requires energy. Energy is related to matter since energy either creates a potential to do work or it enables matter to do work. The close association of energy and matter was discovered by Albert Einstein who showed that in certain cases energy could be converted into mass and *vice versa.*

Potential energy is energy stored in an object as a result of its position, chemical composition or stretching and compression of elastic material. This stored energy is shown by the release of energy when the restraints maintaining the objects position or chemical makeup is removed. EXAMPLES: Stored water in a dam being released by a floodgate, release of a stretched elastic band, breaking down of food to produce body energy.

When energy is in action or is performing work it is known as kinetic energy. EXAMPLES: Flow of blood, chemical manufacture and muscle contraction. Both potential energy and kinetic energy can be converted into the opposite kind of energy.

APPLICATION: STRETCHING AND RECOIL OF THE AORTA

Stretching of the elastic walled aorta results from blood being ejected from the heart during ventricular contraction or systole. The stretched aorta recoils to its original size during ventricular relaxation or diastole. In terms of energy the stretched aorta represents a potential to do work or potential energy. When the walls recoil, blood is squeezed along the artery thus doing work in the form of kinetic energy.

Potential energy and kinetic energy can be subdivided into various forms of energy, each of which has its own properties. The energy forms used in body function and health care are: electrical, radiant, thermal, sonic, chemical and mechanical. These forms of energy can only occur in the presence of matter, apart from radiant energy (e.g. light and X-rays).

5.2 Conservation of energy

Generally energy is neither created nor destroyed but is converted from one form to another. This principle of the interconversion of energy enables humans to obtain energy in the form of chemical energy stored in food, and to transform it into forms that can be utilized by the body for the growth and maintenance of normal body function. The ingested chemical energy is transformed by the body into the following forms:
1. chemical energy for cell growth and function,
2. electrical energy for nerve and muscle function,
3. mechanical energy for body movement and the movement of objects by the body,
4. thermal energy for body temperature homeostasis.

The body also converts sound energy and radiant energy (light) into the electrical energy of nerve function.

The total energy in any system is constant. The total amount of energy taken into the body is either stored or removed from the body as chemical energy (faeces, urine, sweat etc.), mechanical energy or thermal energy.

APPLICATION: DRIP SET FUNCTION

The principles of constant energy within a system and the ability of energy to be converted from one form to another is shown in the functioning of a drip set. The liquid contained in a raised one litre bottle contains one hundred per cent potential energy and zero kinetic energy. When the bottle is sixty-seven per cent full, the bottle will contain sixty-seven per cent of the initial potential energy. The potential energy of the liquid that has left the bottle was converted into kinetic energy as the liquid flowed out of the bottle and down the tube. Likewise, when the bottle is fifty per cent empty, the bottle will contain only fifty per cent of the inital potential energy. Fifty per cent of the original potential energy has therefore been converted into kinetic energy of the infusing liquid. When the bottle is empty, all the potential energy of the liquid has been converted into kinetic energy.

5.3 Electrical energy

Electrical energy results from a flow of charged particles. In the body, information is conducted along nerves in the form of electrical energy and is then converted into muscular movements (mechanical energy) or other energy forms. Electrical energy is easily converted into radiant energy as occurs in light bulbs, X-rays and short-wave radiation. Since the body relies on electrical energy for a major part of its internal communication and for muscular function, these processes can be monitored by machines which measure electrical activity in various body tissues. EXAMPLES: The electrocardiograph (ECG) for the heart, the electroencephalogram (EEG) for the brain and the electromyogram (EMG) for peripheral nerve injuries and muscle disease.

ELECTROLYSIS

When opposite charges exist in nature, the charged particles are attracted to each other in an attempt to attain a zero charge. Remember, unlike charges attract and like charges repel. (See 2.5.)

When ions are added to a solution, they move freely throughout the liquid so that they are randomly distributed. If two rods of opposite charge are placed into a liquid containing ions, negatively charged ions or anions are attracted to the positive rod and repelled by the negative rod. Positive ions or cations are attracted to the negative rod and are repelled by the positive rod (Figure 5.1). These rods or electrodes are named according to their attraction of ions. Anions are thus attracted to the anode (positive electrode) and cations to the cathode (negative electrode). Electrolysis is the movement of ions to oppositely charged electrodes. Note that these ions are no longer randomly distributed in the liquid. Substances that move to oppositely charged electrodes when placed in a liquid or when molten are called electrolytes.

As most metals are cations, they will be attracted to the cathode. This results in the cathode being coated by the metal and this process is referred to as electroplating. EXAMPLE: Gold plating.

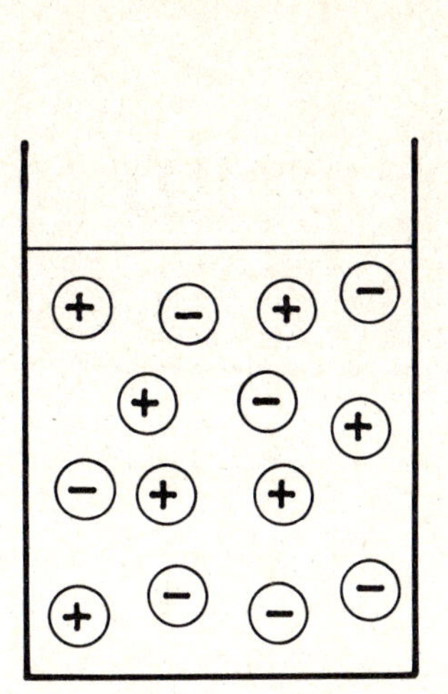

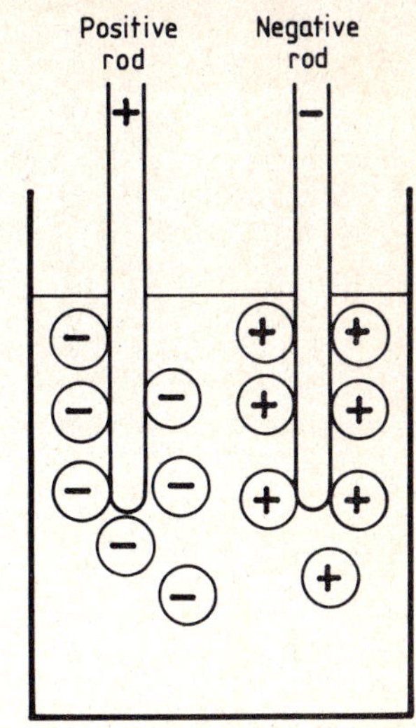

Figure 5.1 (a) Ions are randomly distributed in solution, (b) when two oppositely charged rods are placed in the solution there is a non-random distribution of ions, negatively-charged ions are attracted to the positive rod and positively-charged ions to the negative rod.

POTENTIAL DIFFERENCE

An analogy of water stored in a dam often assists in the understanding of potential difference. Water stored in a dam possesses potential energy. This water has the potential to flow down to the river below due to the influence of gravity. The potential energy is converted to kinetic energy when the dam's spillway is open, that is, the barrier that maintained the potential energy has been removed.

In a similar way to the dam, potential energy exists in a liquid between two separated regions of oppositely charged ions. This can be seen when a vessel containing a liquid is separated by a barrier into two compartments, one of which contains positively charged ions and the other contains negatively charged ions. If a positive electrode or anode is placed in the compartment with the positive ions and a negative electrode or cathode in the other, the barrier prevents the movement of ions to the oppositely charged electrode (Figure 5.2a).

The dam with a closed spillway whose water contains potential energy is analogous to oppositely charged ions being separated by a barrier. These ions possess potential energy. Removal of the barrier will cause a flow of ions and the potential energy being converted into kinetic energy (Figure 5.2b).

These differences in charge between the two regions are referred to as a difference in potential, that is, a potential difference. The greater the difference in charges between two regions, the greater is the attraction of the oppositely charged particles to each other. Potential difference is therefore a form of potential energy, since a

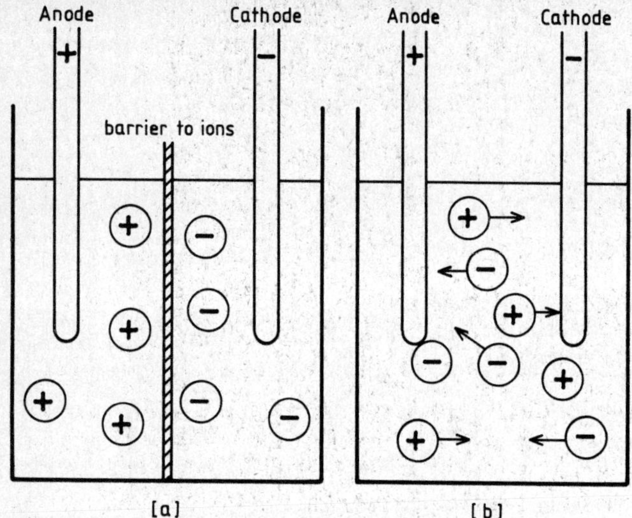

Figure 5.2 (a) A barrier separates two oppositely charged regions of a solution. The ions in these regions possess potential energy since removal of the barrier (b) results in the flow of ions and the conversion of potential energy into kinetic energy.

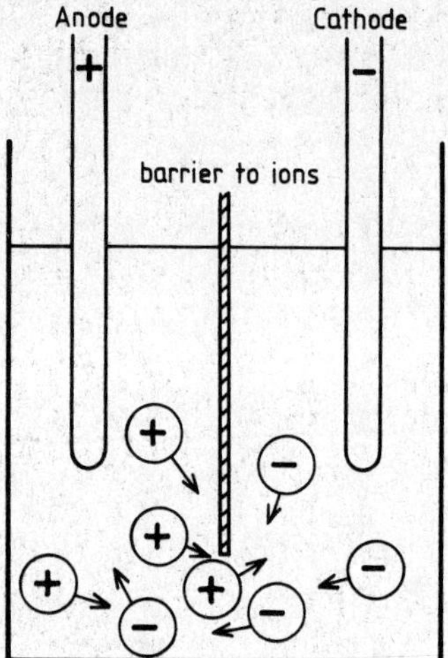

Figure 5.3 The flow of ions or current is reduced by the presence of a restriction or resistance. Complete removal of the barrier would result in a greater flow of ions as in Figure 5.2b.

potential exists for the movement of charged particles.

A dam higher than another dam will possess more potential energy that the lower dam, since water from the first dam flows a greater distance. As a result, the higher dam's water is able to generate a greater amount of kinetic energy. The greater the height that water is raised the greater is the potential energy of that water. With respect to ions, the greater the difference in charges between two regions, the greater is the potential difference or voltage.

The greater the potential difference between two regions, the greater is the amount of stored energy. Potential difference is measured in volts. The greater the voltage, the greater is the attraction of oppositely charged ions to each other. This potential for the movement of charges is only fulfilled if there is a medium present that allows ion movement to the differently charged regions.

The conversion of potential energy into kinetic energy results in a gradual decline in the potential difference. As the potential difference decreases, the flow of ions also decreases. A situation eventually develops whereby there is no difference in charges remaining, that is, zero potential difference and kinetic energy exists.

In the body, a potential difference exists across cell membranes. The charges are distributed so that the inside of the cell is negative and the outside is positive. The membrane acts as the barrier which maintains the potential difference. This potential difference across membranes plays an important role in muscle and nerve function, along with determining the distribution of electrolytes in the body. Since the potential difference is small, a unit one thousandth of a volt is used. This S.I. unit is known as the millivolt and is symbolized as mV.

FLOW OF ELECTROLYTES IN LIQUIDS

Potential difference is only converted into kinetic energy if a medium known as a conductor is present, which enables charged particles to move to their oppositely charged regions. In a liquid the flow of ions is referred to as an electric current. Convention states that current flows from a positive region to a negative region. The S.I. unit for the measurement of current, that is, flow of charge is known as the ampere, which is symbolized as A. The ampere is often abbreviated to amp.

The greater the potential difference, the greater is the amount of available energy that can be converted into kinetic energy. When no restrictions to the flow of ions exist, potential difference is converted into kinetic energy in the form of an electric current. Generally, the greater the potential difference, the greater is the quantity of current that can be generated. EXAMPLE: A potential difference of 240 volts can generate more current than a potential difference of 110 volts for the same medium.

The existence of a restriction to ion flow reduces the flow of ions, that is, current (Figure 5.3). The potential difference only represents the potential for ions to flow and not the actual quantity of ions that flow.

When considering the current that results from a given potential difference it is necessary to consider any restrictions, that is, resistance to ion flow. The amount of resistance present determines the quantity of current that actually flows for a given voltage. The greater the resistance for a given potential difference, the lower the amount of current that flows. The unit which is used to measure resistance is the ohm, which is symbolized as Ω. When the resistance to ion flow is absolute, a barrier occurs which results in no flow of current. In this case the potential difference is maintained until the barrier is removed.

APPLICATION: NERVE IMPULSES

In nerve cells, the positive ions sodium and potassium flow between the outside and inside of these cells. This flow of positive ions or current constitutes a nerve impulse or action potential. Nerve action potentials are extremely important in body homeostasis, since they act as units of information which are necessary for the coordination and regulation of body function. Action potentials can be divided into four stages:

1. resting potential,
2. depolarization,
3. repolarization,
4. redistribution of sodium and potassium ions.

Resting Potential

The potential difference between the contents of a nerve cell and the liquid surrounding the cell is approximately 70 to 90 mV. The inside of a cell is negative with respect to the outside of a cell and therefore this potential difference is denoted as -70 to -90 mV. This voltage represents a nerve cell "at rest" and is referred to as the resting potential.

The distribution of ions between inside and outside a nerve cell determines the resting potential (Figure 5.4). The membrane acts as a barrier to the flow of large negatively charged proteins and the large positively charged sodium ions (Na^+). The smaller ions, potassium (K^+) and chloride (Cl^-) are able to move freely across the membrane. The resting potential occurs as a result of the presence of large negative charges such as proteins and amino acids inside the cell and the high concentration of Na^+ ions outside the cell.

The large negative ions are manufactured within the cell and are incapable of passing through the nerve membrane. These negative charges are retained within the nerve cell during any nerve activity and are therefore referred to as fixed negative charges.

The negative charge within nerve cells attracts Na^+ and K^+ ions. Sodium ions are unable to move to the negative region as the membrane structure acts as a barrier to these ions. Potassium ions can move into the cells but there is an insufficient quantity of K^+ ions to overcome the excess negative charge and therefore the cell contents remains negative with respect to the outside of the cell.

Depolarization

A stimulus causes the nerve membrane to be momentarily altered so that the barrier to Na^+ ions

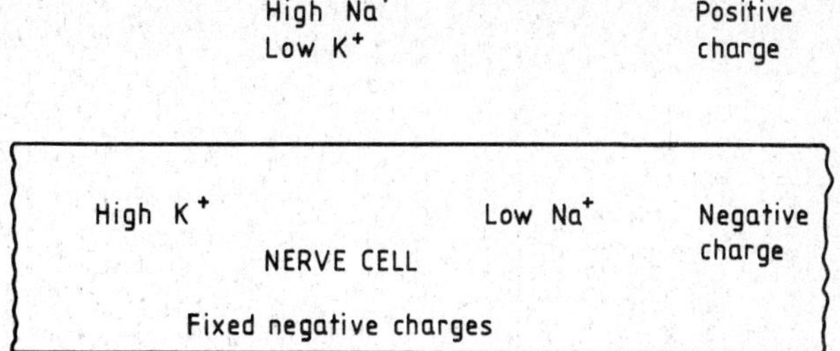

Figure 5.4 The distribution of ions inside and outside a nerve cell "at rest".

is removed. The negative charge within the cell attracts a large flow of Na^+ ions into the cell and a current results. This flow of positive ions causes the inside of the cell to become more positive, that is, less negative with respect to the outside of the cell (Figure 5.5a). The decrease in potential difference is referred to as depolarization and is the first stage of a nerve impulse or action potential.

During depolarization, the rapid flow of positive ions or current results in the potential difference across the membrane firstly approaching zero and then becoming positive. This overshooting of Na^+ ions results in the potential difference attaining a voltage of approximately + 45 mV. Notice that now the cellular contents are positive with respect to the outside of the cell. If this change in potential difference is measured during the depolarization stage a graph similar to Figure 5.5b would be observed.

Repolarization

At the completion of depolarization the nerve cell membrane has returned to acting as a Na^+ ion barrier. Since the inside of the cell is now positive, K^+ ions flow to the negative region outside the cell, that is, current flows to the region outside the cell.

The difference in the amount of K^+ between the outside of the cell and the inside also assists the flow of K^+ from the cell (Figure 5.6a). Relatively high concentrations of K^+ ions inside the cell compared with the outside results in a flow of these ions to lower concentration by a process called diffusion. The relationship of differences in the amount of ions to ion flow is described in Section 17.2.

The loss of K^+ ions from the cell results in the cellular contents becoming less positive, that is, more negative with respect to the outside of the cell. The potential difference returns to zero and then decreases to the original resting potential difference. The loss of K^+ ions from the nerve, and subsequent increase in negative charge within the cell is known as repolarization. Measuring potential difference across a nerve membrane during repolarization will result in a graph similar to Figure 5.6b.

Redistribution of Na^+ and K^+ ions

At completion of repolarization, the nerve has a similar potential difference to the resting potential

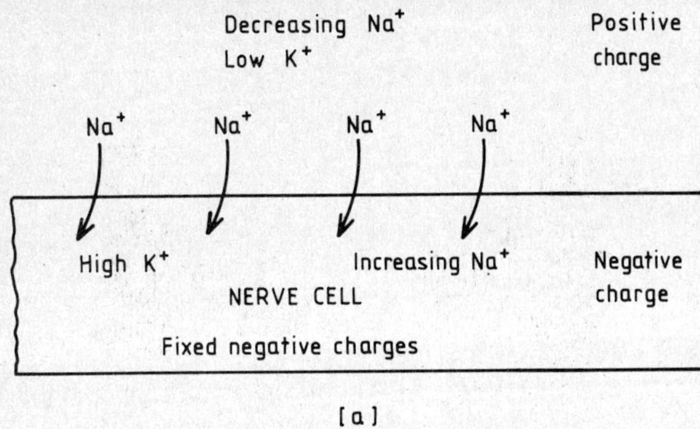

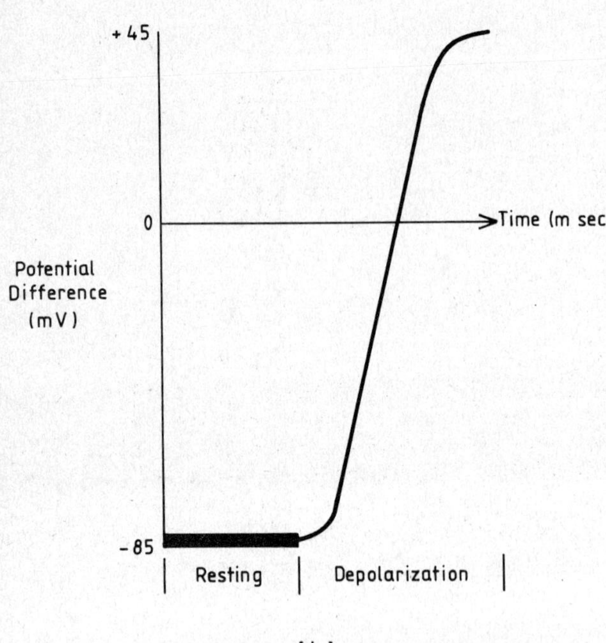

Figure 5.5 (a)A stimulus momentarily removes the barrier to Na+ ions resulting in the flow of these ions into a nerve cell; this flow of Na+ causes the inside of the cell to become more positive.
(b) Measuring the potential difference across a nerve membrane prior to, and immediately following, a stimulus, shows a rapid change of potential difference from -85 to +45 mV.

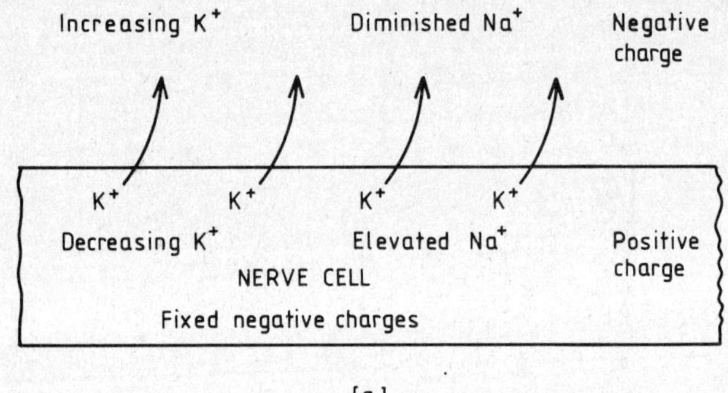

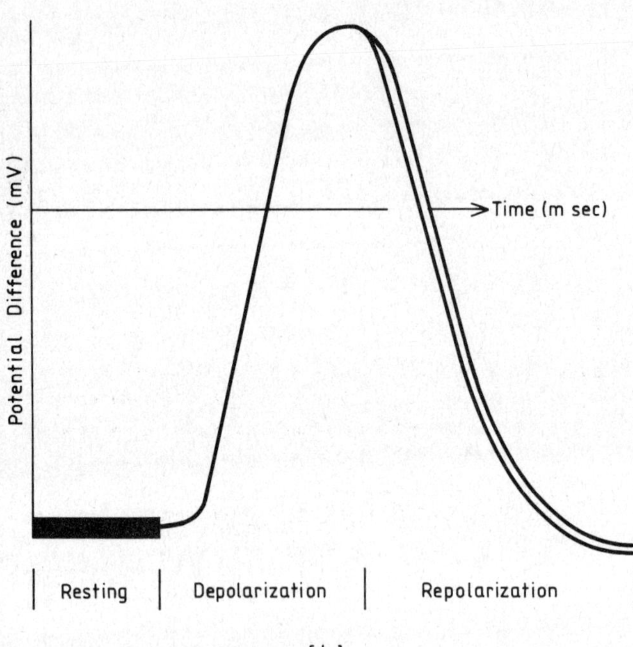

Figure 5.6 (a) K+ ions flow from the positively-charged nerve to the negatively-charged region outside the nerve.
(b) Measuring this flow of K+ ions will show a changing potential difference across the membrane from +45 to -85 mV(=).

but the cellular contents of Na+ and K+ ions are different. For another action potential to occur the distribution of these ions has to return to their original resting potential distribution. The removal of Na+ from the cell and the transport of K+ into the cell is performed by the use of a carrier that uses energy to carry Na+ out of the cell and K+ into the cell. The mechanism of this carrier is described in 17.3.

MAINTENANCE OF A CONSTANT POTENTIAL DIFFERENCE

To ensure a constant flow of ions it is necessary to maintain a constant potential difference. Without the maintenance of a constant potential difference, the flow of ions will decrease the potential difference. A decrease in the flow of current will ensue.

A constant potential difference can be maintained by the use of specific chemical reactions which replace ions that are lost as a result of current flow, and this process occurs in batteries. The quantity of current that flows is dependent upon the potential difference developed by a battery. The strength of a

Figure 5.7 (a) The ability of metals to donate electrons enables them to pass electrons from one metal atom to the next, that is, conduct a current. (b) The inability of non-metals to donate electrons stops the flow of electrons.

battery is measured by the potential difference generated by the battery and is thus measured in volts.

FLOW OF CURRENT THROUGH A SOLID

The potential difference between two regions can be decreased by connecting a piece of metal to the two regions. Unlike ions in solution, it is very difficult to move positive or negative ions through a solid and therefore a different mechanism is used for decreasing the potential difference. The small mass of electrons enables them to flow through certain solids from a negative region to a positive region. Since the flow of negatively charged electrons decreases the potential difference between two regions, this flow is also referred to as electric current. Substances that carry current are known as conductors, whereas substances incapable of carrying current are insulators or non-conductors.

Conductors are generally composed of metals whereas insulators are composed of non-metals. Metals are able to conduct electricity as they are electron donors, that is, metals lose their outer orbit electrons to nearby atoms. This enables electrons to jump from one atom to another along a metal wire, from a negatively charged region to a positively charged region. The net flow of electrons is also called a current (Figure 5.7). Insulators are unable to release electrons because they require electrons to complete their outer shell (Figure 5.7). As electrons do not move freely a current cannot occur.

ELECTRIC CIRCUITS

Completing Electric Circuits

Electrons only flow through metal when it connects two oppositely charged regions. If a wire is broken, electrons are unable to jump across the non-metal gap as this region acts as an insulator. A complete piece of metal wire is required for current to flow from one region to another, that is, a completed electrical circuit is required.

Short Circuits

When oppositely charged regions are connected by several pathways, most electrons will flow along the path which offers the least resistance.

An active wire carries current from the power source to the electrical components of a piece of equipment. Contact between an active wire and another wire or a conducting medium can result in most current flowing along the latter path especially if the latter offers less resistance to the flow of electrons. When the current flows via an unintended, alternative pathway a short circuit occurs.

Short circuits can occur when a loose wire comes

in contact with the metal casing of a piece of equipment or with a frayed electrical cord. This casing is now "live" and contact with it can cause an electric shock.

The presence of a large quantity of electrolytes within the body enables the body to conduct electricity. Metals conduct electricity better than the human body and this property is used in preventing or decreasing the intensity of an electric shock in a process called earthing.

Earthing

Earthing is the process whereby a wire connects the metal casing and other areas of equipment to the earth.

A large potential difference exists between live wires and the earth. This potential difference enables electrons to flow to the earth whenever a short circuit occurs.

APPLICATION: ELECTRIC SHOCK

When a person touches a piece of earthed equipment they will receive only a minor shock as most of the current will flow through the low resistance earth wire (Figure 5.8a). A short circuit in equipment that is not earthed results in a person conducting the entire current as no alternative pathway exists (Figure 5.8b). They therefore receive an electrical shock.

Electric shock can also be minimized by breaking the conduction path between the body and the earth by wearing non-conducting or insulating material such as rubber gloves or rubber shoes. When a person is receiving an electric shock, it is essential to isolate the person from the main power supply by turning it off or by dragging the person away from the source of supply. The rescuer involved in moving a person that is receiving an electric shock needs to avoid receiving a shock by using insulated materials such as thick rubber gloves or rope.

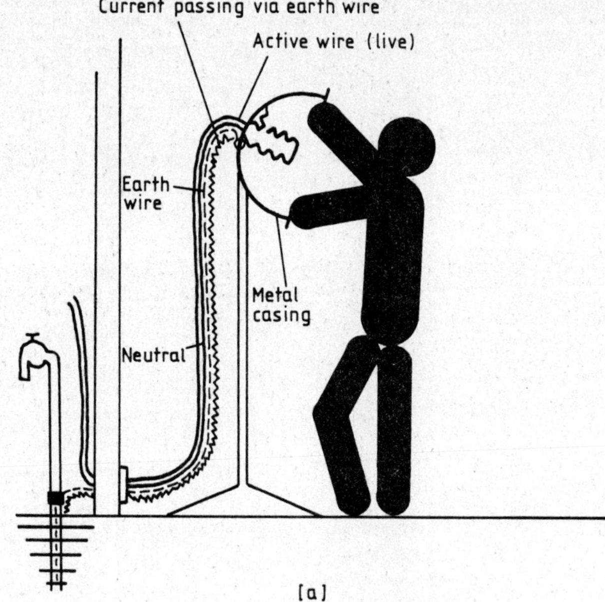

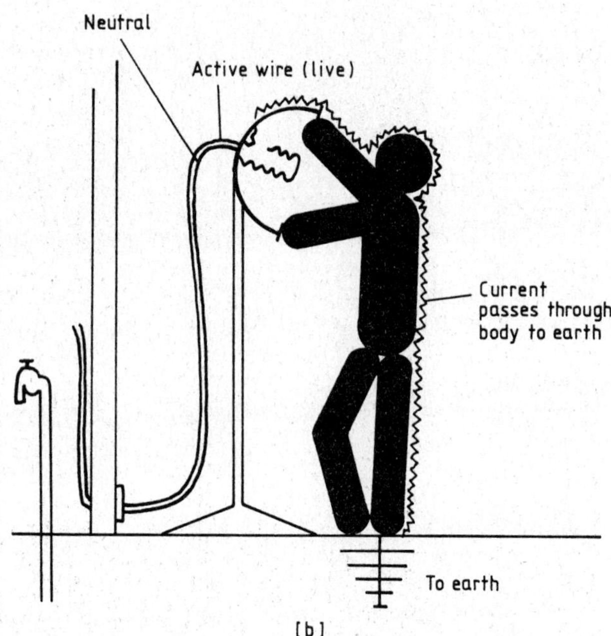

Figure 5.8 A faulty lamp leads to the active wire connecting with the outer casing.
(a) When the casing is earthed the current flows via this low resistance pathway to the earth.
(b) When no earth is present the current flows through the body resulting in the person receiving an electric shock.

Switches

A switch is a device which closes or opens a circuit and therefore either allows current flow or stops current flow respectively (Figure 5.9). Switches are designed to regulate the flow of electricity and thus the function of electrical equipment.

Mains Wire

Electrical circuits involving hospital equipment originate from power stations which cause the flow of electrons and create a potential difference between the power station and the hospital power points. The power station usually maintains a potential difference of 260 to 250V with electrical equipment.

A completed electrical circuit between a power station and a piece of equipment consists of:

1. A live or active wire carrying current from the power station; the wire is covered by brown or sometimes red insulating material,
2. A neutral or return wire from the equipment; the wire is covered by blue or sometimes black insulating material,
3. An earth wire which usually connects the equipment case to the earth; the wire is insulated with a green/yellow or sometimes green material.

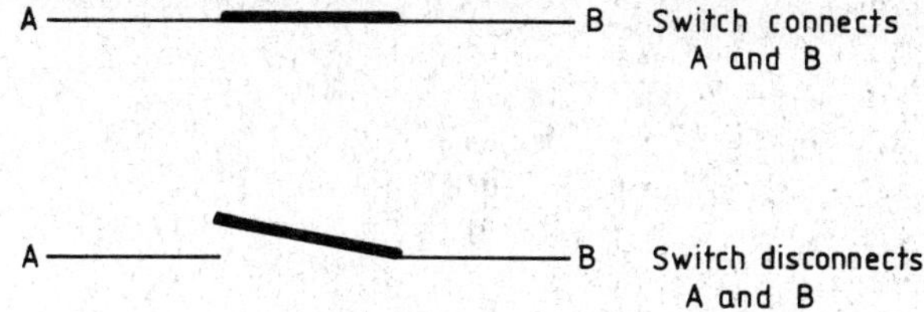

Figure 5.9 A closed switch allows current to flow between A and B, whereas an open switch prevents current flow.

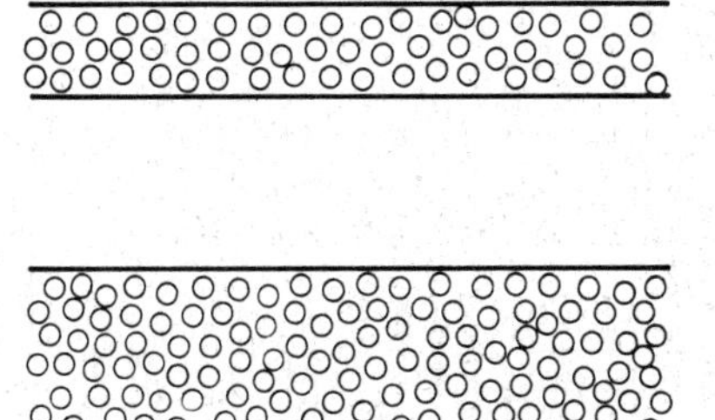

Figure 5.10 The flow of water through a smaller diameter pipe is less than that through a larger diameter pipe.

Resistance in Wires

The amount of electrons flowing along a piece of wire will depend upon certain features of the wire such as wire diameter and the type of metal, both of which alter the resistance and thus the flow of current. A thin piece of wire can only conduct a small number of electrons whereas a larger diameter wire can carry a greater number of electrons. ANALOGY: The quantity of water that can flow through a pipe is determined by the pipe's diameter. A small pipe has a greater resistance to water flow than a larger diameter pipe which results in less water being able to flow through the smaller diameter pipe at any given moment (Figure 5.10).

Metals differ in their ability to conduct electrons, with poor conducting metals carrying a smaller quantity of electrons or current than high conducting metals. The above characteristics of metal wires are used in controlling the amount of current that can flow in electrical equipment.

APPLICATION: FUSES

A fuse is a device that can be put into an electrical circuit and which breaks when an unduly high current passes through it. A fuse consists of a short length of wire which is selected according to its particular resistance to current. The greater the resistance, the less is the quantity of current that can be carried before the fuse wire overheats and breaks. The breakage of the fuse wire prevents excessive current overloading and damaging electrical components.

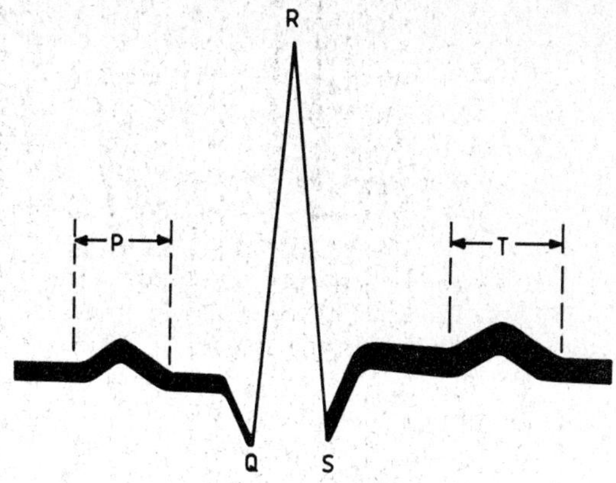

Figure 5.11 A normal ECG.

RECORDING BODY ELECTRICAL EVENTS

Electrical currents and potential differences exist throughout the body and are essential for body function. The electrical activity of the heart, brain and muscle can be measured by placing external electrodes on the body's surface. Detection of various electrical events associated with the function of these organs can assist in determining a patient's condition.

APPLICATION: ELECTROCARDIOGRAPH

During contraction of the heart, potential differences develop between different regions of the heart. The resultant electrical currents spread into tissues surrounding the heart, and a small proportion of these spread to the surface of the body. By placing leads at different points on the body's surface, a comparison can be made as to whether the current is flowing towards or away from a particular lead. There are various arrangements of electrocardiograph leads, each of which produce a different shaped wave of electrical activity known as electrocardiograms or ECGs. Figure 5.11 shows the most common ECG which is composed of:

1. a P wave—due to atrial electrical events,
2. a Q.R.S. complex—due to ventricular electrical events,
3. a T wave—due to ventricles returning to the resting potentials.

5.4 Radiant energy or electromagnetic radiation

The sun is the major source of electromagnetic radiation in this solar system. Without this source of energy life could not exist on earth. Energy from the sun is converted into other forms of energy such as chemical energy in plants. Humans rely on the

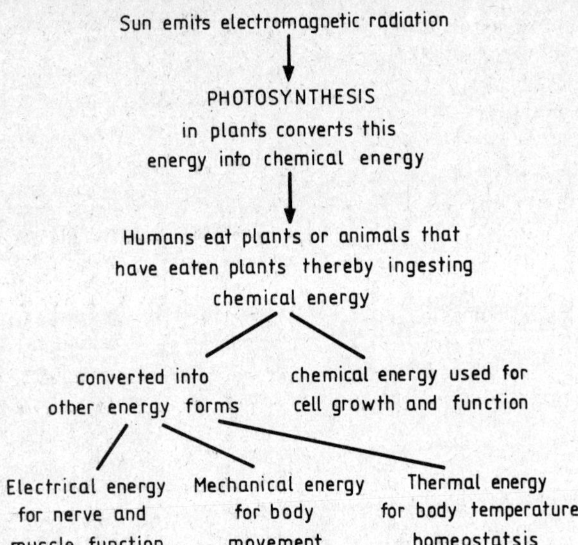

Figure 5.12 Inter-relationship of body function to the sun's emission of electromagnetic radiation.

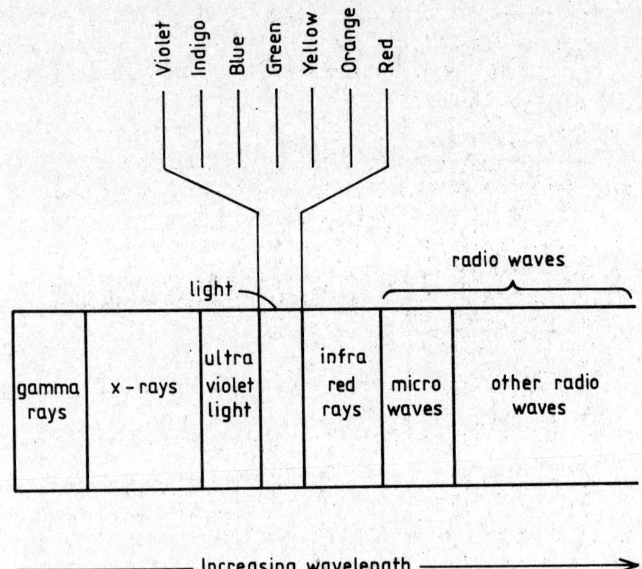

Figure 5.13 The electromagnetic spectrum.

intake of this energy for maintaining body function. (Figure 5.12).

When the electrons of atoms gain energy from sources such as heat, the electrons jump briefly to a higher energy level. On return to their original level they emit energy in the form of waves or rays which are referred to as electromagnetic radiation. Microwaves, heat (infra-red rays), light, ultraviolet light, X-rays and gamma rays are examples of electromagnetic radiation.

The different properties of various forms of radiation is due to differences in the nature of their energy waves. Wave characteristics are determined by the degree of energy that has been absorbed and released by electrons. The electromagnetic spectrum demonstrates the range of wavelengths and thus the properties of the various forms of radiation (Figure 5.13).

Radiation exists in the form of short discontinuous bundles of waves known as photons, and not in the form of a continuous wave. A wavelength is composed of one complete wave or one cycle, for example, the distance between the highest point of one wave and the same point on a following wave (Figure 5.14a). The frequency of a wave is the number of waves passing through a point in one second.

The speed for all electromagnetic waves is constant at about 300,000 kilometres per second in a vacuum. The shorter wavelength wave requires more waves to travel the same distance, at the same speed as a wave with a longer wavelength. Shorter wavelengths thus have a higher frequency than longer wavelengths (Figure 5.14b). The shorter the wavelength, the greater is the amount of energy of those waves. The wavelength and energy of photons is therefore directly related to the frequency of a wave.

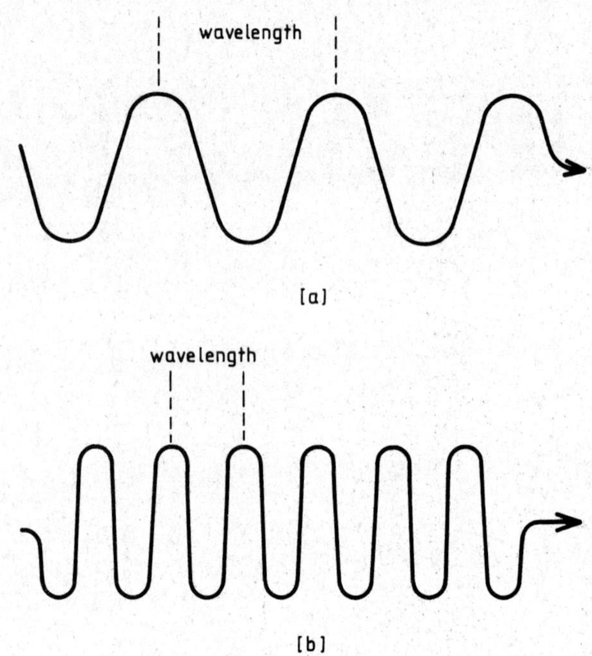

Figure 5.14 The nature of a photon. The wavelength of the photon in (a) is larger than in (b). The greater the number of waves for a given distance or time, the greater is the frequency. The photon in (b) thus has a greater frequency than the photon in (a).

Electromagnetic rays produce their effects at the point at which the rays are absorbed. This is achieved by changing from radiant energy into another form. EXAMPLE: Heat can be converted to chemical energy.

MICROWAVES

Microwaves are the shortest wavelength group of a range of rays that are known as wireless or radio rays. Wireless rays have wavelengths that are greater than infra-red rays. Microwaves can penetrate three to five centimetres of tissue. These waves heat tissues, especially those tissues with a high liquid content. There is an appreciable heating of tissues such as muscle, but less heat is produced in those with a low liquid content, such as fat. Care must be taken to prevent burns.

INFRA-RED RAYS

These rays have a relatively long wavelength and are emitted by hot bodies. As the temperature of an object increases the wavelength of the emitted rays will become shorter and contain more energy. The absorption of these rays by matter produces heat or thermal energy.

APPLICATION: INFRA-RED THERAPY

The low energy content of infra-red rays restricts the penetration of these rays to the outer regions of skin tissue. The effect of infra-red irradiation is a local rise in tissue temperature. A mild increase in temperature appears to have a sedative effect on sensory nerves, thus enabling infra-red irradiation to be a useful method of relieving pain. This heat will also lead to an increase in the diameter of superficial blood vessels which allows an improved blood supply that assists in the healing of wounds and superficial infections.

LIGHT

What is Light?

Everything that is seen by the human eye results from the eye detecting light waves. Light is electromagnetic rays that stimulate the retina of the human eye to produce images in the brain. Light consists of electromagnetic waves that contain a smaller wavelength than infra-red rays but a greater wavelength than ultraviolet rays.

Differences in the colour of light are due to distinctive wavelengths within the range of visible light. The sequence of these distinctive wavelengths forms the pattern of a rainbow which is known as the visual spectrum. The spectrum consists of seven colours arranged in the following order of descending wavelength size: red, orange, yellow, green, blue, indigo and violet (Figure 5.13). White light is a mixture of all of these wavelengths and black is the total absence of visible light. When a single colour exists, it represents a narrow wavelength and is referred to as monochromatic light.

Light is next in the spectrum of wavelengths to infra-red radiation (Figure 5.13). When an object that is emitting infra-red radiation absorbs more heat, the resulting electromagnetic radiation will consist of waves with a shorter wavelength. As the wavelength continues to shorten with the addition of energy the colour of the object changes from a dull red to white, emitting visible light. Light can be generated in this manner by heating material such as metal wire to a level where light is given off. EXAMPLE: Light bulbs.

Interaction of Light with Matter

A green object owes its colour to the ability of that piece of matter to absorb all other wavelengths apart from green. The colour of matter is due to the wavelengths of light that are reflected by individual pieces of matter. A black object does not reflect any light and therefore absorbs all light falling upon it.

The penetration of matter by light depends upon the nature of the matter. Matter can be described as: transparent, translucent and opaque according to light penetration. Firstly transparent material allows light to move freely through it, so that a clear image emerges from the material. EXAMPLE: glass. Translucent matter alters the path of light and causes light to emerge in a diffuse non-ordered manner so that a clear image cannot be seen through it. EXAMPLE: Frosted glass. Opaque matter absorbs all light waves and therefore does not emit light. EXAMPLE: This book. Some matter can also reflect all light from its surface, that is, turn back light from the surface. EXAMPLE: Mirrors.

Lenses

The size and shape of transparent matter influences the movement of light through it. This property is used in the functioning of curved

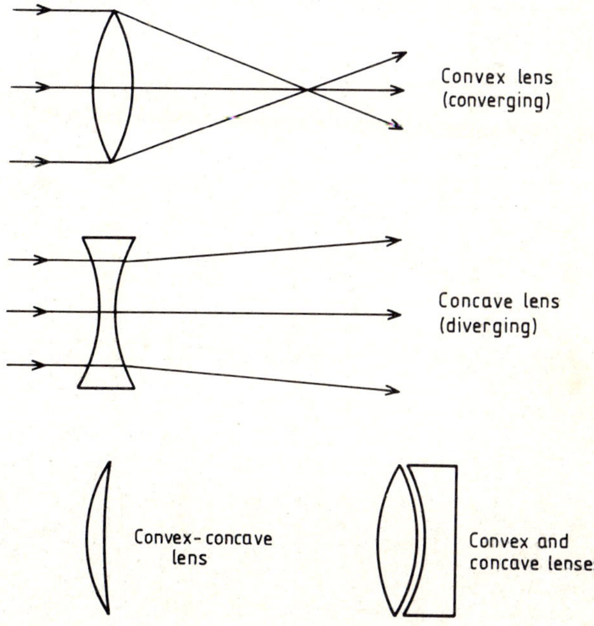

Figure 5.15 Examples of a convex and concave lens and a combination of these.

transparent objects known as lenses. A lens consists of either a convex or a concave surface or a combination of both (Figure 5.15). They are used to form a proportionately smaller or larger visual replica of an object. This replica is called an image. The image occurs at the point where the light rays from an object meet after passing through a lens (Figure 5.16).

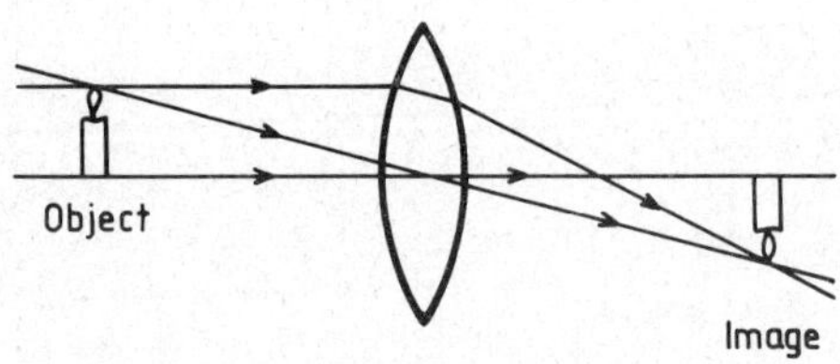

Figure 5.16 Formation of an image by a lens.

A convex lens is able to converge light rays to a point or focus where an image is formed. These lenses are known as converging lenses and they can be used to form an image that is smaller than the object. EXAMPLE: Camera lens. They can also be used to form images that are larger than an object. EXAMPLE: Magnifying glass. The ability of a convex lens to form a reduced or magnified image depends upon the nearness of the object to the lens. Generally, the closer the object is to the lens the larger the resultant image.

Lenses that consist of concave surfaces diverge or spread light rays further apart. These lens are referred to as diverging lens and they can only produce reduced images.

APPLICATION: THE EYE

The eye consists of a large convex lens which magnifies or decreases the size of an image. It forms the image on a layer of photoreceptive cells known as the retina (Figure 5.17). The retina consists of cells that convert the image into nerve impulses which are interpreted in the brain as an image. The colour characteristics of an image are determined by three types of cells that absorb either red, green or blue rays. The colour of an object is determined by the relative number and combination of these cells that are stimulated. An equal stimulation of the three cell types results in the sensation of white light. A loss of one or more types of the cell types results in colour blindness. The lack of one cell type causes an inability to distinguish some colours from others.

Fibre Optics

Fibre optics is the area of study attributed to the movement of light through optic fibres. Optic fibres are flexible glass rods that are capable of conducting light within the rod from one end to the other. The surface of each fibre is coated so that light is trapped within the core of the fibre. Since the light rays are trapped within the fibre they can only emerge from the fibre's core at the opposite end to which they

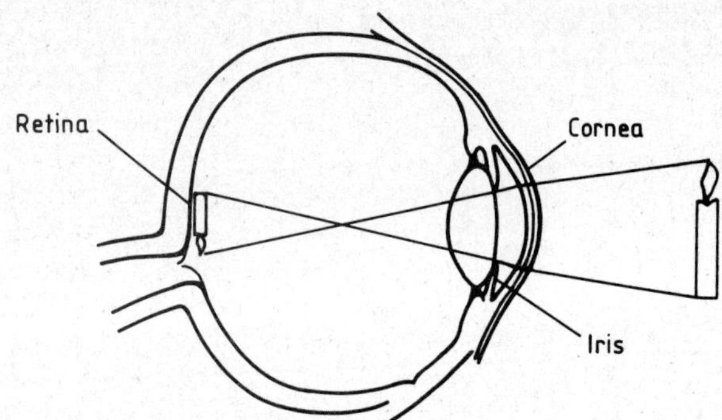

Figure 5.17 In the normal eye, light rays converge to form an inverted image on the retina.

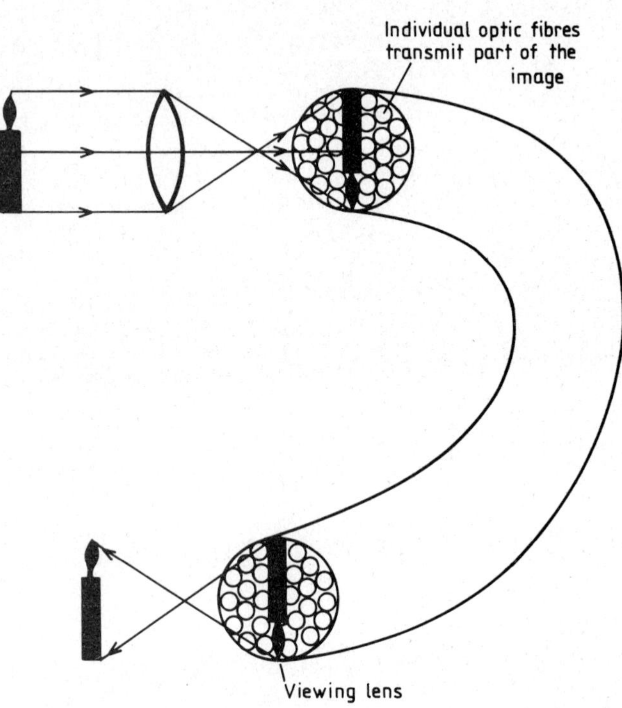

Figure 5.18 An endoscope shows the principle of fibre-optics.

have entered. A group of parallel fibres can transmit a pattern of light which is displayed at one end (image) by each fibre transmitting light from a small area of the pattern to the other end. At the viewing end the image is reassembled (Figure 5.18).

The advantage of instruments that use optic fibres over other optical systems is their ability to reach into previously inaccessible regions.

APPLICATION: ENDOSCOPES

Endoscopes are instruments that examine hollow organs or internal cavities such as the bladder, lung bronchi, gastrointestinal tract. These instruments consist of a long fibre of optically transparent material with a lens system at one end and a viewing head for the observer and/or camera use (Figure 5.18).

The internal examination of patients has been revolutionized by the development of various endoscopes such as the cystoscopes for bladder examination, bronchoscopes for lung bronchi, laproscope for inspection of the peritoneal cavity and the gastroscope for stomach inspection. The addition of small surgical instruments to the end of an endoscope enables biopsy material to be collected without operating on the patient.

ULTRAVIOLET RAYS

These rays contain a shorter wavelength and thus a higher frequency and energy than that of light waves. Ultraviolet rays cause chemical changes when absorbed by certain matter. EXAMPLES: Formation of vitamin D and melanin deposition in the skin, fading of certain dyes. The relatively high energy of these rays can cause damage to any tissue, especially sensitive tissues such as the retina. Most of the ultraviolet rays emitted from the sun are absorbed by ozone in the earth's atmosphere.

APPLICATION: SUNBURN

Excessive exposure to a specific band of ultraviolet rays called ultraviolet B produces blisters and burns on the skin which are commonly referred to as sunburn. Overexposure to ultraviolet lamps will also result in skin burns. Sunburn lotions are available which contain substances that specifically absorb the harmful portion of ultraviolet radiation. The blocking out of these rays enables the remaining rays to develop a person's suntan.

X-RAYS

X-rays have a shorter wavelength than ultraviolet rays and are capable of penetrating most matter including human tissue. This property is used both diagnostically and therapeutically.

APPLICATION 1: DIAGNOSTIC USE

X-rays are absorbed by dense tissues such as bone, whereas tissues such as skin absorb only minimal amounts. X-ray machines emit X-rays over a particular part of the body. Apart from dense tissue such as bone, most X-rays will penetrate the tissues and emerge on the opposite side. These emerging X-rays are recorded on a photographic plate.

Fractures in bones may therefore be detected, since X-rays can pass through the gap formed by the fracture and be recorded on a photographic plate. This X-ray film will contain a slightly different picture when compared with a normal X-ray film of the same region.

The size, shape and function of some soft tissue organs can be determined by administering a dense X-ray absorbing metal such as barium. EXAMPLE: Barium sulphate known as a barium meal is used to outline the gastrointestinal tract.

APPLICATION 2: THERAPEUTIC USE

Exposure of tissue to high doses of X-rays results in chemical changes which may cause cell death or damage to cells, which may lead to tumours or genetic abnormalities in offspring. The ability of X-rays to destroy cells can be used to advantage in X-ray therapy. High doses of X-rays are fired in a fine beam at a tumour and so kill the tumour cells and any healthy neighbouring cells. X-ray therapy has an advantage over radioisotope therapy as the former can be used in regions where an implant is not practicable.

APPLICATION 3: ASSOCIATED HAZARDS

All cells can be damaged or destroyed by X-rays, the most sensitive of which are rapidly dividing cells such as cancer and foetal cells. Pregnant women should not work in areas where a chance of X-ray exposure exists.

As a result of the hazards of X-ray exposure, it is necessary for people who work in such areas to have effective protection. High X-ray absorbing materials such as lead and concrete are used as a protective barrier in order to minimize unnecessary exposure.

Staff who work in areas where X-rays are used need to wear a photographic film which is frequently checked to determine the amount of X-ray exposure. In the ward, mobile X-ray machines should be handled with great caution to ensure that patients or other staff are not in the line of the X-ray emission.

GAMMA RAYS

Gamma rays are high energy, short wavelength waves emitted from radioactive isotopes. These rays are similar to X-rays in that they are dangerous to man. Their property of causing cell death is used in cancer therapy (see 2.10).

5.5 Heat

Heat or thermal energy results from the vibration of atoms and molecules. Any increase in the vibration of atoms results in an increase in the generation of heat energy by these particles. Since vibration of particles is energy of movement, the greater the thermal energy of particles, the greater is the kinetic energy of these particles.

MEASUREMENT OF HEAT

Temperature is a measure of the intensity of heat in an object, that is, the frequency of vibration of particles. Temperature indicates the degree by which particles are vibrating, it does not specify the number of particles that are vibrating. Temperature is only a measure of heat intensity and not the quantity of heat. EXAMPLE: The intensity of heat in a relatively small object such as a match can be identical to a log fire. If in both the match and the log fire, all of the particles are vibrating at the same rate, then the intensity of heat will be identical for the match and the log fire although the quantity of heat will be different.

Temperature is measured in units referred to as degrees. Three scales have been devised to measure temperature: Farenheit, Celsius or Centigrade and Kelvin. Since temperature is a measure of particle vibration, zero degrees should indicate a temperature where no vibration occurs. The more recent Kelvin (K) scale uses this basis for determining zero degrees. Unfortunately, the older scales of Celsius (°C) and Farenheit (°F) base zero degrees on the zero vibration of pure water and water saturated with salt respectively. Substances such as oxygen are still vibrating at these temperatures and thus these scales do not represent the temperature at which there is no vibration of particles in all substances. The temperature at which no particles vibrate is referred to as absolute zero which is recorded as 0 K, -273°C and -460°F. The relationship of these scales is described in 3.2. The units used to measure the quantity of heat are the kilojoule (S.I. unit) and the kilocalorie.

TRANSFER OF HEAT

Heat flows from matter with a higher temperature to matter with lower temperatures. Thermal energy causes particles to increase their rate of vibration, thus generating further thermal energy. The transfer of heat to other matter occurs by either conduction, convection or electromagnetic radiation.

Conduction
This is the transfer of heat by particles interacting with nearby matter that contains less thermal energy. The heat spreads to the adjacent particles causing these particles to increase their vibrations and thus their thermal energy. EXAMPLE: Element on a stove heating pots.

Convection
In convection, heat is transferred by the flow of a liquid or gas with a higher thermal energy to a colder region of matter. EXAMPLE: A hot wind.

Electromagnetic Radiation
Matter with high thermal energy can emit energy in the form of electromagnetic radiation. Particles that come in contact with electromagnetic rays absorb energy from these rays and this results in the particles vibrating more frequently. Heat is thus transferred from the rays to the absorbing matter. Electromagnetic radiation can transfer heat over large distances. EXAMPLE: The sun transfers heat to the earth via electromagnetic radiation.

5.6 Sonic energy

Sonic energy or sound is the vibration of particles in the form of waves through matter. Sound waves can only be transmitted through substances that allow their particles to vibrate, thus they cannot be transmitted through a vacuum. Sound travels in a similar manner to a wave that has been generated by throwing a rock into still water.

Whenever matter vibrates sound is produced. The ability to detect this sound depends upon the nature of the vibration wave. The number of vibrations per second that pass a given point is termed the frequency. The unit of frequency is the hertz which is symbolized as Hz.

The ear is able to convert sound waves into nerve impulses in the range of 20 to 20,000 Hz. The conscious perception of any particular frequency is termed the pitch which may not be the true sound frequency. Pitch varies from the true frequency at very high and very low frequencies. Loudness is the amount of energy that is detected by the ear for a given frequency. This is a subjective term, that is, what is loud to one person may not be to another. Over-stimulation of the ear with excessive sound results in pain which may lead to temporary or even permanent deafness.

Extremely high frequencies can be used for diagnosis and therapy. These frequencies can be up to 800,000 Hz and are referred to as ultrasound.

APPLICATION 1: ULTRASOUND DIAGNOSIS

Sound waves travel at different speeds in different tissues. This property is used in composing pictures of sound that have been reflected by different tissues. These pictures are produced onto television screens or can be photographed for a permanent record. Ultrasound is used for the examination of foetal positions, the heart and the brain. The use of ultrasound in determining the position of the foetus and the placenta is an important diagnostic aid since ultrasound does not damage the sensitive foetal tissue.

APPLICATION 2: ULTRASOUND THERAPY

Ultrasonic beams can penetrate skin to a depth of three to four centimetres. The adsorption of these waves by the skin results in a heating of the tissue, and a decrease in swelling and pain. It is thus useful for a variety of traumatic, inflammatory and muscular conditions.

5.7 Chemical energy

Energy is required to bind atoms together. This energy is known as chemical energy. Some atoms require large amounts of energy to be linked to other atoms and so form high energy bonds. Breaking these bonds results in the release of large amounts of energy, since a large quantity of energy was used to form them. When these bonds are broken the energy released can be converted into other forms such as mechanical energy in muscular contraction.

All of the energy used in the body is derived from the chemical energy stored in chemical bonds of food. The breaking of bonds releases energy which is then either used to form other bonds or is converted to other forms of energy. Metabolism is the body process whereby energy is transferred from the breakdown and formation of chemical bonds. In Unit 8 you will read about metabolism and how most substances can be broken down to produce energy for body functions such as muscular contraction, nerve function and the manufacture of other substances.

5.8 Mechanical energy

Usually energy associated with moving parts is known as mechanical energy. EXAMPLE: The movement of bones by muscles is a form of mechanical energy. This energy and related terms are described in Chapter 7.

Summary

Matter is anything that occupies space and has some mass. To move matter requires energy.

Potential energy is energy stored in an object as a result of its position, chemical composition or stretching and compression of elastic material. Kinetic energy is energy of action. These two kinds of energy can be converted into each other, and be manifested in different forms. The common forms that are involved in nursing, body function and diagnostic and therapeutic equipment are:

1. Electrical energy results from the attraction and movement of charged particles. It is involved in electrolyte balance, nerve function, instrument function and the recording of body electrical events.
2. Radiant energy or electromagnetic radiation consists of the movement of waves that are known as photons. The wavelength of these waves varies from very small to very large. The properties of electromagnetic radiation are dependent upon the wavelength, and it is for this reason that electromagnetic radiation has been categorized according to wavelength. Categories of waves are named in descending order of length as follows: radiowaves, infra-red rays, light, ultraviolet rays, X-rays and gamma rays.
3. Heat or thermal energy results from the vibration of atoms and molecules.
4. Sonic energy or sound is the vibration of particles in the form of waves through matter. It is involved in hearing, ultrasound diagnosis and therapy.
5. Chemical energy is required to bind atoms. The breaking of these bonds can release energy for body function.
6. Mechanical energy is associated with moving parts such as bones and muscles.

Care must be taken when dealing with all forms of energy since exposure to excessive amounts in any form can be harmful, even lethal.

Chapter 6

Physical states of matter

Objectives

At the completion of this chapter the student should be able to:
1. relate kinetic energy to the different physical states of matter,
2. describe the properties of solids, liquids, gases and fluids,
3. explain how matter changes from one physical state to another,
4. describe surface tension.

6.1 States of matter

Matter can exist in three states or phases which are referred to as solids, liquids and gases. The amount of relative movement of atoms or groups of atoms determines the atom's state. Atoms with the lowest kinetic energy are solids whereas those with the greatest amount of kinetic energy are gases. Solids, liquids and gases each contain various properties which can be either similar or different to the other states. These properties are determined by the amount of kinetic energy that is contained within an atom or group of atoms.

6.2 Solids

In solids, the atoms are maintained in a relatively fixed position close to adjacent atoms. These atoms are held in position by attractive forces operating between adjacent atoms. EXAMPLE: Ions are held close together by the force of attraction of opposite charges that exists between cations (positive charge) and anions (negative charge). The distance between atoms in solids is minimal, thereby preventing individual atoms from being compressed or pushed closer to adjacent atoms. Solids are therefore incompressible.

In solids, the attractive forces between atoms or groups of atoms are greater than the kinetic energy of these atoms. This difference results in atoms being unable to move with respect to other adjacent atoms, thus resulting in a well-defined shape for a solid (Figure 6.1).

The degree of attraction between different atoms or groups of atoms varies according to each element. Distinctive structures, such as crystals, result from the particular attractive forces of the elements of which they are composed.

When energy such as heat is added to a solid, the energy is converted into kinetic energy. A liquid is formed when the kinetic energy of atoms is greater than some of the weaker forces that link the atoms. This process of imparting sufficient energy into a solid to form a liquid is known as melting. The rapid addition of large amounts of energy can result in a solid forming a gas. This process is called sublimation. EXAMPLE: When solid carbon dioxide (dry ice) is exposed to the relatively high room temperature this solid is converted directly into gas.

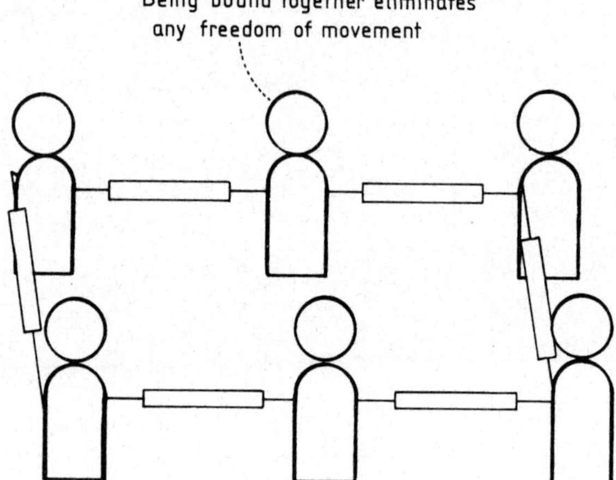

Figure 6.1 For a solid the distance between each atom is fixed and therefore a fixed shape results. People in the figure are held by bars and can therefore only be moved as a total group, with no difference occuring between their relative positions.

6.3 Liquids

The atoms of a liquid contain enough kinetic energy to enable limited movement. The available kinetic energy is insufficient to allow all atoms

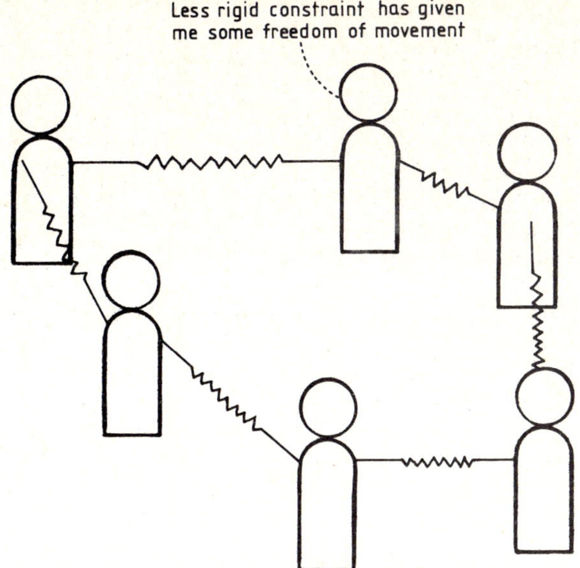

Figure 6.2 In a liquid the distance between each atom may vary in a manner similar to people being joined or linked by large elastic bands (these elastic bands can be broken and reformed).

complete freedom of movement. The quantity of kinetic energy is able to break the weakest attractive forces so that limited movement can occur (Figure 6.2). The variation in position of atoms results in liquids having no characteristic shape. The prevention of atoms escaping from their neighbours causes liquids to maintain a constant volume. The restricted flexibility of a liquid enables it to take the shape of the container in which it is placed.

The close proximity of atoms in a liquid results in a low compressibility, that is, a liquid is relatively incompressible when compared with gases.

SURFACE TENSION

Surface tension is the pull of the surface particles of a liquid into the body of the liquid. Atoms within a liquid are surrounded by a balance of attractive forces. Atoms on the surface of a liquid do not have these attractive forces surrounding them. This results in the greater attractive forces pulling these atoms towards the body of the liquid than those forces between the liquid and gas.

Liquids attempt to contract to the smallest possible area. This can be seen when drops form as liquids come in contact with solids. EXAMPLE: Rain on a windscreen. In gases, a liquid will either form bubbles or droplets. EXAMPLE: Soap bubbles and droplets from a dropper respectively.

The strong attractive forces of solids causes liquids to adhere to a solid's surface, since the attractive forces of solids is greater than the attraction of other

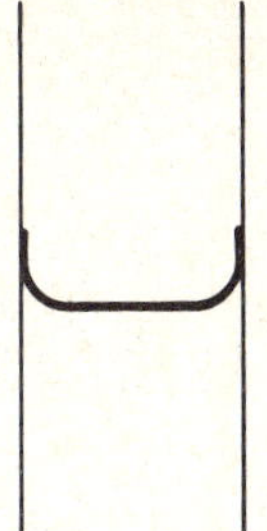

Figure 6.3 A meniscus is formed by the wall's strong attractive forces for the liquid.

liquid atoms. In a narrow piece of tubing this results in an indentation being formed on the surface of the liquid (Figure 6.3). This indentation is known as a meniscus and results from the surface of the liquid being pulled along the walls of the tubing. This movement of a liquid along the walls is often referred to as capillarity.

APPLICATION: SURFACTANT

A surfactant is any substance that reduces the surface tension of a liquid. In the lungs, a pulmonary surfactant is released to reduce the surface tension of the pulmonary liquids, thereby contributing to the elastic properties of lung tissue. A deficiency of pulmonary surfactant results in a collapse of the sacs within a lung.

In premature infants the amount of surfactant present in the lungs may not be enough to enable correct lung function. Lack of pulmonary surfactant in newborn babies is referred to as respiratory distress syndrome or hyaline membrane disease.

6.4 Gases

Gases are formed by imparting more energy into a liquid by a process termed evaporation. The kinetic energy of atoms is now at a level where the attractive forces between atoms or groups of atoms are broken. The high kinetic energy results in the independent movement of atoms or groups of atoms in a random motion (Figure 6.4). Boiling is a special form of evaporation where liquid is converted into a gas via the formation of bubbles within the liquid.

Gases thus have neither a defined volume or shape and occupy the entire space available in a container. Particles continue to move within the container by deflecting off the walls of the container (Figure 6.4).

The collision of gas particles with the wall of a container results in pressure being exerted on the walls of the container. Pressure is also exerted as a result of gas particles trying to escape from a container. If the number of particles in a given volume, or the kinetic energy of particles is increased, the frequency with which these particles

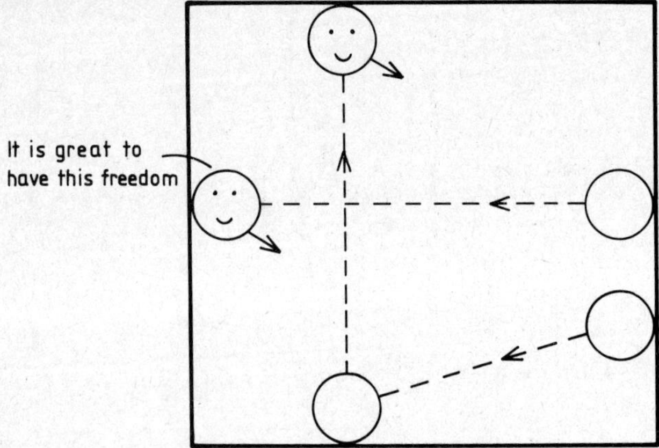

Figure 6.4 In a gas the distance between atoms is random, as is the location of any particular atom in a container. The high kinetic energy of gas atoms enables them to completely occupy the container by deflecting off the walls of the container.

collide with the walls of the container will be increased thus exerting a greater pressure.

6.5 Comparison of the physical properties of solids, liquids and gases

The main differences and similarities between solids, liquids and gases are summarized in Table 6.1. Note that gases and liquids contain many similar properties. The large number of similar properties require a collective term for liquids and gases. A fluid is a substance that takes the shape of its container and can be a liquid or a gas.

6.6 Changes of state

A substance can change from one state to another by gaining or losing kinetic energy. The gaining of kinetic energy by atoms results in a change from either a solid to a liquid, a solid to a gas or a liquid to a gas. When atoms lose kinetic energy they change from either a gas into a liquid, gas into a solid or a liquid into a solid.

A change of state is usually reversible so that no chemical alteration occurs with a substance when it changes from one state to another, that is, the amount of kinetic energy is the only variable. EXAMPLE: Heating ice causes water to be formed whereas cooling water results in the formation of ice.

Table 6.1 Comparison of physical properties of solids, liquids and gases

Property	Solid	Liquid	Gas
Fixed volume	Yes	Yes	No
Compressible	No	No	Yes
Rigidity	Yes	No	No
Characteristic shape	Yes	No	No
Takes the shape of its container	Sometimes	Yes	Yes
Completely fills the container	No	No	Yes
Flows along tubes	Sometimes	Yes	Yes
Can be poured	Sometimes	Yes	Yes

Sometimes a change of state is irreversible since the chemical structure of a substance is altered. EXAMPLE: Heating wood can result in ash and gas or smoke. Cooling smoke and ash will not reform wood.

Summary

Matter exists as solids, liquids or gases. These states of matter differ by the amount of movement of atoms or groups of atoms. This relative movement of atoms is due to the quantity of kinetic energy present.

1. In solids the atoms contain small amounts of kinetic energy and thus they remain in a relatively fixed position close to adjacent atoms.
2. In liquids the atoms contain a greater amount of kinetic energy which enables the atoms to have a limited movement with respect to adjacent atoms. This movement results in liquids being able to pour and flow, that is, exhibit properties of a fluid.
3. Gases contain the greatest amount of kinetic energy. The high kinetic energy results in the independent movement of atoms or groups of atoms. Gases are also fluids since they can flow and be poured.

The difference in the attractive forces of solids with liquids and gases with liquids, that is, surface tension, results in the liquid attempting to contract to the smallest possible area.

Chapter 7

Mechanical properties of matter

Objectives

At the completion of this chapter the student should be able to:
1. explain the terms associated with motion,
2. describe the term force and relate force to nursing and body function,
3. explain the actions of gravity,
4. explain the terms friction, work power, inertia, momentum and elasticity.

Mechanics is a branch of science that is concerned with the motion and the maintenance of equilibrium of objects. The human body relies upon the principles of mechanics for: body movement, the movement of substances into and out of the body and the movement of substances within the body.

7.1 Terms used to describe motion

VELOCITY

Velocity is the distance that something has travelled divided by the time taken. Velocity is a measure of motion in a specified direction. Speed is the motion of an object without reference to any direction and is usually used when describing how fast something is moving.

Velocity is an example of a vector since it has both magnitude and direction. As speed has only magnitude, it is classified as a scalar.

ACCELERATION

When velocity is altered this change is referred to as acceleration. Acceleration is the change of velocity divided by the time taken. A particle with an acceleration of ten metres per second per second means that the particle will alter its velocity by ten metres per second every second. Acceleration has direction and therefore requires a description of direction when referring to it. As acceleration has magnitude and direction it is a vector.

7.2 Forces

A force is any influence that either changes the position, the state of rest or the motion of an object. Force is measured in the S.I. unit called the newton (N).

VECTORS

The net force acting on an object is determined by taking the sum of all of the individual forces. It is necessary to account for the direction and magnitude of each force since these factors can alter the influence of the net force. Force is another example of a vector, that is, it has both direction and magnitude.

The relative influence of individual forces on the magnitude and direction of a net force or the resultant force can be determined by drawing the individual forces to scale.

APPLICATION 1: LIFTING A PATIENT ONTO A BED

Lifting a patient from a bedside chair to a bed requires two main forces. Firstly, a horizontal force is required to move the patient from the chair to the bed (Figure 7.1a). A second force is required to lift the patient to the bed height (Figure 7.1b). The net force developed by the individual forces or the resultant force is determined by placing the individual forces into a "nose to tail" diagram (Figure 7.1c) and then joining the tail of one force to the nose of the last force (Figure 7.1d).

APPLICATION 2: RESULTANT FORCE OF MUSCLES

Individually, muscles are limited in the direction of the force that they can produce. However, when their actions are combined, a large number of resultant directions of muscle action are now possible. A desired movement can be achieved by the body summing muscle actions, the combined action produces the desired resultant force in both magnitude and direction.

FORCE OF GRAVITY

Gravity is the universal force which results in the attraction of all objects to each other. Those objects with a greater mass pull those with a lesser mass towards them. The large mass of the earth exerts a

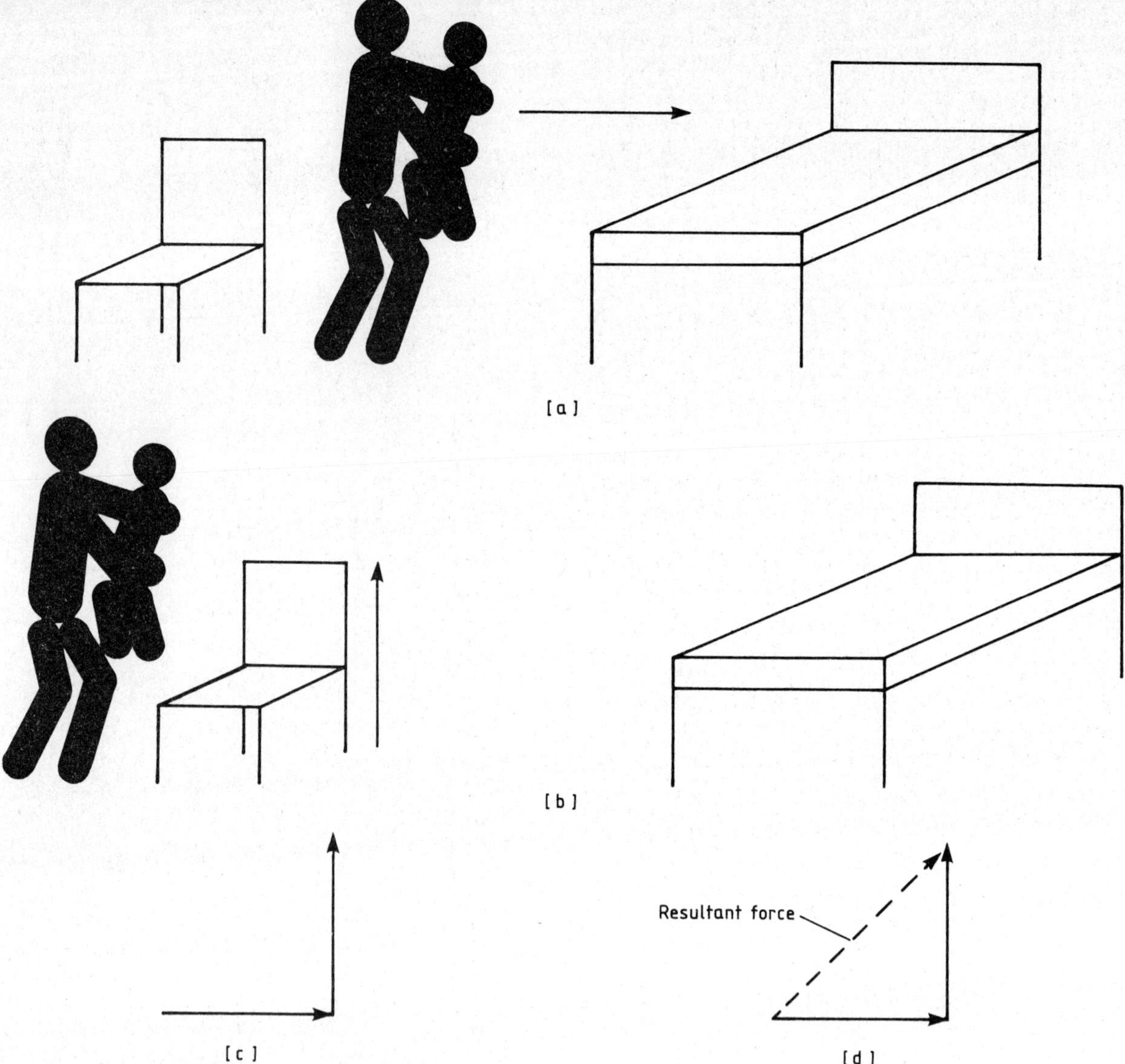

Figure 7.1 (a) A horizontal force is necessary to shift the patient from the chair to the bed. (b) A vertical force is needed to elevate the patient to the height of the bed. (c) These forces can be represented as arrows drawn nose to tail. (d) The resultant force is determined by joining the tail of one force to the nose of the other.

large force attracting matter to the earth's centre. The relationship between mass and its force of attraction to the earth is expressed as weight.

weight = mass × acceleration due to gravity (Eqn 7.1)

The unit of mass is the kilogram which is usually also synonomous with weight.

The movement of matter away from the earth requires a force to oppose gravity, whereas the movement of matter towards the earth will be assisted by gravity.

APPLICATION: GRAVITY AND BLOOD CIRCULATION

Gravity exerts a uniform effect on the blood circulation of a person lying in a horizontal position. If this person now moves to an upright position, gravity will exert different influences on the blood circulation. The regions of the body above the heart can now only receive blood from the heart if it is moved against the force of gravity. To enable blood to reach these portions of the body an increased force is developed by the heart. This force has to be of sufficient strength to overcome the effect of gravity. Failure of the heart to develop the necessary force to supply sufficient blood to the head and brain can result in fainting, and is known as postural hypotension.

Blood flows to the lower parts of the body with the assistance of gravity. Gravity however, interferes with the return of blood from the lower extremities to the heart. This is evident by the appearance of fluid and thus swelling in the feet of people who stand for long periods of time.

The effect of gravity can be used to advantage in order to decrease bleeding. A cut hand raised above the head will receive less blood than when it is located below the heart. This is due to the heart only pumping with sufficient force to overcome the effect of gravity for the distance to the head, and not the additional distance to the raised hand.

CENTRE OF GRAVITY

Since gravity acts on all matter, an object will have this force acting on all its constituent particles. The resultant of the gravitational force on each individual particle is termed the centre of gravity. In a uniformly shaped object with a uniform mass, the centre of gravity is the centre of that object (Figure 7.2a). In objects that contain either varying amounts of particles or which are not uniform in shape, the centre of gravity will be located towards the area of greatest mass (Figure 7.2b). This position of the resultant or net force exists because more gravitational forces are acting in the region of greatest mass.

The position of the centre of gravity is important since it determines the position of the necessary forces that are required to oppose an object falling from the influence of gravity. The net effect of gravity on an object is exerted through a vertical line that passes from the centre of gravity to the earth (Figure 7.3). In Figure 7.3 the resultant force of gravity is exerted onto the opposing force located at position 2. This object will remain balanced or in equilibrium at this position if the magnitude of the force equals the force of gravity. Equilibrium exists in any object when all the forces and vectors total zero and it appears that no force is acting on the object. The position of the force at 1 results in the object falling to the right, whereas the force at position 3 results in the object falling to the left.

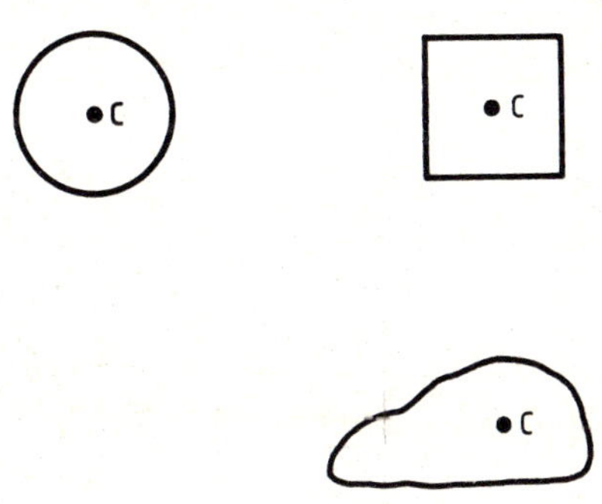

Figure 7.2 (a) In objects with uniform shape and mass the centre of gravity (c) is located at the centre of the object. (b) In non-uniform objects the centre of gravity (c) is located towards the area of greatest mass.

Figure 7.3 The ball is only balanced at position 2 where the force directly opposes the net force of gravity. At position 1 the ball falls to the right and at position 3 the ball falls to the left.

Stability

An object can move horizontally and not fall, as long as an opposing force acts in line with the centre of gravity. The ability of an object to move and not fall is restricted to the boundaries of action of that opposing force. EXAMPLE: A ball will not fall off a table until it moves to the boundary or edge of the table where no opposing force to gravity exists. The region of support for holding an object in a balanced position is known as the base of support.

An object has stability when it resists falling. This occurs when the base of support for that object is below or through the centre of gravity (Figure 7.4). The wider the base of support the greater is the stability of an object, since a small base of support can only support an object that has a small degree of movement.

Figure 7.4 A book will remain on a table as long as the centre of gravity (c) is over the base of support.

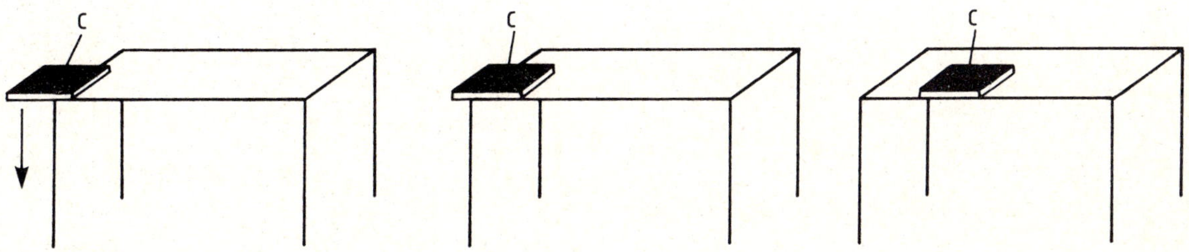

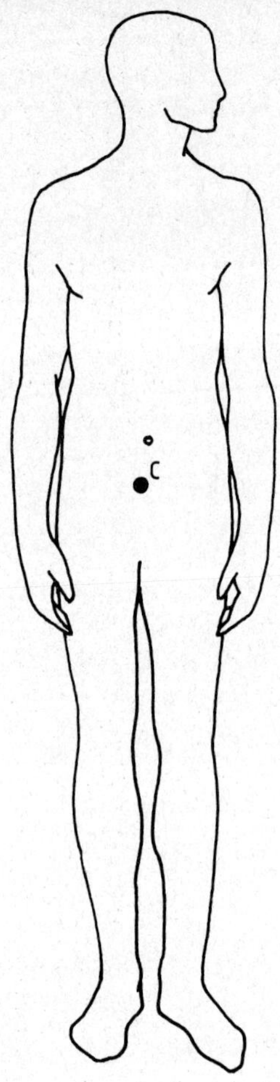

Figure 7.5 Location of the centre of gravity in humans.

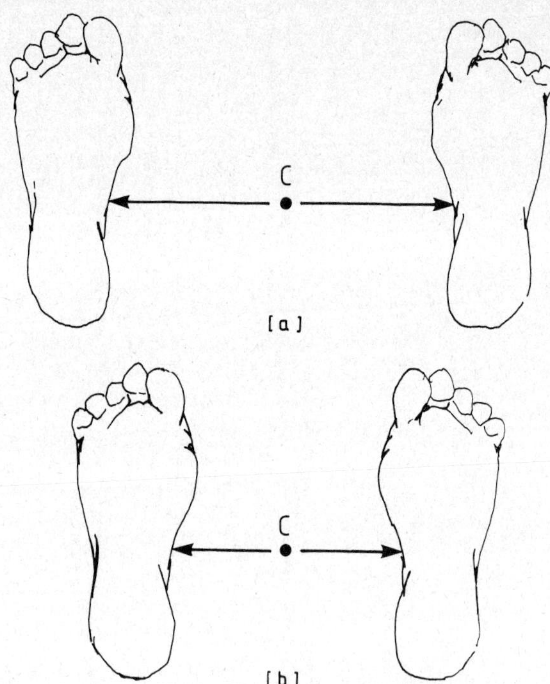

Figure 7.6 A person (a) with a wide base of support can move their centre of gravity a relatively large distance when compared with a person (b) with a narrow base of support.

Centre of Gravity and Man

Generally, the centre of gravity in man is located in the pelvic region near the base of the spinal column. When sitting or standing with an erect posture, the centre of gravity is in line with the spinal column and in the pelvic region (Figure 7.5). The location of the centre of gravity in this region results in the least strain on the body muscles. The spinal column and large thigh muscles are used to oppose the action of gravity.

Greatest stability occurs when a person has a wide base of support. A person who has their feet separated by a greater distance than another person has the ability to shift their centre of gravity to a greater extent, that is, the former can have a greater variety of postural changes without falling over (Figure 7.6).

Changes of posture can alter the centre of gravity. EXAMPLE: When bending forward the centre of gravity shifts forward according to the degree of bending. Other factors that can alter the centre of gravity and require alterations in the placement of feet are pregnancy, ascites (accummulation of fluid in the abdominal cavity), loss of a limb and carrying a heavy object.

APPLICATION: LIFTING

When lifting any object the powerful thigh muscles should be used rather than the back muscles. The object to be lifted should be held close to the body so that the centre of gravity does not shift substantially from the normal upright position. Bending the back excessively or not lifting objects close to the body will result in the relatively weak back muscles being used to oppose the displaced centre of gravity. This results in back strain which would not have occurred if the object had been lifted close to the body with the back straight and the knees bent. In this position, the thigh muscles are placed to oppose most of the extra effects of gravity.

SPECIFIC GRAVITY

Specific gravity of a substance is the weight of a certain volume of substance compared with the weight of an equal volume of water. Specific gravity can also be defined in terms of density, where density is the amount of mass in a given volume. A substance with a greater density than another substance will contain a greater amount of mass than the latter substance in the same volume. Specific gravity of a fluid is the density of a fluid compared with an equal volume of water.

Water is selected as the reference substance since

the relationship between the volume and weight of water, i.e., density, is one. EXAMPLE: A volume of ten millilitres of water weighs ten grams:

$$\text{density} = \frac{\text{weight}}{\text{volume}} = \frac{10}{10} = 1 \quad \text{Eqn (7.2)}$$

Specific gravity is calculated as follows:

$$\text{specific gravity} = \frac{\text{Density of weight of substance}}{\text{density or mass of water}}$$

EXAMPLE:

$$\text{Urine} = \frac{1.010}{1.000} = 1.010$$

APPLICATION: VARIATIONS IN URINE SPECIFIC GRAVITY

Urine consists of mainly water and a few substances such as urea and sodium chloride. The normal range of the specific gravity of urine is between 1.002 to 1.040 (1002 to 1040). Note that the specific gravity of pure water is 1.000, and thus the closer the reading is to 1.000, the more diluted is the fluid. In the same way, the further that the reading is away from 1.000, the greater is the amount of other substances present in the water. A reading lower than 1.002 (1002) indicates a dilute urine and may be due to diuresis. An abnormally high urine specific gravity indicates a concentrated urine. This may be due to dehydration causing excessive water retention or from the excessive excretion of a particular substance.

FRICTION

Friction is a force that resists the movement of one object over another. The amount of friction depends upon the nature of the surfaces and the weight of the moving object. A relatively rough surface such as concrete will exert greater friction than a smooth surface such as ice. An object requires a greater force to move over soil than over ice, that is, a greater force is needed to overcome the friction. More friction results from heavy objects moving over a surface than lighter objects over the same surface.

Two types of friction can be encountered when moving a stationary object. Static friction results from the rough edges of an object interlocking with surface irregularities of another object (Figure 7.7). Static friction exists between most objects since very few objects are perfectly smooth. Kinetic friction is the friction exerted by a surface on an object that is moving. This friction is less than static friction as a greater force is required to move an object out of an interlocked position (static friction) than to "bounce" an object along from one minute ridge to the next (kinetic friction).

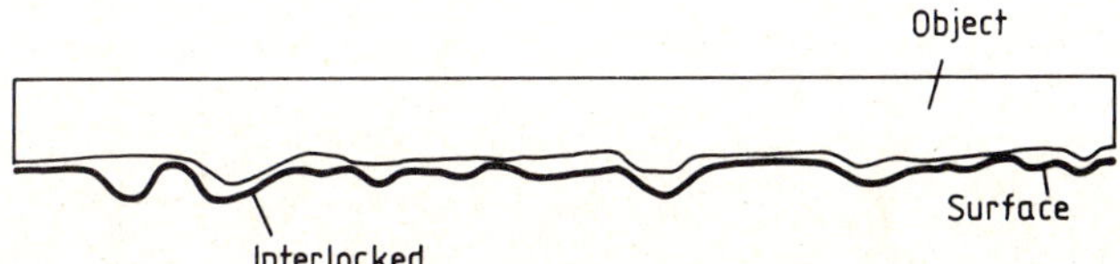

Figure 7.7 At rest an object settles in a position where the surface of the object interlocks with another object's surface.

APPLICATION: INSERTION OF A NASOGASTRIC TUBE

In the process of inserting a nasogastric tube, far less force is required when the tube is inserted in a continuous motion than when the tube is inserted in a series of discontinuing steps. Since less force is required for a continuous motion, the patient will experience less discomfort.

Friction and Humans

Humans use friction to walk, as sliding results from little friction. Try walking on ice or a highly polished floor! The movement of blood through the vascular system is opposed by friction. This friction can be varied (see 13.3) to control blood supply to different regions of the body as the greater the friction or resistance to flow the slower the speed of blood.

Friction can also exert harmful effects on people. EXAMPLE: Abrasions. The body relies upon bones moving against each other for body movement. The friction between bones is overcome by the use of a low friction substance known as synovial fluid.

TURNING FORCES

The measure of the tendency of a force to rotate an object about a point is referred to as a moment. The magnitude of a moment depends upon the size of the force and its distance from the pivot. The point around which rotation occurs is termed the fulcrum or pivot. Moments are involved in body movements to convert mechanical energy developed

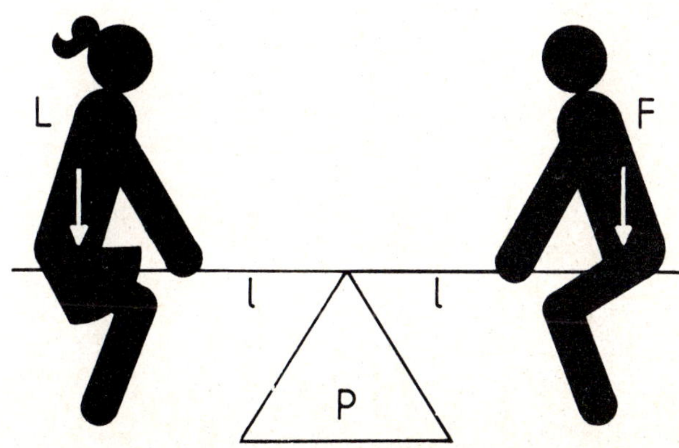

Figure 7.8 Representative diagram of a moment showing the force applied (*F*) to a level (*l*) whereby a load (*L*) is moved about a pivot (*P*).

from muscle–bone combinations, into energy that is used to alter posture and move, or exert a force on other objects.

A lever is a bar which connects an object and a force. There are three different ways of arranging the forces and pivot with respect to each other on a lever. Each of these types of lever is represented by bone–muscle combinations in the body.

All moments consist of a force applied to some point along a lever whereby a load is moved about a pivot (Figure 7.8).

Type 1 lever

The force and load are always located on opposite sides of the pivot. A seesaw is an example of a type 1 lever (Figure 7.8). The tendency of the boy to move the girl and *vice versa* is determined by multiplying the mangitude of the force, that is, their weight and their distance from the pivot. The boy and the girl will remain in a balanced state or equilibrium if the following exists:

boy's weight × distance from pivot = girl's weight × distance from pivot

If the boy exerts a greater moment the girl is forced upwards.

A greater moment on the side of the force, rather than the side of the load, is required to move an object. When the magnitude of a force is sufficient to cause the load to rotate, the moments are known as torque.

Type 2 lever

The load is located between the force and the pivot (Figure 7.9).

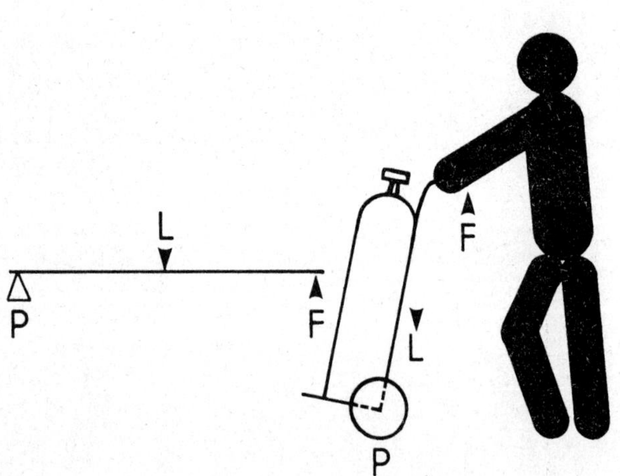

Figure 7.9 Type 2 levers have the load located between the pivot and force. When pushing a gas cylinder the wheel acts as the pivot and the gas cylinder as the load.

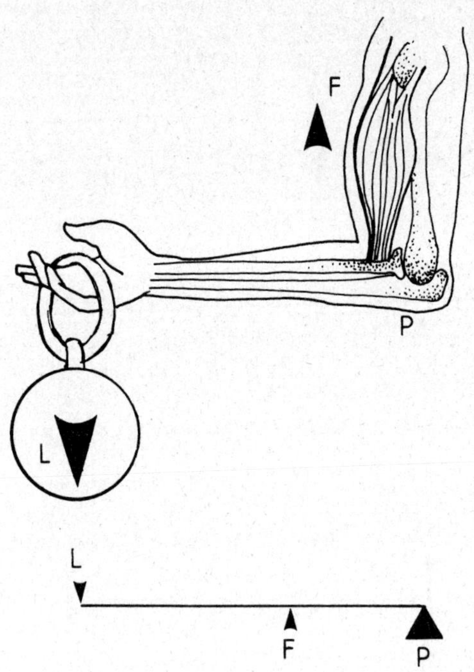

Figure 7.10 Type 3 levers have the force located between the load and the fulcrum. When holding a weight the biceps exerts a force between the weight and the pivot or elbow.

Figure 7.11 The greater the distance of a load from a force the greater is the effectiveness of that force.

Type 3 lever

The force is located between the load and the pivot (Figure 7.10).

With all three types of levers, the greater the distance of the force from the load, the greater the load that can be moved. EXAMPLE: Using a crowbar (Figure 7.11).

7.3 Work

Work is the transfer of energy from one object to another. It is defined as the distance that an object is moved in the direction of a force, multiplied by the magnitude of that force:

work = force × distance moved (Eqn 7.3)

Note that time is not associated with work. A person who pushes a wheelchair from one end of ward to another in several minutes will perform the same amount of work as a person who takes longer.

The S.I. unit for work is the joule (J).

The amount of energy used in performing work does not equal the amount of work done, since some of the energy is converted into heat or other energy forms. Energy may also be used to overcome other forces such as friction. EXAMPLE: When pushing a wheelchair, additional energy is required to overcome friction. The greater the friction between the floor and the wheelchair, the greater is the extra energy required to move the wheelchair a given distance.

Efficient work is only performed when the force exerted on an object is in the same direction as the desired direction the object is to be moved. When the force is not in the desired direction an object is to be moved, some of the force that is being exerted will not contribute to the work being done. EXAMPLE: Pushing a wheelchair with a downward force (Figure 7.12).

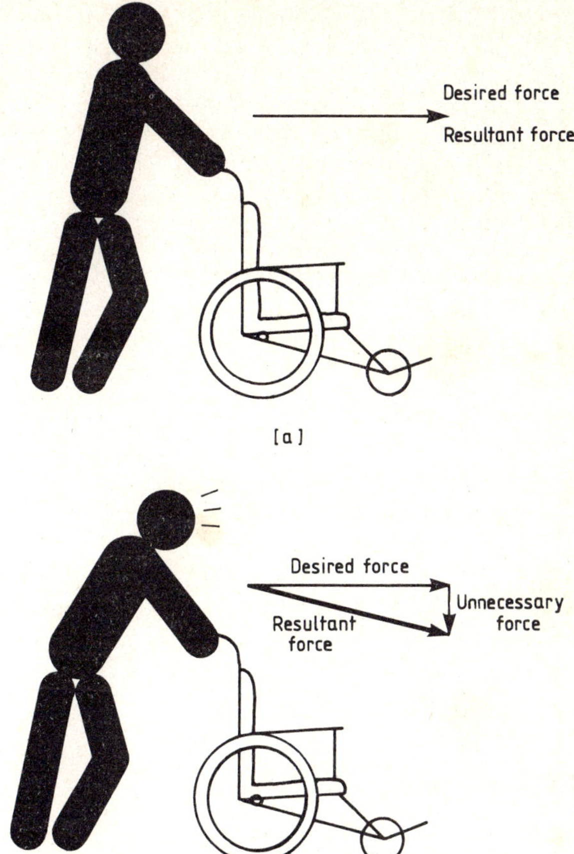

Figure 7.12 (a) The movement is in the same direction as the applied force. (b) The pushing of the wheelchair with a downward motion results in the applied force containing a horizontal and a vertical vector. For a given distance more energy is used in (b) compared with (a).

7.4 Power

Power is the rate of doing work and is therefore different from both work and energy. The units for power are joules per second or watts. Since power is the rate of work being performed, the amount of work done per second determines the power. The faster a given amount of work is performed, the greater is the power.

7.5 Newton's laws of motion

Sir Isaac Newton studied the movement of objects and determined three laws of motion. These laws explain how an object moves and what effect that object has when it collides with another object.

FIRST LAW OF MOTION

This law can be expressed as follows: matter will continue to be either in a state of rest or will continue to move in a straight line with the same speed unless it is acted upon by a net external force.

When an object is at rest or is moving in a straight line at constant speed the object is in equilibrium. Most moving matter has forces such as gravity and friction acting upon it. These forces slow down the motion of matter, eventually bringing the matter to rest.

A heavy object requires a greater force to bring it to rest or change its direction than a lighter object. This relationship between mass and velocity is explained in the terms inertia and momentum.

Inertia

An object with a large amount of mass requires a greater force to initiate motion than an object with a smaller mass. Inertia is the reluctance of a body to start moving after a force has been applied to it and the reluctance of the body to stop once it has begun to move. According to Newton's first law, a moving object's reluctance to move and stop moving is directly related to the object's mass. The greater the mass of a body, the greater is the inertia of that body and the greater is the force needed to alter the motion of that body. Mass is therefore an important factor in initiation of motion and the stopping of motion in any matter.

Momentum

Moving objects with a large intertia require large forces to slow them down. This property is important when considering a collision between two objects and the effect that the collision has on each object's motion. The two factors of mass and velocity are important when considering the effect of a collision between two objects. In any collision it is the combined effect of the mass and the velocity of each object that determines the outcome of the collision. Momentum is the motion of a body as a result of the mass and velocity of the body:

$$\text{momentum} = \text{mass} \times \text{velocity} \qquad \text{(Eqn 7.4)}$$

Newton used the term momentum to explain how much force was required to alter the equilibrium of any piece of matter.

SECOND LAW OF MOTION

Newton's second law can be expressed as follows: the net force acting on a body is equal to the rate of change of momentum of that body. A large car travelling at a low speed can contain the same amount of momentum as a small car travelling at a high speed. A large truck requires a greater breaking force to stop it than a car when both are travelling at the same speed. When two objects collide with each other, the total momentum of the collision is the same, that is:

$$\underset{\text{object 1}}{\text{mass} \times \text{velocity}} = \underset{\text{object 2}}{\text{mass} \times \text{velocity}} \qquad \text{(Eqn 7.5)}$$

This equation shows that if the mass of each object is unchanged in a collision, the velocity is the only factor that can be altered. An object with a large inertia will lose a small amount of velocity in preference to an object with small inertia. The latter has less reluctance to a change in its motion and thus tends to bounce off the object with the larger mass at a greater velocity.

THIRD LAW OF MOTION

This law is expressed as follows: for every action there is an equal and opposite reaction exerted. EXAMPLE: A patient lying on a bed exerts a force on the bed and in turn the bed exerts an equal but opposite force on the patient.

7.6 Pulleys

A pulley consists of a grooved wheel on an axle, with a cord of rope passing along the groove. Pulleys are used to change the direction at which a force is acting and are used in the process of traction. This is the process in which the fractured ends of a bone are kept in position by a system of pulleys and weights (Figure 7.13). The force of gravity on the weights equals the body's opposition to the weights. This balance is important, and any adjustments should only be made after specific instructions have been given.

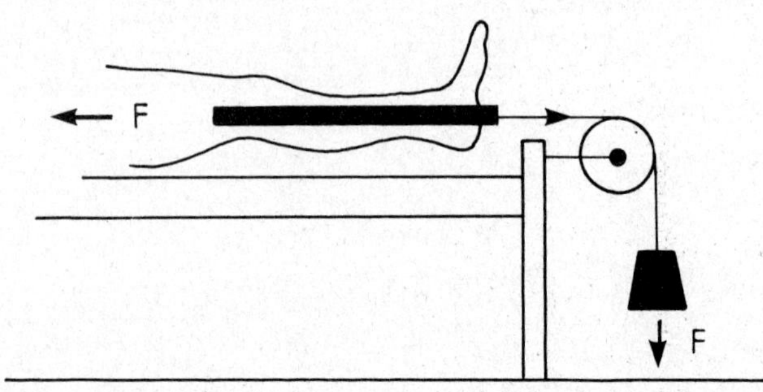

Figure 7.13 Traction.

7.7 Elasticity

Elasticity is the property of matter that returns it to its original shape after a force has altered the shape. The body uses elasticity to recoil stretched lungs and arteries. There are four types of elasticity:

1. elasticity of compression—EXAMPLE: rubber ball,
2. elasticity of extension—EXAMPLE: rubber bands, elastic arteries, thoracic cage,
3. elasticity of torsion—EXAMPLE: the twist in a coiled spring, limited twisting of long bones,
4. elasticity of flexion—EXAMPLE: bending a strip of steel, limited bending of long bones.

There is a limit to the elasticity of any piece of matter, and past this limit the material breaks. The limit depends on the individual elastic properties of the particular material. EXAMPLE: A piece of steel can be bent to a greater extent than an equal length of bone before it breaks.

Summary

Mechanics is the branch of science that is concerned with the motion and the maintenance of equilibrium of objects. The motion of objects in a specified direction is termed velocity. Acceleration occurs when velocity alters with time.

A force is any influence that changes the position, the state of rest or the motion of an object. Since force has direction and magnitude it is an example of a vector. Several forces may constitute a net force. The net or resultant force is determined by adding the individual forces.

Gravity is a force of attraction between objects. The large mass of the earth pulls objects to it. Gravity exerts its net effect on an object at the centre of gravity. Equilibrium exists when an opposing force acts through the centre of gravity. The greater the boundaries of action or base of support of the opposing force, the greater is the stability of the object.

Friction is a force that resists the movement of one object over another.

The measure of the tendency of a force to rotate an object about a point is referred to as a moment. Three arrangements exist by which a force can rotate an object. These are known as type 1, type 2 and type 3 levers.

Work is the transfer of energy from one object to another. It equals the force applied by the distance moved. The rate of performing work is termed power.

The relationship between moving objects is explained in Newton's laws of motion.

The direction that a force is acting upon can be changed with the use of pulleys. This process is used in traction.

Elasticity is the property of matter that returns it to its original shape after a force has altered its shape.

Unit Four

This unit describes how atoms combine and relates this information to the properties of biologically important organic compounds.

Chapter 8

How atoms combine

Objectives

At the completion of this chapter the student should be able to:

1. describe how the structure of atoms relates to the formation of compounds,
2. explain the differences between ionic and covalent compounds,
3. relate the structure of polar molecules to their ability to form dipole–dipole interactions,
4. explain the difference between structural and molecular formulae,
5. determine whether a chemical equation is balanced,
6. define the terms mixture, compound, molecule, polar molecule, radical, valence and chemical equilibrium.

8.1 Mixtures

Mixtures contain two or more elements that are not bound to each other and which may exist in variable proportions. EXAMPLE: The constituents of air such as oxygen, nitrogen and carbon dioxide are not bound to each other and the proportion of these elements in a sample of air would vary in different environments. A mixture's properties result from a combination of the mixture's constituents, whereas a compound's properties are different to the component elements. EXAMPLE: Carbon dioxide is a compound which does not exhibit the properties of carbon or oxygen but has its own particular properties.

8.2 Compounds

A compound is a combination of atoms in a fixed proportion which confers definite physical and chemical properties. EXAMPLES: Water (H_2O), carbon dioxide (CO_2). Compounds can be broken down chemically whereas elements cannot. A molecule is the smallest combination of atoms that can combine together and still exhibit the properties of a compound. EXAMPLE: One molecule of water consists of two hydrogen atoms and one oxygen atom, that is, H_2O.

IONIC COMPOUNDS

Ionic compounds are formed by the exchange of electrons between an electron donor or cation and an electron acceptor or anion. Electrons are donated or accepted in order that the fundamental force in nature of completing the outer orbit of electrons is satisfied. The donation and acceptance of electrons results in the formation of oppositely-charged ions. The difference in potential developed between the ions causes the oppositely-charged ions to combine as a result of the fundamental force of opposite charges attracting. When atoms combine because of differences in electrical charge ionic bonds are formed. The formation of ionic bonds combines the positive and negative charges so that no difference in charge remains, that is, the compound has zero charge. EXAMPLE: In sodium chloride, sodium donates one electron to chloride. The positive charge of the sodium combines with the negative charge of chloride to form sodium chloride which has no net charge (Figure 8.1).

COVALENT COMPOUNDS

Atoms are able to satisfy the force of completing the outer electron shell without the necessary number of electrons being present one hundred per cent of the time. This property allows electrons to be shared between two atoims. The shared electrons spend a proportion of their time under the influence of each of the atoms.

Covalent compounds consist of atoms that complete their outer orbit by sharing electrons with other atoms. Elements with four outer orbit electrons such as carbon find it easier to share electrons than attempt to either gain or lose four electrons. The bonds that are formed by sharing electrons between two atoms are termed covalent bonds. These bonds can be formed between atoms of the same element such as oxygen, or atoms of

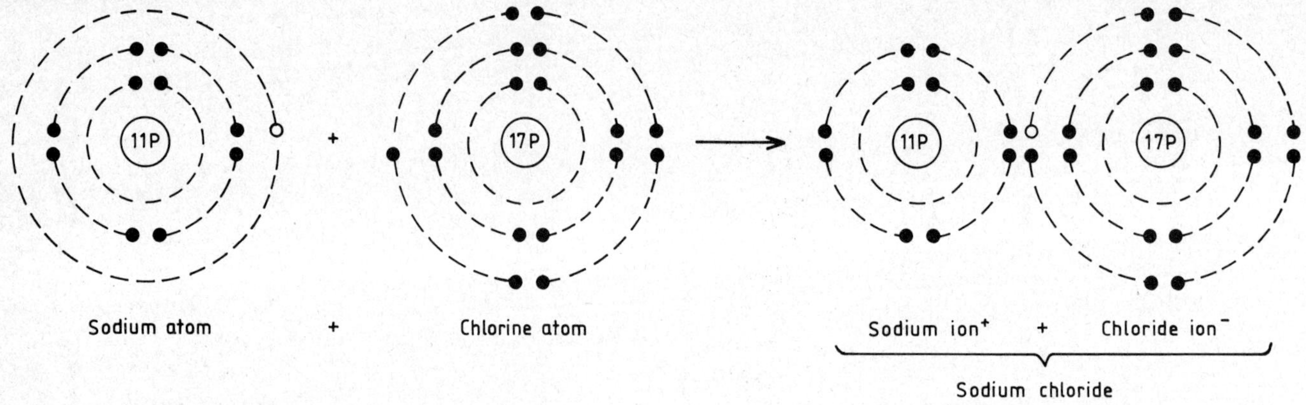

Sodium atom + Chlorine atom → Sodium ion⁺ + Chloride ion⁻ — Sodium chloride

Figure 8.1 Formation of an ionic bond. A sodium atom donates an outer shell electron (o) to a chlorine atom, resulting in a positively-charged sodium ion and a negatively-charged chloride ion. These two ions contain complete outer orbits. The oppositely charged ions are attracted and held together to form an ionic compound with no net charge.

different elements such as carbon and oxygen which form carbon dioxide. Generally covalent compounds are formed from elements with four, five, six or seven electrons in the outer orbit, as these elements attract electrons. The metals with one, two or three electrons in the outer orbit tend to lose electrons and thus are generally unable to share electrons and form covalent compounds. Hydrogen has unique properties and is able to gain or lose its electron or share it with other atoms.

Three forms of covalent bonds can be formed between two atoms. These forms depend upon the number of electrons that are shared between two atoms.

Single Covalent Bonds

In a single covalent bond, one pair of electrons is shared between two atoms. EXAMPLE: Two hydrogen atoms (Figure 8.2a).

Double Covalent Bonds

This bond involves the sharing of two pairs of electrons between two atoms. As a result the outer electron shell of each atom is complete for a proportion of time. EXAMPLE: Two oxygen atoms (Figure 8.2b).

Triple Covalent Bonds

Some elements such as nitrogen, are capable of sharing three electrons with another atom in order that the outer shell is complete for a proportion of time. A triple bond is formed when two atoms share three pairs of electrons between them. EXAMPLE: Nitrogen (Figure 8.2c).

A covalent compound may consist of a mixture of single, double and triple bonds. Each individual atom can form any permutation of covalent bonds as long as the total number of shared electrons does not exceed eight electrons in the outer orbit. EXAMPLE: A carbon atom contains four electrons in the outer orbit and therefore requires four electrons for a proportion of time for the outer shell to be complete. The carbon atom can achieve this by any permutation of covalent bonds as long as an additional four electrons are shared. The possible covalent bond permutations for carbon are:

	4 single bonds	= 4 electrons
1 double bond plus	2 single bonds	= 4 electrons
1 triple bond plus	1 single bond	= 4 electrons
	2 double bonds	= 4 electrons

POLAR MOLECULES

Polar molecules result from an unequal sharing of electrons between atoms. These molecules lie between the extremes of equal sharing of electrons (covalent bonds) and ionic bonds. Some atoms that share electrons with other atoms will have a greater affinity for the shared electrons. The affinity is insufficient to cause the electrons to spend one hundred per cent of their time under the influence of these atoms but is of sufficient strength to hold the electrons in their influence for more than fifty per cent of the time. These electrons are disproportionately shared but not donated to these high electron affinity atoms. Atoms can contain a complete outer shell for less than fifty per cent of the time and still satisfy the force of completing outer orbits.

An atom that gains one electron forms an anion with a charge of 1^-, since each electron has a charge of 1^-. An atom that gains a number of electrons to complete the shell for fifty per cent of the time and loses an equal number of its electrons for the other fifty per cent of the time has a net charge of zero. An

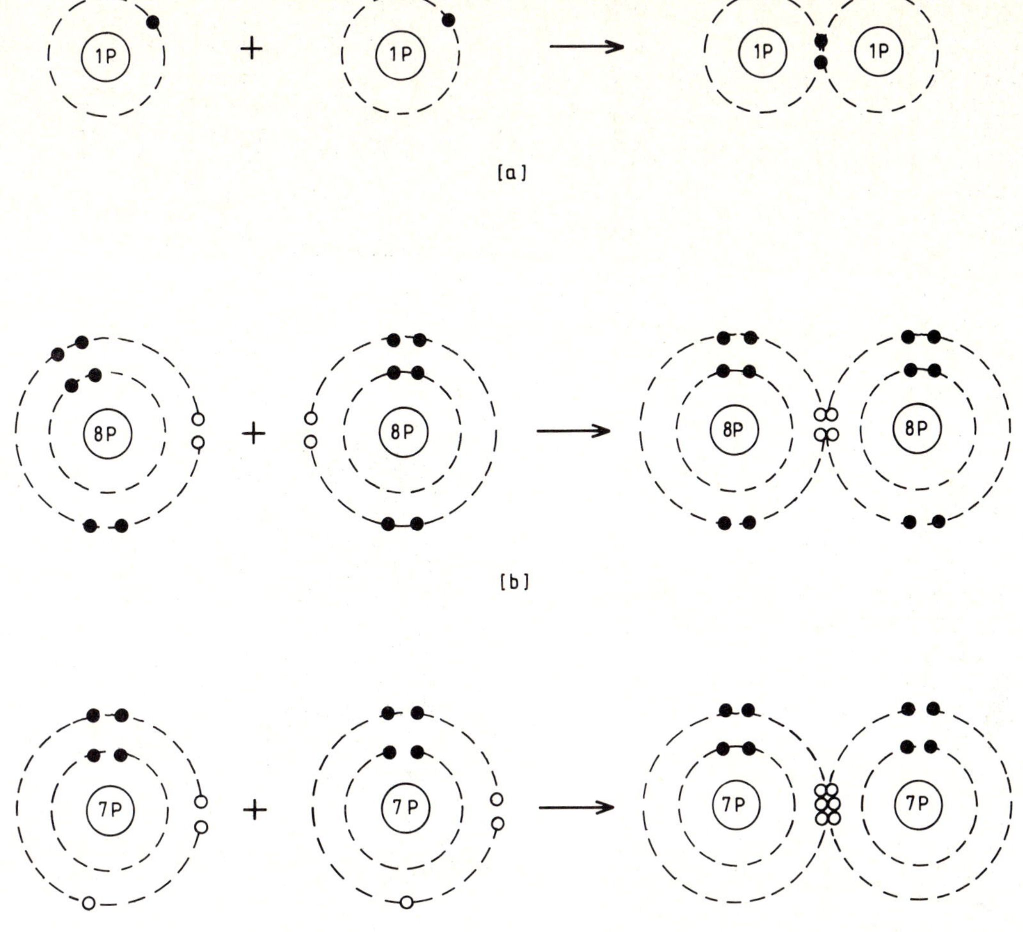

Figure 8.2 Formation of covalent bonds. (a) Two hydrogen atoms combine to complete their outer orbits by sharing their electrons to form a single covalent bond which is symbolized as H· ·H or H–H. (b) A double covalent bond is formed between two oxygen atoms when a pair of electrons from each oxygen atom is shared with the other oxygen atom to form eight electrons in the outer orbit; symbolized as O: :O or O=O. (c) In a triple covalent bond three pairs of electrons are used to complete the outer orbit; symbolized as N⁝⁝N or N≡N.

atom that contains the correct number of electrons for greater than fifty per cent of the time but less than one hundred per cent of the time must have a charge that is smaller than zero but less than 1^-. These atoms contain a slightly negative charge which is represented as δ^-. Similarly an atom that contains a completed outer shell for less than fifty per cent of the time has a slight positive charge which is represented as δ^+.

Polar molecules contain atoms that share electrons disproportionately and therefore contain a separation of charge. This separation of charge is termed a dipole.

Highly polar molecules have a greater difference in electron sharing than molecules with a low polarity. EXAMPLE: In hydrogen fluoride, the electron from the hydrogen atom spends a greater proportion of its time (about seventy per cent) in the sphere of influence of the fluoride atom (Figure 8.3). The greater the difference in sharing between two atoms, the greater is the dipole and thus the polarity of the molecule.

Non-polar molecules contain atoms that share electrons for approximately fifty per cent of the time and thus lack a dipole. All molecules that consist of the same element are non-polar. EXAMPLES: Oxygen (O_2), hydrogen (H_2), chlorine (Cl_2) (Figure 8.3). Sometimes different elements have a similar electron affinity which also results in the formation of non-polar molecules. EXAMPLE: Carbon and hydrogen (Figure 8.3).

The degree of polarity of a substance determines the ability of a substance to dissolve in water. Generally polar substances dissolve in water, which is a polar molecule. Non-polar substances are

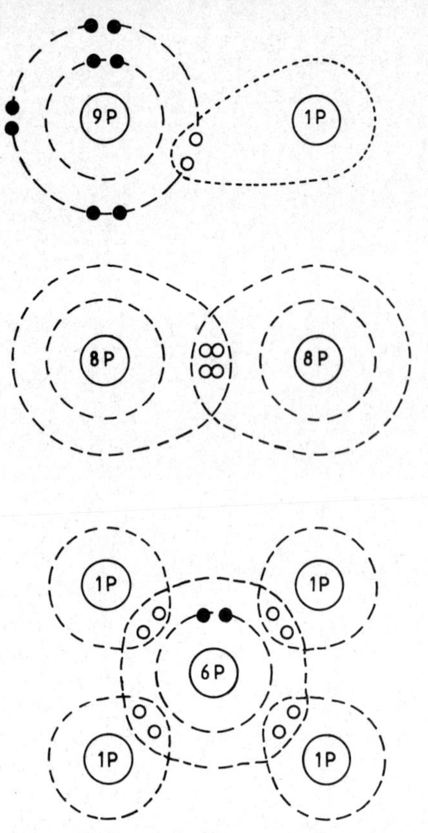

Methane

Figure 8.3 Hydrogen fluoride shares electrons disproportionately and is highly polar. Molecular oxygen and methane share electrons approximately equally and are non-polar.

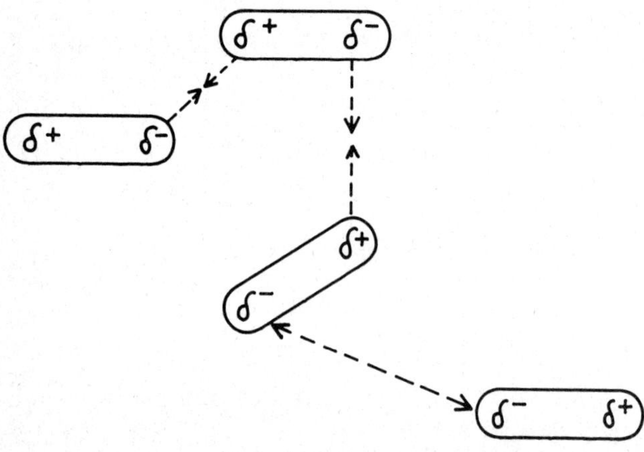

Figure 8.4 Dipole–dipole interaction occurs between oppositely charged portions of different polar molecules. Like charged regions between polar molecules are repelled.

unable to dissolve in water. Non-polar substances however will dissolve in non-polar solvents such as acetone (see 10.2).

Polar molecules are able to form weak bonds with other polar molecules. These weak bonds or loose linkages are known as dipole-dipole interactions. Since opposite charges attract, then the small opposite charges from different polar molecules can be attracted to each other (Figure 8.4). This ability of oppositely charged parts of different polar molecules to form dipole-dipole interactions is of great importance in determining the structure and function of substances such as enzymes, antibodies, genetic molecules and pharmacological agents. When hydrogen is one of the atoms involved in the dipole-dipole interaction, the linking is referred to as a hydrogen bond.

8.3 Radicals

A radical is a group of atoms which acts as a unit within a molecule, and when that molecule undergoes a chemical change the radical remains as an intact unit. A radical cannot exist by itself since it contains either a net positive or negative charge and thus is required to combine with other atoms to neutralize its charge. A radical therefore acts as if it were an individual ion. EXAMPLE: The bicarbonate radical (HCO_3^-) always acts as a separate entity. When sodium bicarbonate ($Na^+HCO_3^-$) is involved in a reaction with hydrochloric acid (HCl), the bicarbonate ion combines with the H to form carbonic acid (H_2CO_3):

$$Na^+ \; HCO_3^- + HCl \rightleftharpoons H_2CO_3 + NaCl \quad \text{(Eqn 8.1)}$$

The importance of radicals in body function is summarized in Table 8.1 The symbols in Table 8.1 will be described in 8.5.

Table 8.1 The main radical constituents in the body

Name	Formula and charge	Function
Ammonium	NH_4^+	Breakdown chemical of amino acids and is used in the regulation of urine acidity
Bicarbonate	HCO_3^{2-}	Important in acid/base balance and carbon dioxide transport in the blood
Hydroxide	OH^-	Important factor in alkaline solutions and in neutralizing acids
Phosphate	PO_4^{3-}	Important acid control mechanism in blood cells and the kidneys

8.4 Valence

A system is required to determine which element will combine with another element to form a compound. In addition, a system is needed to determine the proportion of atoms of an element that are required to combine with atoms of another element to form a particular compound. Valence refers to an atom's ability to combine with other atoms. It therefore relates to an atom's attempt to achieve a completed outer electron orbit. Atoms achieve a full outer electron shell by sharing electrons or by either gaining or donating electrons.

In ionic compounds, the valence of an atom is the number of electrons gained or lost by the atom in order to complete the outer electron shell. An atom with seven electrons in the outer orbit requires one electron to complete the outer shell and therefore has a valence of 1^-. The minus sign is due to the atom gaining one electron. Similarly an atom with six outer shell electrons has a valence of 2^- and an atom with five outer electrons has a valence of 3^-. The valence of an atom that donates its electrons to form a cation, with a completed outer shell, is determined by the number of outer orbit electrons. An atom with one outer shell electron has a valence of $1+$ as one electron has been lost. In the same way an atom with two outer orbit electrons has a valence of $2+$ and an atom with three outer electrons has a valence of $3+$.

A compound has no net charge and thus the number of electrons gained and lost must be equal within a compound. EXAMPLE: An element with a valence of 2^- requires either two atoms of $1+$ valence or one atom of $2+$ valence to form a stable compound with no net charge.

In covalent compounds, the valence of the constituent atoms is determined by the number of shared electron pairs that are present. An atom with seven electrons in the outer orbit can achieve its eight electrons by forming a single covalent bond, that is, one electron pair. This atom has a valence of 1. Similarly an atom with six outer orbit electrons has a valence of 2 since two electron pairs are necessary to complete the outer orbit. An atom that requires three electron pairs to complete the outer shell has a valence of 3. Note that no minus or plus sign is used since electrons are shared and not gained or lost. When determining the valence required to combine two atoms to form a covalent compound, it is necessary to match numerically the valence of each atom. EXAMPLE: An atom with a valence of 2 requires either two atoms with a valence of 1 or one atom with a valence of 2 to form a covalent compound.

Throughout this book a system of diagrams is used to show how valence relates to the ability of an atom to combine with other atoms (Figure 8.5). Each atom, or in some cases a radical group, is represented by a square. The addition of each electron to an atom is represented by an individual peak. EXAMPLE: Two peaks indicates an anion which contains two additional electrons. The removal of each electron from an atom is shown by a triangular indentation. EXAMPLE: Two indentations indicate a cation which has a valence of $+2$. The sharing of each electron pair between two atoms is indicated by a triangular indentation in each atom.

Indicates the gain of 1 electron, each peak represents 1 electron

Indicates the loss of an electron, each trough represents 1 electron

Indicates the sharing of 1 electron pair, that is, a covalent bond

Indicates an element that has a valence of 2 as 2 electrons have been gained

Indicates that the elements have a valence of 2 since 2 pairs of electrons are being shared

Indicates that the elements have a valence of 2 since 2 electrons have been either gained or lost, that is, an ionic bond exists

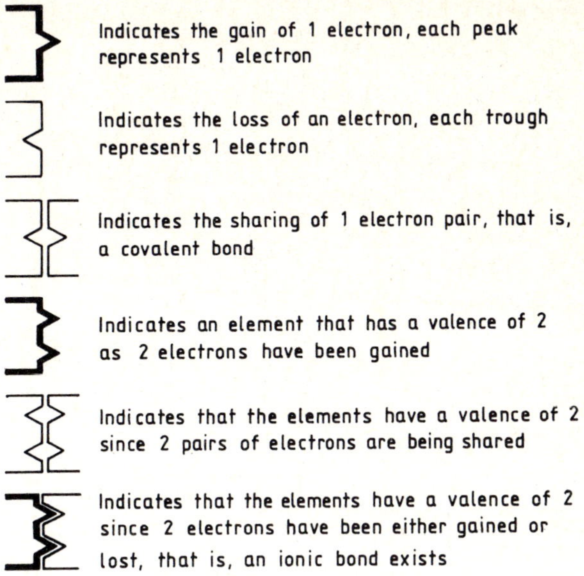

Figure 8.5 Schematic representation of valence. This system is used throughout this book to show how valence relates to the ability of an atom to combine with other atoms. An anion is identified by (-) outline where each peak represents one extra electron. A cation has a (-) outline and each lost electron is represented by a triangular indentation. Where covalent bonds exist, each shared electron pair is indicated by a triangular indentation in each atom.

8.5 Expressing chemical combinations

CHEMICAL FORMULAE

When atoms are combined, it is necessary to use a system of symbols that indentifies the combination of atoms. A formula conveys information about the structure and composition of any specific combination of atoms. An understanding of how formulae are used is necessary for understanding how substances are formed and broken down in the body, along with their chemical and physiological action. The two forms of chemical formulae used in this book are the molecular formula and the structural formula.

Molecular Formulae

A molecular formula states the actual number of atoms of each element present in either a molecule or group of atoms in a radical. The number of atoms of an element is represented as a subscript. EXAMPLE: H_2O indicates that two hydrogen atoms are present and one oxygen. When no subscript is present, such as with oxygen, only one atom is present. The molecular formula for glucose is $C_6H_{12}O_6$. This formula informs us that there are six carbon atoms, twelve hydrogen atoms and six oxygen atoms present in every molecule of glucose.

In a radical such as the bicarbonate ion, which is represented as HCO_3^-, the $_3$ refers to the number of oxygen atoms and the minus indicates that the radical group contains a charge of -1. Similarly, the hydroxyl ion or OH^- is a radical with a charge of -1. The minus refers to the OH and not the H alone.

A number in front of a molecule indicates the quantity of molecules present. EXAMPLE: Two water molecules would be represented as $2H_2O$ and five water molecules would be $5H_2O$.

In many cases the molecular formula represents a specific structure since the constituent elements can only be combined to form one unique structure. EXAMPLES: H_2O, CO_2, NH_3 (ammonia). In the case of $C_6H_{12}O_6$, there are sixteen different ways that the atoms can be arranged. These sixteen different structures all have unique chemical and physiological properties. An alternative formula is required for indicating which substance is being represented.

Structural Formulae

A structural formula shows the actual number of atoms of each element present along with the structural composition of the molecule. The molecular formula $C_6H_{12}O_6$ is represented by sixteen different structural formulae. Substances that have the same molecular formula but a different structural formula are termed isomers. EXAMPLE: The two different sugars glucose and galactose are isomers of $C_6H_{12}O_6$. The difference in chemical and physiological action in the body of these two sugars is due to the relative difference in the position of an H and an hydroxyl ion (OH) about a single carbon atom, with respect to the other atoms in the molecule (Figure 8.6). A structural formula also indicates the type of bonds that are present within a molecule, that is, the presence of single, double and triple covalent bonds.

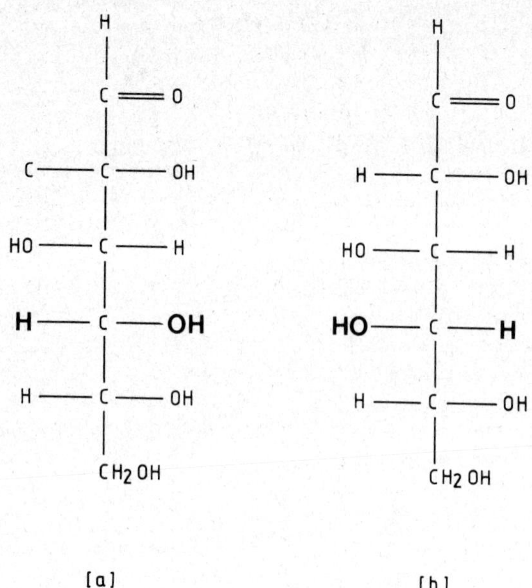

Figure 8.6 The molecular formula $C_6H_{12}O_6$ is represented by sixteen isomers. The structural formulae of (a) glucose, and (b) galactose are shown. Glucose and galactose differ only in the relative position of H and OH about a single carbon atom.

CHEMICAL EQUATIONS

These are used to show which substances are used (reactants) and which substances are formed (products) in a chemical reaction. An equation assists a person in understanding the nature of a chemical reaction. The reactants and products can be separated by either a single arrow →, a double arrow ⇆, or an equal sign.

A single arrow indicates that the reaction is predominately a one way reaction, that is, products are formed in the direction of the single arrow.

A double arrow indicates that the conversion of reactants into products is always incomplete. These are reversible reactions which can operate in either direction. The relative amount of each substance present is maintained in a fixed proportion to the other substances in the equation, that is, a state of chemical equilibrium exists.

APPLICATION: RATIO OF CARBONIC ACID TO BICARBONATE

Carbonic acid and bicarbonate exist in a proportion of 1:20 in our body. This chemical equilibrium is shown in Eqn 8.2

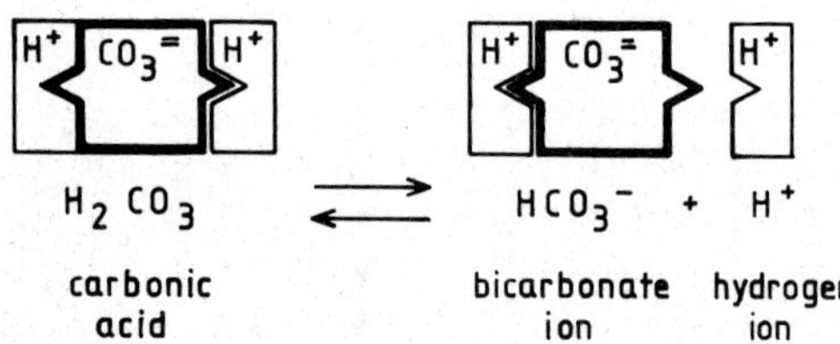

An elevation of carbonic acid would cause an alteration to the fixed proportion. This results in a proportionate amount of HCO_3^- and $H+$ being formed from the H_2CO_3. A ratio of one carbonic acid to twenty bicarbonate is maintained. Similarly, an alteration to the amount of bicarbonate will cause a proportionate change in the amount of carbonic acid. This maintenance of a ratio of one to twenty is important in maintaining the acid/base balance in blood (see 11.5).

An equal sign between reactants and products is often used instead of double arrows when chemical equilibrium exists.

A chemical equation has to be consistent. Elements cannot be changed into other elements by chemical means. If charges are involved, the net charge on both sides of a chemical equation must be equal. All equations must be balanced, with the amount of reactants and their charges being equal to the amount of products and their charges.

Balancing an equation may require altering the quantity of molecules, as the ratios of elements within a molecule are fixed for a specific molecule. EXAMPLE: In Eqn 8.3a, the first point to note is that two Cl^- are required to combine with the Zn^{2+} so that all compounds have a net charge of zero. Eqn 8.3a is rewritten to form Eqn 8.3b. In Eqn 8.3b an excess of Cl^- ions now occurs on the right hand side of the equation. An excess of H^+ ions also occurs on the same side. To balance the equation an increase of H^+ and Cl^- is required on the left hand side. This balance can only be achieved by increasing the quantity of HCl. Increasing the quantity of Cl^- or H^+ in the molecule will result in a different compound being formed. Eqn 8.3c shows the balanced equation for this reaction. The amounts of each substance and their charges are equal on both sides of the equation.

$$Zn^{2+}O^{2-} + H^+Cl^- \longrightarrow Zn^{2+}Cl^- + H_2^+O^{2-} \quad \text{(Eqn. 8.3a)}$$

zinc oxide + hydrochloric acid ⟶ zinc chloride + water

$$Zn^{2+}O^{2-} + H^+Cl^- \longrightarrow Zn^{2+}Cl_2^- + H_2^+O^{2-} \quad \text{(Eqn. 8.3b)}$$

zinc oxide + hydrochloric acid ⟶ zinc chloride + water

$$Zn^{2+}O^{2-} + 2H^+Cl^- \longrightarrow Zn^{2+}Cl_2^- + H_2^+O^{2-} \quad \text{(Eqn. 8.3c)}$$

zinc oxide + hydrochloric acid ⟶ zinc chloride + water

A balanced equation can be used to determine:

1. the amount of reactants required to form a specific amount of product,
2. the chemical result of combining two substances,
3. the valence of an element or radical.

In a balanced equation the number of charges on each side of the equation are equal. A compound contains no net charge and therefore the valence of most elements and radicals can be determined by knowing the valence of only one element or radical. Since the valence of many elements and radicals can be determined from a small number of elements, it is therefore only necessary to remember the valence of a small number of common elements and radicals!

In Eqn 8.4 the valence of all the elements and the radical OH is determined from knowing that hydrogen has a valence of 1+. Firstly, only one Cl is required to form a zero net charge with H^+ on the left hand side. This indicates that Cl has a valence of 1^-. Similarly, the radical OH must have a valence of 1^- since one H^+ combines with one OH. The valence of Na must be 1+ as only one Na ion is used to combine with the Cl^- ion.

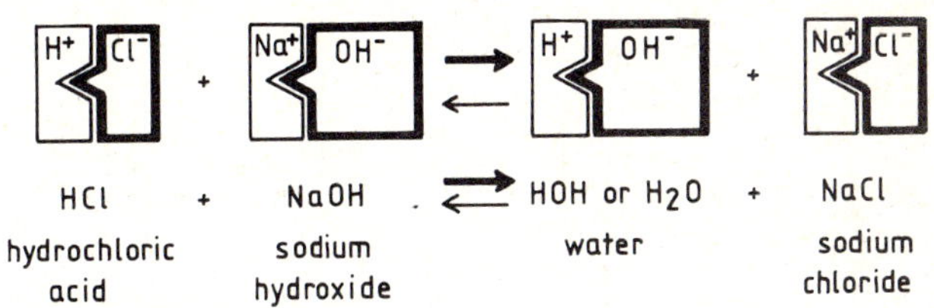

Summary

Compounds are combinations of atoms in fixed proportions whereas mixtures are combinations in variable proportions. The smallest combination of atoms that can exhibit the properties of a compound is called a molecule.

Compounds are formed in order that atoms can satisfy two fundamental forces of nature, that is, that atoms can attain a neutral charge and that they have a completed outer orbit. Two mechanisms exist by which atoms can form compounds.

1. Ionic compounds result from the combination of oppositely charged ions. These ions are formed by atoms either gaining or losing electrons in order that the outer orbit is complete. Each addition or loss of an electron results in an alteration of charge by +1 or −1 respectively and a change in valence of 1. In order for a net charge of zero to exist, the number of positive charges must equal the number of negative charges. The bonds that link these ions are called ionic bonds. These bonds can only be formed between anions and cations, that is, oppositely charged ions.
2. Covalent compounds result from the sharing of outer orbit electrons between adjacent atoms. The sharing of electrons enables atoms to complete their outer orbit without the need to donate or accept electrons. Covalent compounds can be formed from either like atoms or different atoms. When different atoms combine, one atom may have a greater attraction for shared electrons than another. In such cases, the shared electrons spend more than fifty per cent of the time with the former atom, which results in this atom becoming slightly negatively charged (δ^-). For a net charge of zero to exist, the other atom must have a slight positive charge (δ^+).

Polar molecules are molecules that are formed from a disproportionate sharing of electrons.

The slight difference in charge or dipole that exists between atoms in a polar molecule enables these molecules to form weak bonds with other polar molecules. When one of the atoms is hydrogen, the loose linking or dipole–dipole interaction is referred to as hydrogen bonding.

A radical is a group of atoms which acts as a unit within a molecule, and when that molecule undergoes chemical change the radical remains as an intact unit.

Chemical combinations can be expressed in a molecular formula and/or structural formula. The former represents the relative proportions of atoms whereas the latter shows the structural relationship of each atom.

When chemical reactions occur, the relative amount of each substance must be the same after the reaction as it was prior to the reaction, that is, a state of chemical equilibrium must exist.

Chapter 9

Biologically important organic compounds

Objectives

At the completion of this chapter the student should be able to:

1. explain how organic and inorganic compounds differ,
2. explain the difference between saturated and unsaturated organic compounds,
3. describe an alkane, alkene and alkyne,
4. describe the difference between straight chain hydrocarbons and cyclic hydrocarbons,
5. define alcohols, phenols, ketones, aldehydes, organic acids, esters, halogenated hydrocarbons, amines, amino acids,
6. describe heterocyclic compounds.

9.1 Organic and inorganic compounds

All chemical compounds are divided into two categories; organic compounds and inorganic compounds. Organic compounds always contain carbon whereas inorganic compounds generally lack carbon. EXCEPTION: Compounds such as carbonic acid (H_2CO_3) and carbon dioxide are classified as inorganic compounds.

Carbon has a valence of four which enables carbon atoms to link with each other to form large and complex covalent compounds. This property of carbon led to a separate branch of chemistry known as organic chemistry. Carbon compounds form ninety-five per cent of the total number of known chemical compounds. Most carbon compounds contain large quantities of hydrogen. The molecular structure of humans is based on organic compounds. EXAMPLES: Proteins, lipids and carbohydrates. Most pharmacological substances and all plastics also contain carbon.

Inorganic compounds are generally smaller than organic compounds. They also serve an important role in the human body by acting independently or with organic compounds. EXAMPLES: Electrolytes (such as sodium and potassium ions), carbon dioxide, and inorganic acids such as hydrochloric acid.

Organic compounds differ from inorganic compounds in the following ways:

1. organic compounds are covalent whereas inorganic compounds can be ionic or covalent,
2. organic compounds are mainly incapable of dissolving in water whereas most inorganic substances can dissolve in water,
3. organic compounds form the main molecular framework of tissue whereas inorganic compounds assist in strengthening skeletal tissues and aid body metabolism.

9.2 Saturated and unsaturated organic compounds

Carbon has a valence of four since the outer electron orbit contains four electrons. Carbon can form single, double or triple bonds with other carbon atoms. The presence of these bonds confers individual properties to a carbon molecule.

When carbon atoms are linked by single bonds the molecule is referred to as a saturated compound. These compounds always contain a chain of carbon atoms that are linked by the sharing of one pair of electrons. The remaining bonds are usually formed with hydrogen (Figure 9.1a) although oxygen and nitrogen can be present. In Figure 9.1a all of the carbon atoms contain four covalent bonds since carbon has a valence of four.

An unsaturated compound contains some carbon atoms that are linked to other carbon atoms by either double or triple bonds (Figures 9.1b, 9.1c). Note that all of the carbon atoms still contain four covalent bonds. A polyunsaturated compound contains many double or triple bonds linking carbon atoms. The word poly is a frequently used prefix in organic chemistry. It simply means "many". EXAMPLE: Most animal fats are polysaturated and thus contain a predominance of single bonds whereas many plants contain polyunsaturated fats.

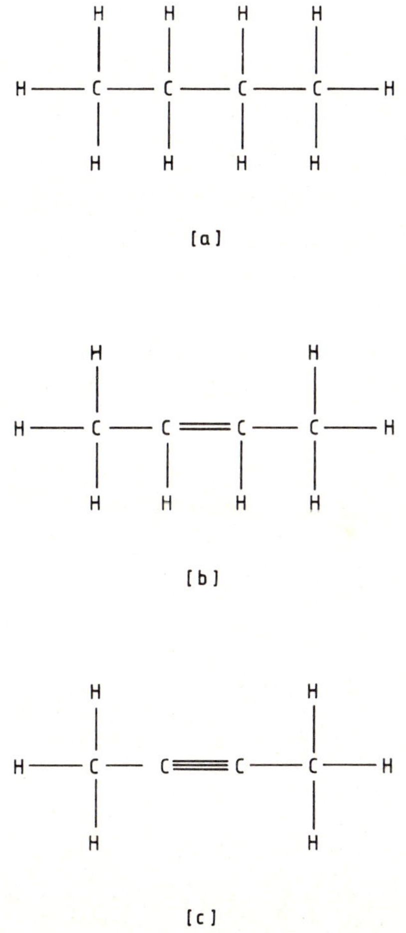

[a]

[b]

[c]

Figure 9.1 (a) Saturated compounds contain carbon atoms linked by single bonds whereas unsaturated compounds contain some carbon atoms that are linked by (b) double bonds or (c) triple bonds.

9.3 Hydrocarbons

WHAT ARE THEY?

Hydrocarbons are compounds that consist solely of carbon and hydrogen. These compounds form the basis of oil and its derivatives such as methane and propane. Carbon has the ability to form long chains of carbon atoms. These atoms can form straight chains, often with a network of side chains or branches (Figure 9.2a). Carbon atoms can also form

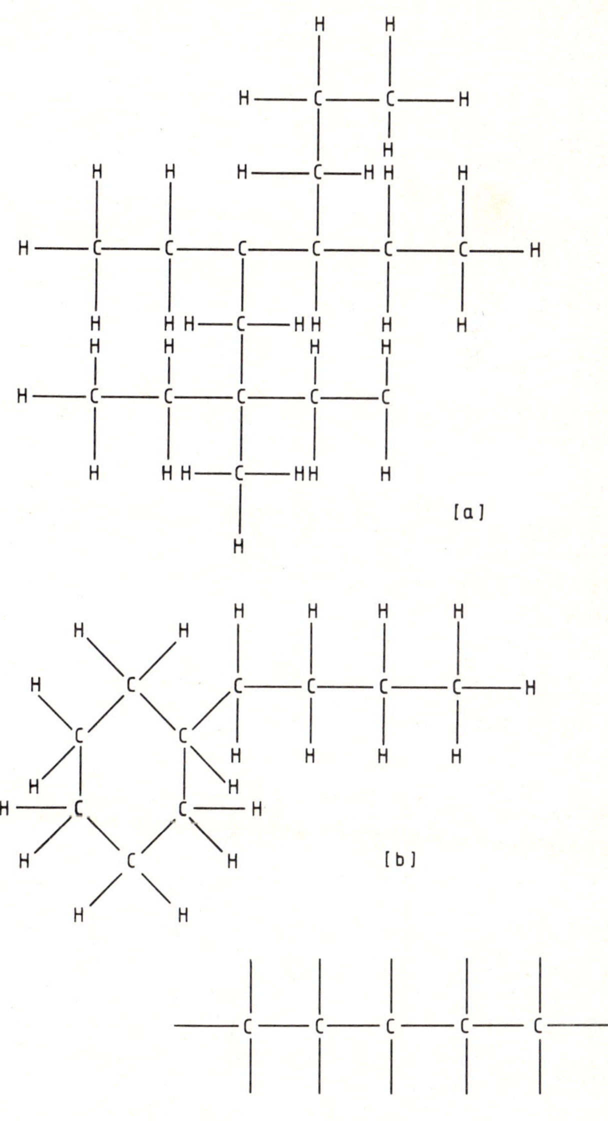

Figure 9.2 Carbon can form either (a) long open ended chains with numerous side chains, or (b) closed carbon chains. (c) The structural formula for carbon compounds is abbreviated by inferring the presence of hydrogen at the end of each incomplete covalent bond.

cyclic chains which may have straight chains attached (Figure 9.2b). The structural formulae of hydrocarbons can be abbreviated by indicating each carbon atom and its covalent bonds, but not the large number of hydrogen atoms. Unless otherwise indicated a hydrogen atom is inferred to exist at the end of any covalent bond (Figure 9.2c).

STRAIGHT-CHAIN HYDROCARBONS

Alkanes

Alkanes are a large group of saturated hydrocarbons that exist in straight chains. All alkane compounds are named in the following manner. Each alkane's name can be split into two parts. The first part indicates the number of carbon atoms in the chain. EXAMPLES: pent- for five, oct- for eight. The second part of the molecular name is -ane. Any substance that contains an -ane at the end of a word is an alkane. The common alkanes are listed in Table 9.1. The structural formula of the alkane propane is shown in Figure 9.3a.

Table 9.1 Names and molecular formulae for the first ten alkanes

Name	Molecular formula
Methane	CH_4
Ethane	C_2H_6
Propane	C_3H_8
Butane	C_4H_{10}
Pentane	C_5H_{12}
Hexane	C_6H_{14}
Heptane	C_7H_{16}
Octane	C_8H_{18}
Nonane	C_9H_{20}
Decane	$C_{10}H_{22}$

Alkenes

An alkene is an unsaturated hydrocarbon with one or more carbon-carbon double bonds. All alkenes are distinguished by a numerical prefix which is identical to the alkanes, and the letters -ene at the end of the word. EXAMPLES: Propene for a three carbon alkene and octene for an eight carbon alkene. The structural formula of the alkene propene is shown in Figure 9.3b.

Alkynes

An alkyne is an unsaturated hydrocarbon that contains one or more carbon-carbon triple bonds. These compounds are distinguished from alkanes and alkenes by the use of -yne at the end of a molecular name. EXAMPLES: Propyne contains three carbon atoms and octyne contains eight carbon atoms. Propyne's structure is shown in Figure 9.3c.

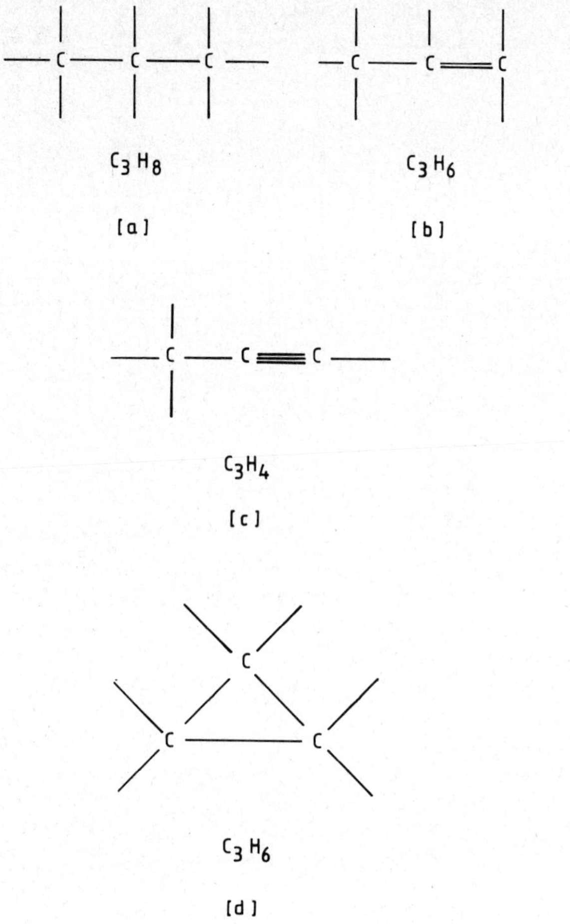

C_3H_8

[a]

C_3H_6

[b]

C_3H_4

[c]

C_3H_6

[d]

Figure 9.3 Structural and molecular formulae of the hydrocarbons: (a) propane, (b) propene, (c) propyne and (d) cyclopropane.

Naming Carbon Chains

The location of double and triple bonds along a carbon chain can vary for the same molecular formula. The longer a carbon chain, the greater the number of positions where a double or triple bond can be located, that is, the greater the number of possible isomers. Remember that isomers have the same molecular formula but different structural formulae. To overcome this problem of only being able to identify organic compounds by their cumbersome structural formulae, an alternative system of identifying compounds has been devised by the International Union of Pure and Applied Chemistry (IUPAC). This system is generally accepted throughout the world.

The IUPAC system is a word picture of a compound's structural formula. The location of double and triple bonds is described by numbering the carbon atoms in a chain from the end that will give the double and triple bonded carbon atoms the lowest possible number. EXAMPLE: 2-Hexene indicates that a double bond is located between the second and third carbon atoms from the end of the

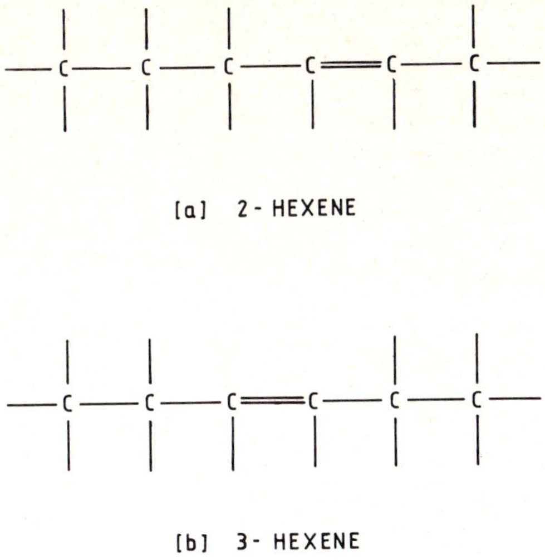

Figure 9.4 The location of the double bonds in two hexene isomers by use of the specific names (a) 2-hexene and (b) 3-hexene.

chain (Figure 9.4a), 3-hexene indicates that a double bond exists between the third and fourth carbon atoms (Figure 9.4b).

The word picture system can result in extremely long and complex words. These words are purely descriptive and if broken down into the constituent parts they enable a person to draw the structure of a compound. It is necessary to have a general understanding of this system in order to study the fundamental components of organic and cell chemistry.

CYCLIC HYDROCARBONS

Cyclic hydrocarbons contain a ring of carbon atoms. These compounds are named by combining the prefix cyclo- to the corresponding alkane, alkene or alkyne name. EXAMPLES: Cyclopropane contains three carbon atoms linked into a ring whereas cyclobutane contains eight carbon atoms in a ring. The structural and molecular formulae of cyclopropane are shown in Figure 9.3d. Note that cyclopropane contains two less hydrogens than propane. Cyclopropane has different properties to propane. Cyclopropane (trimethylene) can be used as a general anaesthetic whereas propane is a natural gas which is used in conjunction with butane as a fuel.

AROMATIC HYDROCARBONS

Aromatic hydrocarbons were originally named because many of them have a spicy or sweet smelling odour. This generalization is misleading as not all aromatic compounds are odorous and not all sweet smelling compounds are aromatic compounds. The name is now used to denote a particular form of six carbon ring. Aromatic hydrocarbons contain six carbon atoms joined together in an alternating sequence of single and double bonds (Figure 9.5). This structure consists of two isomers that are continually changing into each other. An abbreviated symbol is used to indicate the structure of these compounds (Figure 9.5).

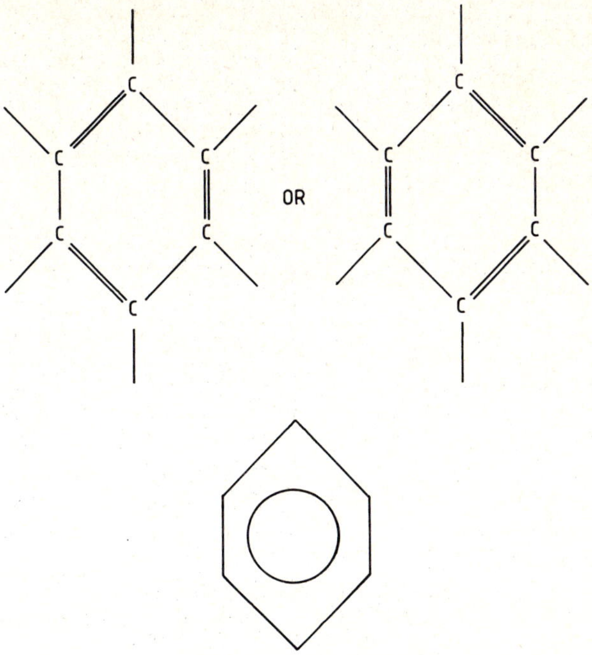

Figure 9.5 The two structural forms of an aromatic compound and the frequently depicted abbreviated symbol.

The simplest aromatic compound is benzene. This compound is represented in Figure 9.5 and it has the molecular formula C_6H_6. Benzene is toxic when ingested and can cause a decrease in red blood cell and white blood cell counts on continued inhalation of its vapour.

Benzene can have straight hydrocarbon chains, cyclic chains or other benzene molecules attached to

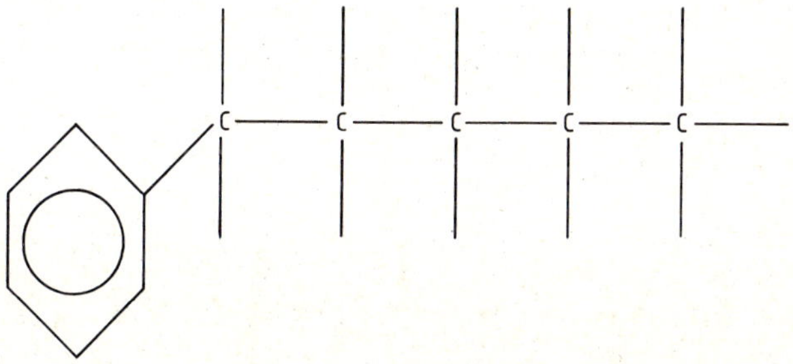

Figure 9.6 The combination of a hydrocarbon chain with a benzene molecule can lead to an infinite number of aromatic compounds.

the ring by simply replacing a hydrogen in the benzene molecule with a carbon atom in a hydrocarbon chain (Figure 9.6). The addition of these chains leads to an infinite number of aromatic compounds with an infinite number of properties. Benzene compounds are used in the body as hormones. EXAMPLE: Steroids. The benzene compound toluene is used as a urine preservative while other benzene compounds are used as drugs. EXAMPLE: Benzodiazepines such as valium, librium and serepax. Some of the multi-benzene ringed compounds can cause cancer. EXAMPLE: 1,2 Benzpyrene which is found in cigarette smoke and automobile exhaust gases.

9.4 Functional groups

A functional group is an arrangement of atoms that will have very similar chemical and physical properties whenever it is found in an organic molecule. The addition of these functional groups to a hydrocarbon chain alters the properties of the chain. Functional groups are the sites of chemical reactions. In the case of chemicals in the body, this means that these groups are the main sites at which the chemical processes that determine body function are located.

For simplicity the hydrocarbon chain can often be replaced by R when referring to the properties of a functional group. The R represents the term radical which refers to the chemically inactive part of a molecule, that is, the molecular part that is not the functional group. EXAMPLE: Alcohols have the general formula ROH. The OH determines the chemical properties of an alcohol and thus R represents hydrocarbon chains such as propane and pentane.

A diverse range of functional groups exist. The more relevant groups involved in body function and their alteration of body function are listed in Table 9.2.

The addition of polar molecules or ions to a hydrocarbon chain alters the previously non-polar properties of the chain. Generally, small organic molecules containing charged functional groups such as alcohols and organic acids are capable of dissolving in water. In the larger-chained molecules, the non-polar hydrocarbon chain exerts a greater influence over the charged region, and therefore the longer the hydrocarbon chain the less able is the molecule to dissolve in water. EXAMPLE: Small-chained alcohols such as ethanol dissolve in water whereas larger-chained alcohols such as octanol have difficulty dissolving in water.

Table 9.2 Important functional groups, their structure and examples of their use

Name of functional group	Functional group structure (attached to a carbon chain)	Use
Alcohol and phenol	R – OH	Alcohol depresses nervous system while phenol is corrosive to tissue; both used as antiseptics and disinfectants
Aldehyde	R – C(=O)–H	Glucose used to produce energy for cell function
Amine	R – N(–H)–H	Stimulant chemicals of the body, nervous function, inflammatory response, protein structure, hereditary molecules; many drugs contain amine groups
Carboxylic acid	R – C(=O)–OH	Fatty acids are a major component of body structure; lactic acid is a by-product of energy production; many drugs contain acid groups
Ester	R – C(=O) – O – R	Body fat stores energy and the ester ATP transfers energy
Halide	R – halide	Anaesthetics, disinfectants and plastics
Ketone	R – C(=O) – R	Organic solvent, sugar fructose is converted into energy, by-product of fat and protein breakdown

ALCOHOLS AND PHENOLS

The addition of a hydroxyl ion OH^- to a hydrocarbon chain results in the formation of an alcohol, whereas if the OH^- is added to a benzene ring it forms a phenol.

Alcohols are named by replacing the "e" at the end of hydrocarbon chains with an -ol. EXAMPLE: A three carbon alcohol is named propanol from the three carbon hydrocarbon propane (Figure 9.7). All alcohols can be identified by the letters -ol at the end of the name. Phenols are usually referred to by their common names. EXAMPLE: Cresol (Figure 9.7).

Alcohols and phenols have different properties. Phenols are more acidic than alcohols and can be corrosive to tissue. Phenols are unable to dissolve in water whereas the common alcohols dissolve easily in water. The term alcohol is often used when referring to ethanol or ethyl alcohol. Ethanol is used for drinking purposes as it has the effect of depressing the nervous system resulting in the

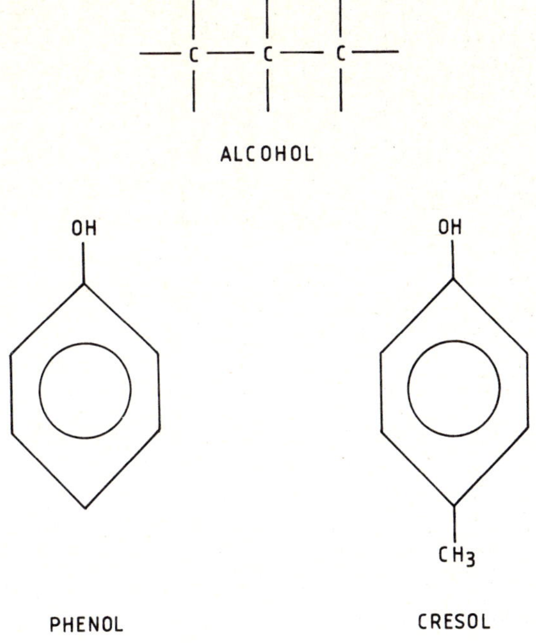

Figure 9.7 The structural formulae of alcohol, phenol and the phenol derivative, cresol.

removal of a person's inhibitions. Methanol is also a commonly used alcohol which is the main constituent of methylated spirits (industrial alcohol or surgical spirit). This alcohol should never be ingested since the intake of small amounts can lead to blindness and paralysis. Both alcohol and phenol have the common property of being able to kill germs.

APPLICATION: ANTISEPTICS AND DISINFECTANTS

A disinfectant is an agent that destroys germs or renders them inert when applied to any surface. An antiseptic is an agent that disinfects the skin and other living tissue surfaces. Many useful chemical disinfectants are too irritating to the skin and mucous membranes and thus are unable to be used as antiseptics. Some alcohols and phenols are used as disinfectants and antiseptics.

Alcohols such as ethanol and propanol have antigerm activity in the presence of water. When using alcohol as a disinfectant it is important to consider the presence of moisture on a surface. Since alcohol dissolves in water, a wet surface will dilute the amount of alcohol present leading to a diminished action.

Oils are non-polar and are unable to dissolve in water whereas they can dissolve in alcohol. This property enables alcohols to be used to cleanse and disinfect skin surfaces by removing oils, cell debris and germs.

When other disinfectant compounds are dissolved in alcohol a tincture is formed. EXAMPLES: Tincture of iodine and tincture of zephiran. Tinctures have better disinfectant properties than either of the constituents along with better removal of oils.

Alcohols and tinctures are usually used on normal skin surfaces as alcohol may have a deleterious effect on traumatized tissue. The use of alcohols as disinfectants is restricted to small areas because of its possible irritating effect on skin.

Phenols are more corrosive than alcohols. The phenol's harmful properties are altered when it combines with soaps or detergents. The derivatives of phenol are used as disinfectants, with the less toxic compounds also being used as antiseptics. Cresol (lysol and sudol) is a phenol derivative which is frequently used for the preparation of theatre and dressing trolleys before use and for disinfecting basins and baths after use. Hexachlorophane is another phenol compound which, when combined with soap or detergents, forms relatively non-toxic antiseptics such as phisohex and dialsoap. Phisohex is used for surgical "scrubs" and hand washing. It can also be used to disinfect and clean walls, floors and surfaces of dressing trolleys and operating tables.

KETONES

Ketones are organic substances whose structure is based on an oxygen atom being attached by a double bond to a carbon atom. This carbon atom must not be at the end of a hydrocarbon chain (Figure 9.8). Ketones are named by using the ending -one. Organic substances readily dissolve in ketones such as propanone which is frequently referred to as acetone (Figure 9.8). This substance is used as nail polish remover and to dissolve the parts of adhesives that adhere to the skin when a dressing is removed.

Ketones are formed from alcohols that have lost the hydrogen from the hydroxyl group. The removal

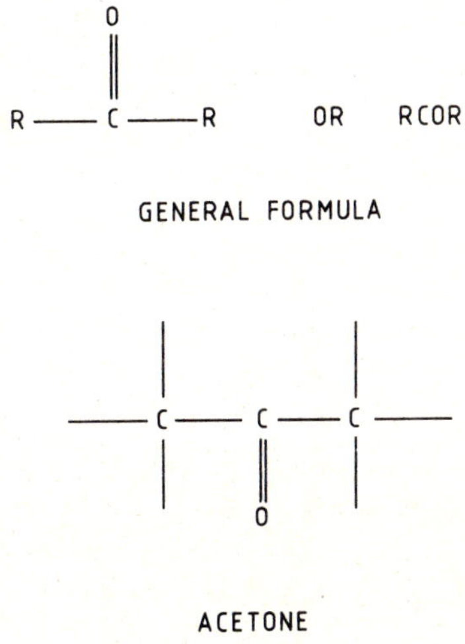

R — C(=O) — R OR RCOR

GENERAL FORMULA

ACETONE

Figure 9.8 The structure of ketones.

of a hydrogen atom means that an electron has been removed from the alcohol group. Oxidation is the process of removing hydrogen. Oxidation can also mean the combination of oxygen with a substance. The process of oxidation is important in the production of most of the body's energy since the lack of oxygen to cells results in very little energy being produced (see 22.5).

APPLICATION: BODY KETONES

Ketones are used by the body to form energy. EXAMPLE: The fruit sugar, fructose, is an important ketone that is digested and converted into energy. The breakdown of fats and protein results in the formation of ketones that are termed ketone bodies. EXAMPLES: Acetone and acetoacetic acid.

In diabetes mellitus, excessive breakdown of fat and proteins results in an excess of ketone bodies in the blood which is referred to as ketosis. The sweet smell of acetone can often be detected in the breath of people with ketosis. The onset of ketosis is associated with malaise and rapid breathing, which if not treated will be followed by vomiting, dehydration, abdominal pain and finally coma.

ALDEHYDES

An aldehyde is a compound with the general formula RCHO (Figure 9.9). Aldehydes are formed from the oxidation of alcohols which contain the hydroxyl group at the end of the hydrocarbon chain. In naming an aldehyde, add -al to the end of the prefix used for the longest hydrocarbon chain which includes the C of the aldehyde group.

A common aldehyde used in hospitals is methanal which has the common name of formaldehyde. This substance readily dissolves in water. Formalin is a mixture of formaldehyde and water. It is used as a disinfectant and is frequently used as a preservative of tissue.

Glucose is an aldehyde which is used by the body, particularly the brain, to produce energy for cell function. The body's supply of glucose comes from either the ingestion of food or from the manufacture of glucose within the body.

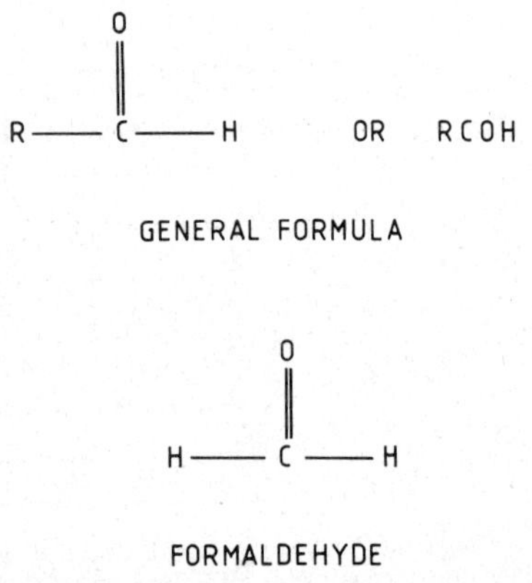

Figure 9.9 The structure of aldehydes.

ORGANIC ACIDS

An acid is a substance that releases hydrogen ions (H^+) into water. The greater the number of hydrogen ions released into water, the greater is the acidity of the liquid (see Section 11.1). An organic acid is a compound with the general formula RCOOH (Figure 9.10). Most organic acids are weak acids. Organic acids are formed from the oxidation of an aldehyde, that is, oxygen is added to the aldehyde functional group. The RCOOH group is called the carboxyl group, and substances with this group are termed carboxylic acids. Many of the carboxylic acids are known by their common names in preference to their IUPAC name.

The IUPAC name for each acid is derived by taking the longest hydrocarbon chain that includes the C of the COOH group and replacing the "e" with the suffix -oic. The word acid follows the -oic. EXAMPLE: Acetic acid consists of ethane that has been oxidized to form the carboxyl group and is thus named ethanoic acid (Figure 9.10). Aromatic compounds can form carboxylic acids by combining a carboxyl group to the aromatic compound. EXAMPLE: The antifungal agent benzoic acid consists of a carboxyl group attached to a benzene ring (Figure 9.11).

Organic acids perform a variety of important roles in body function. Fatty acids are a major component of body structure and function. These organic acids

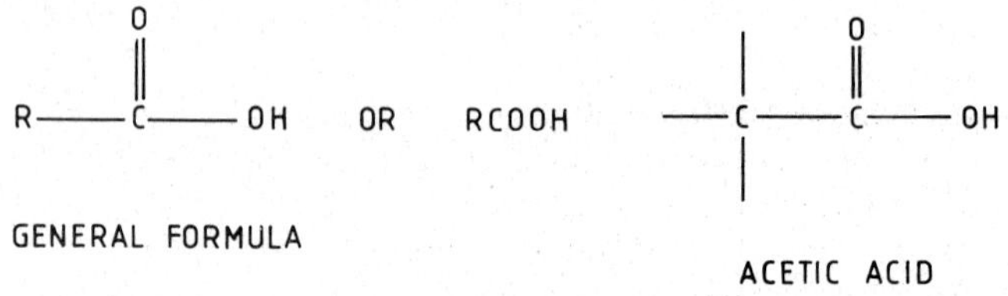

Figure 9.10 The structure of acids.

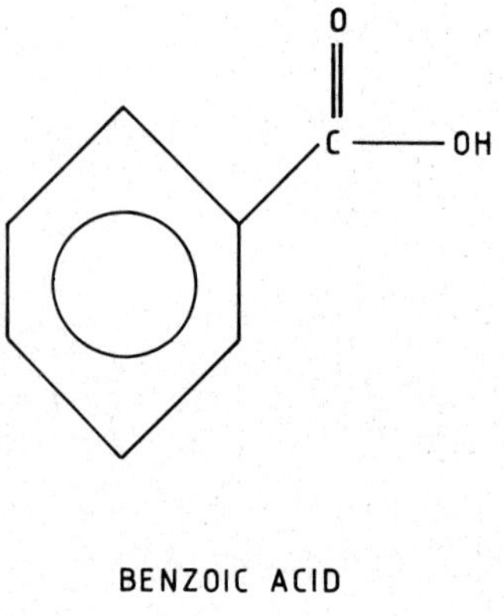

Figure 9.11 The anti-fungal agent benzoic acid consists of a carboxyl group attached to a benzene ring.

are essential for the formation of all body fat (see 23.1). Another important body chemical is lactic acid. This organic acid is formed in tissue when a relative lack of oxygen exists (see 22.5).

Organic acids are also important constituents of many drugs and can be chemically altered to form a wide variety of drugs.

APPLICATION: SALICYLIC ACID

Salicylic acid is an aromatic compound that contains both a carboxyl and an alcohol group (Figure 9.12). This acid exerts remarkable analgesic, antipyretic, anti-inflammatory and anti-rheumatic effects in humans. Compounds formed from salicyclic acid act either by conversion to this acid or by a similar mechanism to its mode of action. These salicylic acid preparations are termed salicylates. EXAMPLES: Aspirin and sodium salicylate.

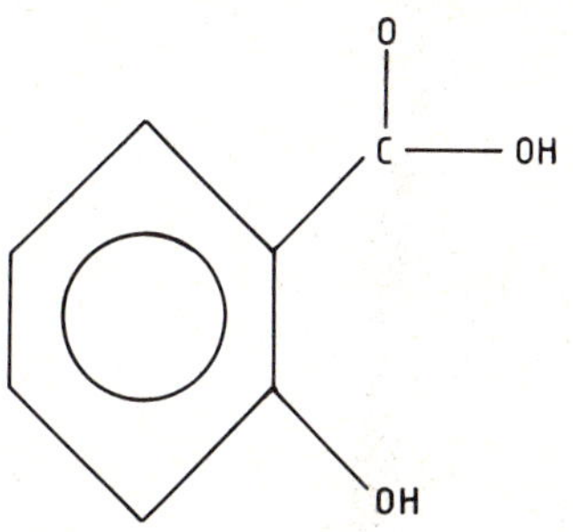

SALICYLIC ACID

Figure 9.12 The structure of salicylic acid.

ESTERS

Esters are substances produced by combining an organic acid with an alcohol (Figure 9.13a). An important example of an ester is body fat. All body fat is formed by the method of combining an organic acid with an alcohol. In the case of fat, the acid is a fatty acid and the alcohol is glycerol.

The general formula for an ester is R_1COOR_2 where R_1 and R_2 can be either the same hydrocarbon chain or different chains.

An ester is identified by two words. The first word is derived from the alcohol by replacing the -ol with a -yl. The second word is derived by replacing the acid -oic with an -ate. EXAMPLE: Methanol is combined with benzoic acid to form methyl benzoate (Figure 9.13b).

The importance of body esters is shown in the formation and breakdown of fat. Briefly, fat is the major source of stored energy for the body. An understanding of these processes is important for gaining an insight into the utilization of fat by the body. These processes will be discussed in detail in 23.1.

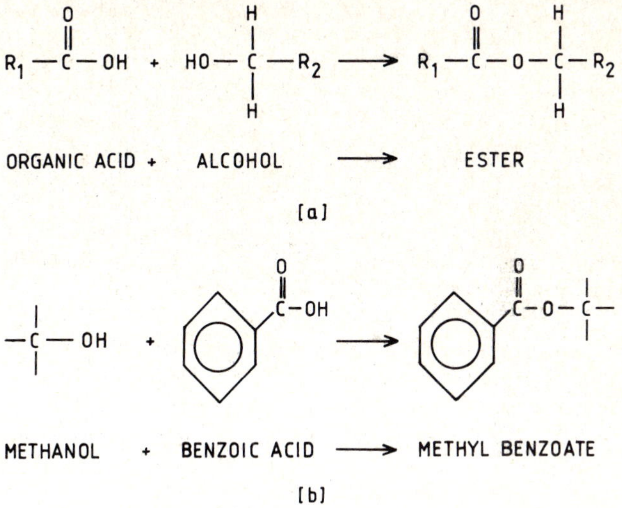

ORGANIC ACID + ALCOHOL ⟶ ESTER

[a]

METHANOL + BENZOIC ACID ⟶ METHYL BENZOATE

[b]

Figure 9.13 (a) The formation of an ester from an alcohol and organic acid. (b) Methyl benzoate is formed by combining methanol and benzoic acid.

Esters are used as local anaesthetics (benzocaine), preanaesthetic medications (scopolamine), and blood vessel dilators (nitroglycerin).

HALOGENATED HYDROCARBONS

Halogenated hydrocarbons consist of a hydrocarbon chain where some of the hydrogens have been substituted by halogens. Halogens are the elements flourine (F), chlorine (Cl), bromine (Br) and iodine (I). In naming these compounds the halogen is specified as fluoro-, chloro-, bromo- or iodo- which are used as prefixes with the hydrocarbon chain. The location of the halogens is determined by the carbon atom numbers where they are attached to the chain. EXAMPLE: The inhalation anaesthetic halothane has the IUPAC name of bromo-2-chloro-1,1,1-trifluoroethane. This long name indicates the specific structure of the anaesthetic (Figure 9.14) but is a cumbersome term and thus the common name is often preferred.

Halogenated hydrocarbons have a variety of uses in hospitals ranging from anaesthetics such as halothane and ethyl chloride to disinfectants such as hexachlorophene.

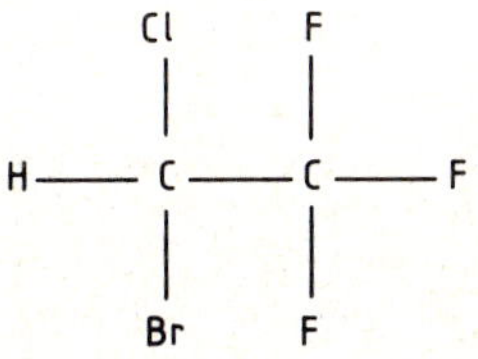

Figure 9.14 Halothane or 2-bromo-2-chloro-1,1,1-trifluroethane.

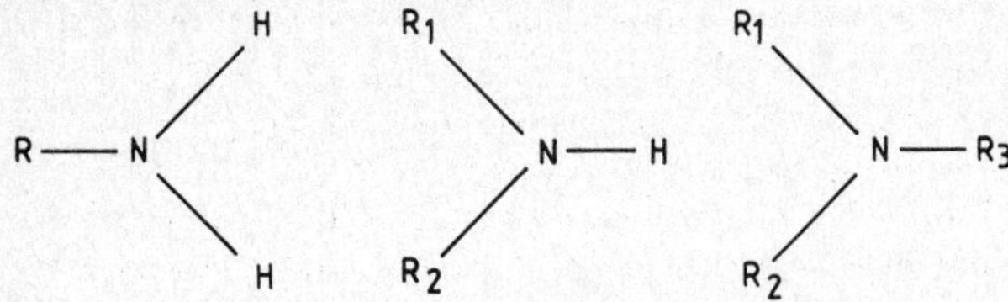

Figure 9.15 General formula of the three forms of amines. R_1, R_2, and R_3 can be different or identical.

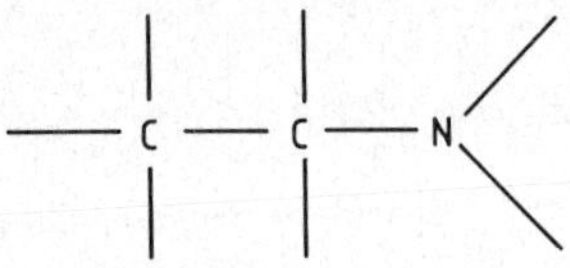

Figure 9.16 Ethylamine.

AMINES

Amines are a group of compounds that are identified by the functional group $-NH_2$. They are derived from ammonia (NH_3). The nitrogen can combine with one, two or three hydrocarbon chains depending on the number of hydrogens that have been displaced (Figure 9.15). An amine is named by adding the suffix -amine to the hydrocarbon name. EXAMPLE: Ethylamine is a two carbon amine (Figure 9.16).

Amines play a vital role in both normal body function and patient care. Amines are used as cardiovascular stimulants (epinephrine), central nervous system stimulants (amphetamines), antihistamines, antiseptics and anaesthetics.

APPLICATION: BODY AMINES

A variety of functions are performed by body amines. The catecholamines are the amines of the nervous system. The main catecholamines are norepinephrine (noradrenaline), epinephrine (adrenalin) and dopamine. The catecholamines are the stimulant chemicals of the body. EXAMPLE: Epinephrine stimulates the cardiovascular system and norepinephrine stimulates parts of the nervous system.

Nitrogen compounds are removed from the body mainly in the form of urea.

Amino Acids

These are small organic acids that contain an amine group (Figure 9.17). Amino acids are the building blocks of proteins. They are also the source of other body nitrogen compounds such as the catecholamines, hormones and the molecules of genetic information. Amino acids will be described in greater detail in 24.1.

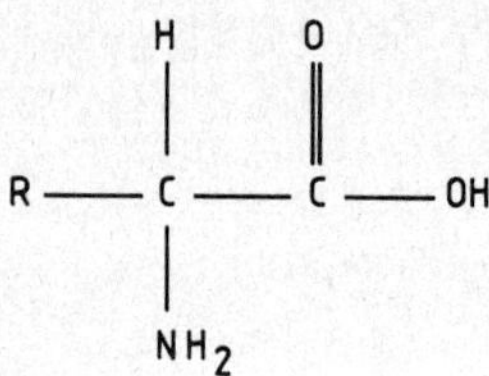

Figure 9.17 General formula for an amino acid.

9.5 Heterocyclic compounds

Heterocyclic compounds are cyclic organic compounds that contain another element within the ring structure. The elements are usually nitrogen, oxygen or sulfur. These compounds comprise a large segment of organic chemistry. The nitrogen based heterocyclic compounds are essential for life. EXAMPLE: The molecules deoxyribonucleic acid (DNA) and ribonucleic acid (RNA) are the key to hereditary information and the maintenance of life. Other important molecules that contain a heterocyclic component are haemoglobin, vitamin B_{12} and the brain chemical, serotonin.

APPLICATION: HETEROCYCLIC DRUGS

Heterocyclic compounds form the basis of a group of drugs used in altering brain function.

Opiates such as morphine, codeine and heroin depress brain function. Codeine is used to depress the cough centre of the brain.

Barbiturates are a group of drugs that have a sedative and sleep-inducing effect when administered in therapeutic doses. Nembutal and seconal are the trade names of two barbiturates that are used as sleeping pills and preanaesthetic sedatives.

Psychedelic drugs or hallucinogens alter sensory perception so that the boundary between real and unreal is lost. LSD appears to act by mimicking the action of the brain chemical serotonin. STP and mescaline are structurally similar to epinephrine and norepinephrine which suggests a possible link between these drugs and body chemicals.

Some heterocyclic drugs have a diverse action on the body. Caffeine stimulates brain nervous activity and acts as a diuretic. Reserpine acts on norepinephrine and is used as a sedative and in the treatment of hypertension.

Summary

Organic compounds generally contain carbon whereas inorganic compounds do not. Organic compounds are covalent and are usually incapable of dissolving in water.

When atoms of carbon are linked by single bonds, the molecule is saturated. An unsaturated compound contains some carbon atoms linked by either double or triple bonds.

Hydrocarbons or compounds consisting of carbon and hydrogen can be arranged in straight chains, cyclic configurations or a combination of both. Straight chained hydrocarbons are termed alkanes, alkenes and alkynes according to the presence of single, double or triple bonds linking respective carbon atoms. When six carbon atoms form a ring by linking the carbon atoms in an alternating sequence of single and double bonds, an aromatic hydrocarbon is formed.

Hydrocarbon chains can have various functional groups, attached to them. Each functional group has its own chemical and physical properties which will alter the properties of a hydrocarbon chain. The site of many body chemical reactions is located at the functional group and thus these groups are important for body function and the maintenance of homeostasis. The main functional groups are alcohols, phenols, aldehydes, amines, carboxylic acids, esters, halides and ketones.

Heterocyclic compounds are cyclic compounds that contain another element such as nitrogen within the ring structure. These compounds play an important role in body function and its alteration.

Unit Five
The composition and properties of fluids

In this unit the properties of fluids are integrated and applied to medications, normal and abnormal body function. You will learn how these properties influence the composition of body fluids.

Chapter 10

Properties of fluid mixtures

Objectives

At the completion of this chapter the student should be able to:

1. describe the differences between solutions, suspensions and colloids,
2. explain why certain substances are soluble in water while others are insoluble,
3. relate the properties of solutions, suspensions and colloids to the administration of medications,
4. relate the properties of fluid mixtures to normal and abnormal body function.

10.1 Composition of fluids

A fluid is any substance that will take the shape of its container. Fluids can thus be either gases or liquids. There are various ways in which substances can be combined in a fluid. These combinations are known as solutions, suspensions and colloids.

10.2 Solutions

GENERAL PROPERTIES

A solution is a fluid that consists of a homogenous mixture of two or more substances. The greater quantity of substance in a solution is called the solvent. All other substances are solutes. An aqueous solution uses water as the solvent. EXAMPLE: Body fluids. Non-aqueous solutions use other liquids such as alcohol as the solvent. EXAMPLE: Iodine tincture contains iodine dissolved in alcohol.

In all solutions, solutes are uniformly dispersed in the solvents by the solutes loosely attaching to the solvent. This loose linking or bonding is called dissolving. When a substance dissolves in a solvent the solute cannot be distinguished from that solvent. Thus when the solution is left to stand the solute does not settle to the bottom of the container unless the solution contains excess solute.

HOW DISSOLVING OCCURS

Solvent molecules are loosely linked together. The ability of a solute to dissolve depends upon the breaking of the interactions that link solvent molecules. A solute will dissolve in a solvent if the formation of loose linkages occurs between the solute and solvent. The forces that form the solute–solvent interactions have to be great enough to break the solvent–solvent linkages. Solutes incapable of breaking solvent–solvent interactions remain undissolved. Generally, polar and ionic solutes (for example, glucose and sodium chloride) dissolve in polar solvents such as water, whereas non-polar solutes such as oil, will not (Table 10.1). A general rule is that like solvents dissolve like substances.

Table 10.1 Nature of a solute and its ability to dissolve in a solvent

Solute	Solvent	Dissolves
Ionic	Polar	Yes
	Non-polar	No
Polar	Polar	Yes
	Non-polar	No
Non-polar	Polar	No
	Non-polar	Yes

Polar substances contain a dipole which consists of one part of a molecule containing an excess of electrons and the other having a deficiency. This difference in charge results from an unequal sharing of electrons between atoms within a polar molecule (see 8.2). The slightly negative part is shown as δ^- and the slightly positive part as δ^+. The delta indicates that a charge exists which is less than the complete acceptance or donation of an electron. Polar molecules dissolve in water by forming dipole-dipole interactions. These interactions or loose-linkages between solute and solvent result from the fundamental force of oppositely charged particles attracting each other (Figure 10.1a).

Ionic substances form ion–dipole interactions with a polar solvent. Ions loosely link with oppositely charged solvent dipoles (Figure 10.1b). In Figure 10.1 both polar and ionic substances have water molecules attached to them. The effective size of the solute has been increased to the size of the solute-water complex. This increase in the size of the solute

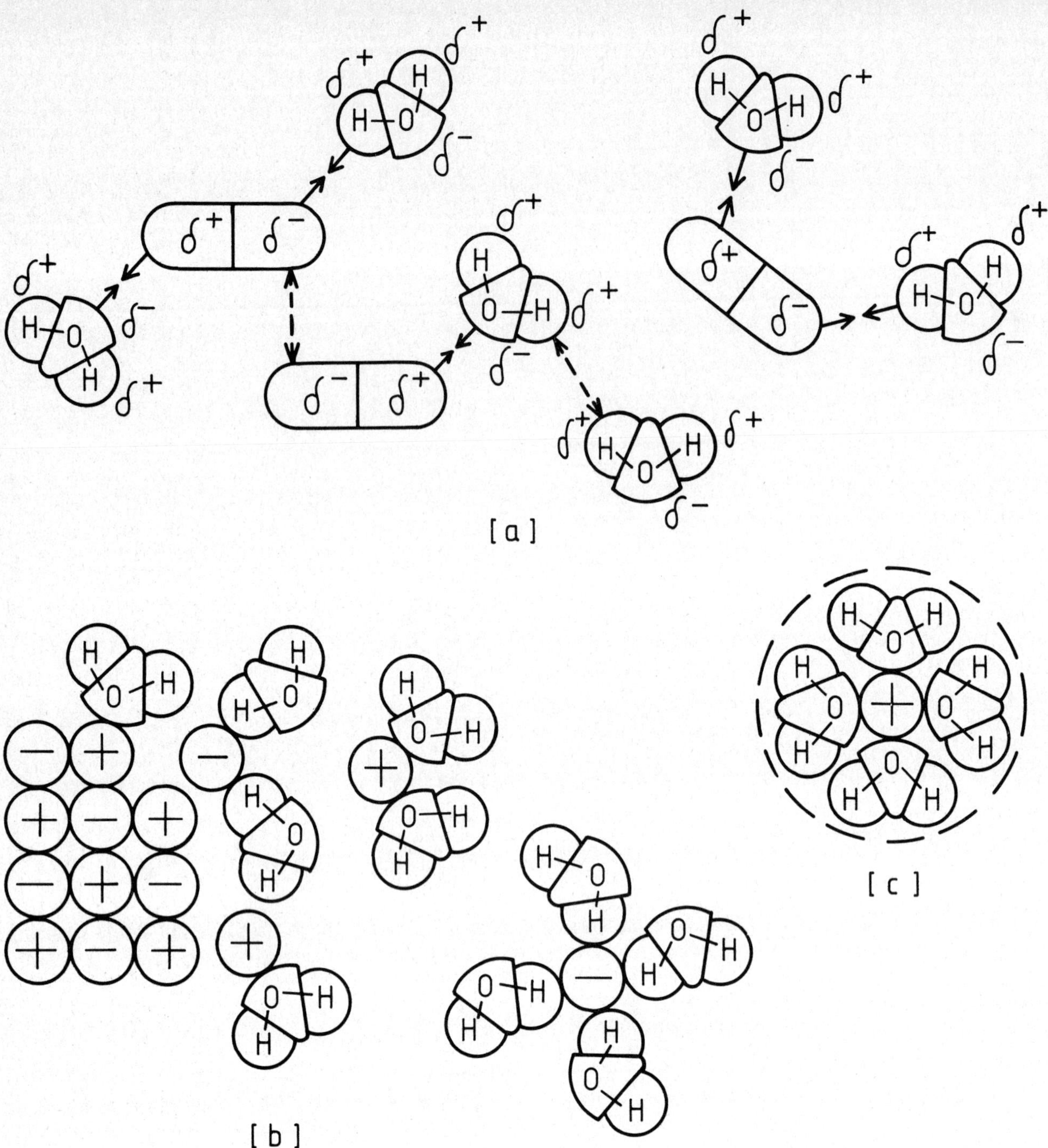

Figure 10.1 The dissolving of solutes in water: (a) Polar substances dissolve by forming dipole-dipole interactions between oppositely charged dipoles (δ+ is slightly positive and δ− is slightly negative). (b) Ionic compounds dissolve by polar solvent molecules pulling the ions apart; the ions form loose linkages with oppositely charged solvent dipoles. (c) The hydrated diameter represents the solute-water complex diameter.

is an important property in determining the movement of substances through a membrane. The solute-water complex is the hydrated form of that substance. The hydrated diameter is the diameter of the solute and water molecule combination (Figure 10.1c).

Non-polar solutes have a low capacity to form linkages with polar solvents and therefore either remain in an undissolved state or only small quantities are dissolved in a polar solvent such as water. EXAMPLE: Oxygen is a non-polar molecule and thus only small quantities can dissolve in plasma. Non-polar solutes are easily dissolved in a non-polar solvent. EXAMPLE: Most adhesives are non-polar and can therefore be removed by being dissolved in a non-polar solvent such as acetone.

SOLUBILITY

Solubility is a measure of the amount of solute that will dissolve in a specified volume of solvent and is thus the relative ability of a solute to form loose linkages with a solvent.

Solutes differ in their ability to dissolve in a given volume of solvent. A solute with a high solubility has the property of a large number of solute particles being able to combine with a given number of solvent molecules. A low-solubility solute can only dissolve a smaller number of particles in the same volume of solvent, that is, a highly-soluble solute can exist in high concentrations whereas a substance with a low solubility can only exist in low concentrations.

SATURATION OF SOLUTIONS

A saturated solution contains the maximum quantity of dissolved solute. Any additional solute remains in an undissolved state and usually settles as a precipitate. A large quantity of a solute that is highly soluble can be dissolved in a given quantity of solvent before saturation occurs. With a low-solubility solute, only a relatively low concentration can exist before saturation develops. Any factor that decreases the ability of a solute to form loose linkages with a solvent or that decreases the number of solvent molecules may effect precipitation of that solute. This means, some of the previously dissolved solute forms a precipitate and thus a saturated solution.

An unsaturated solution has the ability to dissolve additional solute and therefore does not contain precipitate. To determine if a solution is saturated or unsaturated add a small quantity of solute and observe. If it dissolves, the solution is unsaturated.

Several factors can cause an unsaturated solution to change into a saturated solution. This usually results in the formation of precipitate.

Solute Quantity

Increasing the quantity of the solute in a specific volume, that is, increased solute concentration, can cause precipitation. EXAMPLE: Small quantities of sugar dissolve in a cup of tea. The further addition of large quantities of sugar can result in some of the sugar remaining in its undissolved state.

Solvent pH

A substance may have a high solubility at a certain pH and a low solubility at another pH. Thus, pH changes can cause the precipitation of some substances by altering the substances solubility. EXAMPLE: Calcium has a higher solubility in acidic solutions and a low solubility in alkaline solutions.

Solvent Volume

When the number of solvent molecules is decreased the remaining solvent may not be sufficient to form loose linkages with all the solute. The excess solute forms a precipitate. EXAMPLE: Evaporation of water from salt water results in the precipitation of salt crystals.

APPLICATION: RENAL CALCULI

In the kidneys, solutes can be precipitated in one of three ways:

1. increased quantities of serum calcium,
2. alterations in the acidity of the kidney tubule fluid,
3. increased reabsorption of water resulting in a decreased volume of solvent that can pass through the kidneys.

The precipitation of any of the above changes may result in renal calculi (kidney stones). These calculi are usually a combination of calcium and phosphate. The solubility of calcium is greater in acidic conditions than in alkaline conditions, and therefore patients suffering from renal calculi are frequently treated with acidotic diets and drugs.

10.3 Suspensions

GENERAL PROPERTIES

In defining solutions, it was stated that they were homogenous mixtures. Not all mixtures of fluids have this property. EXAMPLE: A container of blood will only be a homogenous mixture after shaking, due to the relatively large red blood cells and white blood cells settling to the bottom of the container upon standing. Such fluids are called suspensions.

Solubility

Suspended particles are insoluble in fluids.

Settle Upon Standing

On standing, the suspended material will settle at the bottom of the container. The heavier the particles, the faster is the settling time, and so the suspended particles will settle in order of weight.

APPLICATION I: ESR (ERYTHROCYTE SEDIMENTATION RATE)

Constant movement of blood maintains erythrocyte (red blood cells) in an even distribution. The sedimentation rate of erythrocytes is altered by various pathological states. Factors such as inflammatory conditions, myocardial infarction, menstruation, pregnancy, pulmonary tuberculosis and septicemia elevate the ESR as a result of the erythrocytes clumping together to form large and heavy aggregates.

APPLICATION 2: MEDICATIONS

Medications that are a suspension contain a label stating "shake before using". It is important that this instruction is followed immediately prior to use, since particles may settle rapidly. Failure to administer a thoroughly mixed suspension will result in the patient receiving an incorrect dosage. In most cases the patient would receive a diluted or less concentrated medication since some of the medication has settled and remained in the container. A stock suspension such as milk of magnesia requires shaking prior to each use. If the initial samples from the bottle are diluted due to an incorrect procedure, then subsequent samples will have a proportionately elevated concentration.

Relative Size
Particles are relatively large. These particles do not pass through ordinary filter paper or cell membranes.

Most liquid medications are suspended in either water or an oil. EXAMPLES of water based suspensions are: milk of magnesia, calamine and benzathine penicillin Gs (L.P.G. injection—a sterile ready-made aqueous suspension). EXAMPLES of oil based suspensions are: epinephrine s. and methenamine mendelate oral (Mandelamine).

10.4 Colloids

GENERAL PROPERTIES

Particles that are too large to be solutes, and too small to form suspensions, form a colloid. The particles are dispersed or distributed throughout a medium called the dispersion medium. A colloid particle can consist of aggregates of molecules or a very large single molecule such as protein.

Solubility
Colloidal particles are insoluble in the dispersion medium.

Settle Upon Standing
Colloidal particles remain in suspension or settle slowly upon standing. The heavier the particle the greater is the chance that the particles will settle over an extended period of time. Colloidal medications therefore require shaking before use.

Relative Size
Particles pass through filter paper but do not pass through cell membranes.

Adsorb Substances
The relatively large surface area of colloid particles can attract and hold specific substances to it. The holding of a substance to a surface is called adsorption.

Most colloids selectively adsorb cations or anions (but not both) onto their surface. This results in the colloid becoming positively or negatively charged. These charged colloids attract either negatively or positively charged particles and other colloids. The addition of positively-charged colloid particles to negatively-charged colloid particles results in the combination of the oppositely-charged colloids. The resulting increase in size causes the colloid complex to act as a suspended particle and settle. This latter process is called coagulation and is used in the neutralization and removal of toxic substances. One form of coagulation is the combination of antagonistic blood antibodies and antigens (see 18.5).

APPLICATION: ADSORBENT MEDICATIONS

Various medications use adsorption to selectively combine with and condense an unwanted substance. Activated charcoal is administered via a nasogastric tube to patients who have overdosed with a polar drug, e.g. barbiturates. The drug is bound to the charcoal and then the colloidal complex is removed by pumping the stomach. Non-polar substances are not absorbed by charcoal. Diarrhoea relief is obtained from colloid medications such as aluminium silicate (kaolin, kaomagma, kaopectate) which act as an adsorbent. Aluminium silicate condenses and holds the irritating substances.

COLLOID CATEGORIES

Colloids are frequently categorized according to the state of matter (solid, liquid or gas) of the colloidal particles and the dispersion medium. The main categories are sols, gels, aerosols and emulsions.

Sols
Sols consist of solid particles dispersed but not dissolved in a liquid. They can be formed by combining small particles or by breaking down large particles. EXAMPLES: Proteins in plasma and intracellular fluid, aluminium silicate.

Gels
Gels consist of solids that are set in a semi-solid that has jelly-like properties. EXAMPLES: agar, gelatine, petroleum jelly, cell membranes.

Gels are semi-rigid and thus do not flow easily whereas sols are easily poured. Small substances can penetrate and move through gels at approximately the same rates as when they flow through liquids. Sols can be converted into gels under certain conditions, that is, sols which can flow easily are converted to gels which do not flow easily. The contents of a cell have been found to alter between sols and gels.

APPLICATION: INTERSTITIAL FLUID

Tissue spaces consist of compartments of sols and gels. In normal conditions the amount of sol occuring in these spaces is negligible. The gel consists of various colloidal particles, the most abundant of which is hyaluronic acid. Fluid in the gel cannot flow readily from one area of tissue to another. The non-mobile properties of the gel hold the interstitial (intercellular) fluid in place and thus prevents the flow of liquid under the influence of gravity from the upper parts of the body to the lower parts. Since only small substances can readily penetrate gels, larger sized particles such as bacteria are prevented from spreading throughout the gel and thus the tissue. This gel barrier assists in restricting the spread of infection in tissues.

Aerosols

Aerosols consist of liquid or solids dispersed in a gas. They are used in inhalation therapy and also in room sterilization. The latter involves the use of the contents of aerosol cans.

APPLICATION I: INHALATION THERAPY

Inhalation therapy may require the use of a device that can propel the aerosol into the trachea, bronchi and lungs. A nebulizer is a device that uses either compressed air or ultrasonic waves to disperse liquids or solids into a medium of gas, thus forming an aerosol. Drugs that can act directly on the respiratory system can be administered using this device. EXAMPLE: salbutamol.

APPLICATION 2: AEROSOL CANS

On standing, the chemicals and liquid settle to the bottom of an aerosol can. When a can that contains a bacteriocidal solution is used without shaking, the resulting spray will consist of mainly the dispersing medium. The liquid containing the bacteriocidal solution will remain at the bottom of the can and not be atomized into the spray. Vigorous shaking ensures a dispersion of the colloid throughout the gas and therefore the spray.

Emulsions

These colloid particles exist as a liquid distributed in small globules or droplets throughout a second liquid, that is, the two liquids do not dissolve in each other. These two liquids are immiscible, whereas liquids that are soluble in each other are miscible.

In some emulsions, the droplets come together when left to stand and form a separate liquid layer from the solvent. Emulsions with this property are called temporary emulsions. EXAMPLE: Mixing oil and water. Other emulsions do not settle or separate into two separate layers upon standing. These are permanent emulsions which exist as homogenous mixtures. EXAMPLE: Mineral oil emulsion (liquid petrolatum) consists mainly of mineral oil, water and alcohol.

The addition of charged particles to dispersed temporary emulsion globules results in each globule resisting recombination with other globules. The repelling of identically charged globules prevents the formation of a liquid layer and thus results in the formation of a permanent emulsion. Substances can form emulsions by the introduction of chemical agents. This process is used in the body to digest and absorb fats.

APPLICATION: FORMATION OF BODY EMULSIONS

In the gastrointestinal system, bile acts as an emulsifying agent in the small intestine. Bile breaks down the ingested fat into droplets. Most of this emulsion is then digested by enzymes and absorbed across the small intestine wall. In addition, some droplets are absorbed as a temporary emulsion in the intestinal cells. These droplets or chylomicrons are converted in these cells into a permanent emulsion by the addition of charged proteins. The identical charges on each droplet prevent the droplets from sticking to each other as like charges repel. Absence of the protein coat, due to poisons or genetic disorders, results in fat accumulating in the intestinal cells.

10.5 Separation of solutions, suspensions and colloids

Solutions, suspensions and colloids can be separated by their individually different properties. A summary of these different properties is listed in Table 10.2.

Table 10.2 Relative properties of solutions, suspensions and colloids

Particle property	Solution	Suspension	Colloid
Solubility	Potential	No	No
Settles upon standing	No	Yes	Yes
Relative size	Small	Large	Medium
Medications require shaking	No	Yes	Yes
Pass through membranes	Yes	No	No
Pass through filter paper	Yes	No	Yes
Adsorb substances	No	No	Yes

Summary

A fluid is any substance that will take the shape of its container, that is gases and liquids. Fluids can exist as solutions, suspensions and colloids.

A solution consists of a homogenous mixture of two or more substances. The lesser quantity of substance is known as the solute and it combines with solvent molecules in the process of dissolving. Generally only polar and ionic solutes dissolve in polar solvents, whereas non-polar solutes dissolve in non-polar solvents. In aqueous solutions water is the solvent, and the size of the solute/water complex is known as the hydrated diameter. A precipitate settles out of a solution when some of the solute has not combined with the solvent. A saturated solution exists when a precipitate occurs. Factors such as the quantity of solute, the solvent pH and the volume of solvent can cause an unsaturated solution to change into a saturated solution.

Suspensions and colloids differ from solutions in that:
1. particles settle upon standing,
2. the size of particles are larger,
3. particles are unable to pass through membranes,
4. they require shaking to form a homogenous mixture.

In addition, colloids can selectively absorb cations and anions. Also, colloids can combine together in the process of coagulation to form suspended particles.

Colloids are categorized according to the type of matter that is mixed with the gas or liquid dispersion medium:
1. sols—solids dispersed in a liquid,
2. gels—solids dispersed in a jelly-like liquid,
3. aerosols—solids or liquids dispersed in a gas,
4. emulsions—liquids distributed in liquids.

Chapter 11

Properties of ions in solution

Objectives

At the completion of this chapter the student should be able to:
1. describe the properties of acids and bases,
2. explain the mechanism of buffer systems,
3. describe the acid-base homeostatic mechanisms,
4. differentiate the various acid-base imbalances,
5. explain how the body compensates for acid-base imbalance,
6. describe a salt and relate its structure to the formation of electrolytes,
7. differentiate between electrolytes and non-electrolytes.

Various ions occur in body fluid. The properties of these ions are important in determining body function. Some of these ions can be formed in the body. EXAMPLE: Hydrogen ion (H^+). Other ions can result from the ingestion of salts. EXAMPLE: Sodium ions (Na^+) and chloride ions (Cl^-) result from table salt.

11.1 Acids

WHAT IS AN ACID?

Acids can be both useful and harmful (Table 11.1), and are continually manufactured in the body as a result of metabolism (Table 11.2). A homeostatic mechanism regulates these acids since abnormal variations in serum concentration of acids will cause a marked alteration of external respiration (breathing). The corrosive property of acids can be both beneficial and harmful. The stomach uses this property beneficially in the process of gastric digestion. The caustic nature of acids can result in harmful burns to tissue that has been exposed to an acid spillage.

An acid is a substance that releases H^+ ions into a solution, the greater the number of H^+ ions the more acidic is that solution. When an acid is placed in solution only a proportion of the H^+ ions are released into the solvent. Dissociation is the breakdown of a substance into ions when that substance is placed in a solvent. A strong acid releases most of its H^+ ions and thus mostly

Table 11.1 Common acids both useful and harmful to patients

Acid/medication	Effect
Acetyl salicyclic/aspirin	Anti-prostaglandin
Benzoic/mycozol	Anti-fungal agent and germicide
Ethacrynic (edecrin or edecril)	Powerful diuretic effective in promoting sodium and chloride secretion
Hydrochloric	Harmful, causes burns
Hypochlorous	Disinfectant and bleaching agent, its salts are used as surgical solutions, e.g. sodium hypochlorite
Nitric	Harmful, causes burns
Sulfuric	Harmful, causes burns

Table 11.2 Summary of acids that are used by the body

Acid	Function
Acetoacetic	Ketone body produced during fat and protein breakdown; excessive formation results in ketoacidosis
Amino	Building blocks of protein
Ascorbic	Promotes many metabolic reactions; may combine with poisons rendering them harmless
Carbonic	Involved in the regulation of pH
Cholic	Most abundant bile acid; forms salts that assist in the digestion of lipids
Deoxyribonucleic (DNA)	Carrier of genetic information
Fatty	Important in triglyceride (neutral fat) and phospholipid structure
Folic	Vitamin that is essential for the normal production of red and white blood cells
Hyaluronic	Important interstitial chemical
Hydrochloric	Important digestive agent in gastric juice
Lactic	Waste product of anaerobic carbohydrate breakdown
Ribonucleic (RNA)	Transmits genetic information from DNA to protein-forming system of the cell
Uric	Breakdown product of nucleus metabolism and causes gout

dissociates into H+ ions and anions (Eqn 11.1).

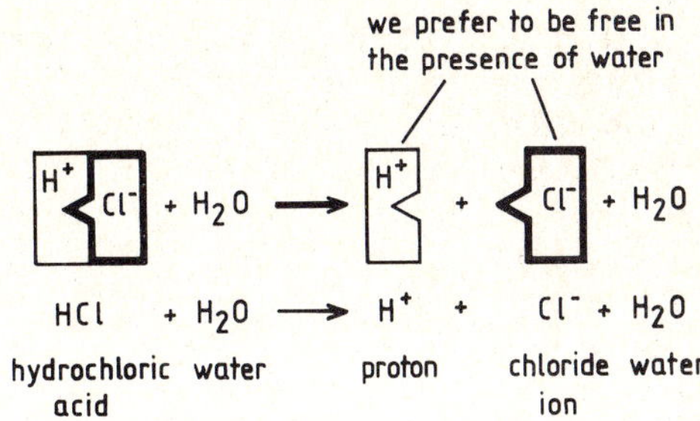

The arrow indicates that most of the HCI has been converted into H+ and Cl⁻. A weak acid only dissociates to a limited extent, with most of the hydrogen atoms being retained by the weak acid (Eq 11.2).

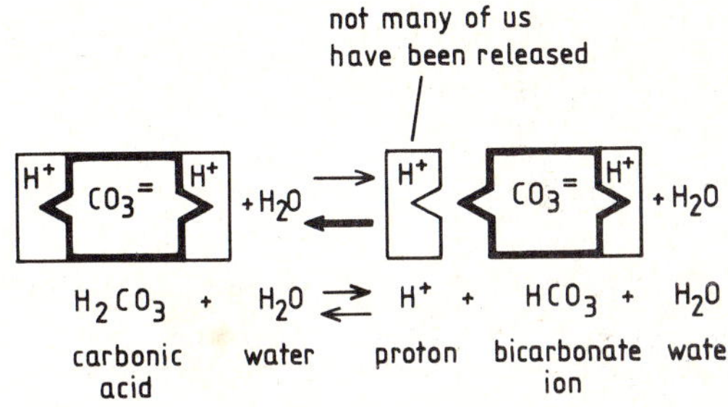

The two arrows indicate that a chemical balance of fixed proportion exists between the concentration of carbonic acid and its dissociation products. The two dissociation products can also be combined to form the weak acid. Any increase in the H+ ion and HCO_3^- ion concentrations will alter the chemical balance or equilibrium. This imbalance is returned to the correct proportions by the formation of the necessary concentration of carbonic acid.

Hydrogen consists of one proton and one electron. When an electron is removed H+ is formed which is a single proton. Since an acid donates H+ ions it therefore, in effect, is a proton donor (Figure 11.1). These free protons are responsible for the properties of an acid.

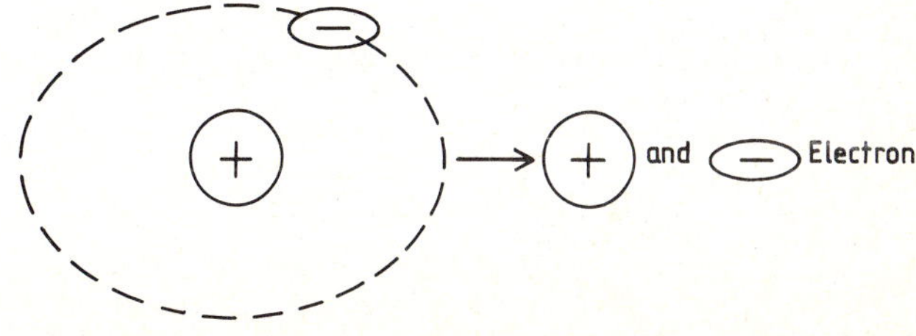

Figure 11.1 Hydrogen consists of a proton (H+) and an electron.

GENERAL PROPERTIES

Taste

An acid has a sour taste. EXAMPLE: Citric acid causes the sour taste in lemons and grapefruit.

Indicator Change

Acids change the colour of substances called indicators. EXAMPLE: Blue litmus indicator changes to red in the presence of an acid.

React with Bases

Acids combine with bases to produce water and a salt. This process is called neutralization. EXAMPLE: Hydrochloric acid combines with sodium hydroxide to produce water and sodium chloride (Eqn 11.3).

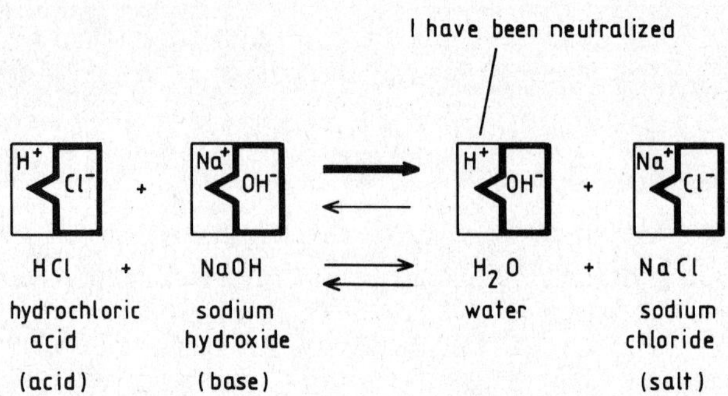

Corrosiveness

Acids are corrosive to body tissue. The stronger the acid the greater is the corrosive effect on tissue.

APPLICATION 1: ACID BURNS

Care must be exercised when handling strong acids such as concentrated hydrochloric acid (HCl), sulphuric acid (H_2SO_4) and nitric acid (HNO_3). Acid burns require a dilution of the spilt acid with copious quantities of water. This dilution is followed by a neutralization of the acid with, for example, sodium bicarbonate.

APPLICATION 2: GASTRIC ULCERS

Excessive production of hydrochloric acid by cells lining the stomach results in the gastric acidity being increased. Over a period of time an elevated gastric acidity can lead to the formation of a gastric ulcer. This is a hole eaten into the stomach wall. These ulcers can also result from a localized loss of the protective mucous barrier that lines the gastric wall.

React with Carbonates

Acids react with carbonates to form carbon dioxide, water and salts.

APPLICATION: GASTRIC ANTACIDS

Sodium bicarbonate (baking powder, bicarbonate of soda) is administered to counteract excess stomach acidity. The combination of sodium bicarbonate and hydrochloric acid produces sodium chloride and carbonic acid (Eqn 11.4).

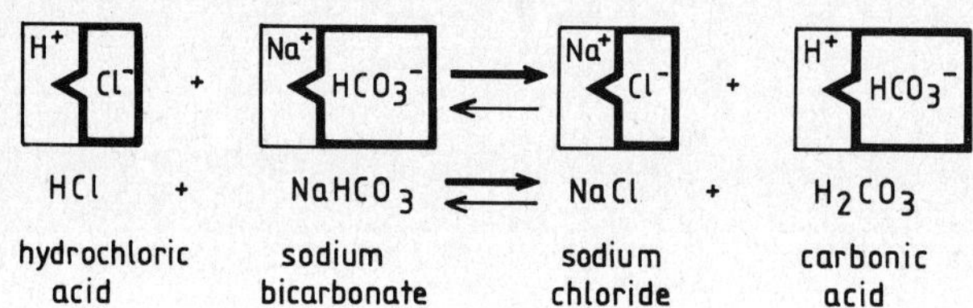

Carbonic acid is a weak acid and thus it releases only a small quantity of H+. In Eq 11.1 a strong acid has been replaced by a weak acid. A decrease in free H+ ions has resulted in a lower acidity. Carbonic acid breaks down to form water and carbon dioxide. (Eqn 11.5). This equation is also reversible with carbon dioxide and water producing carbonic acid.

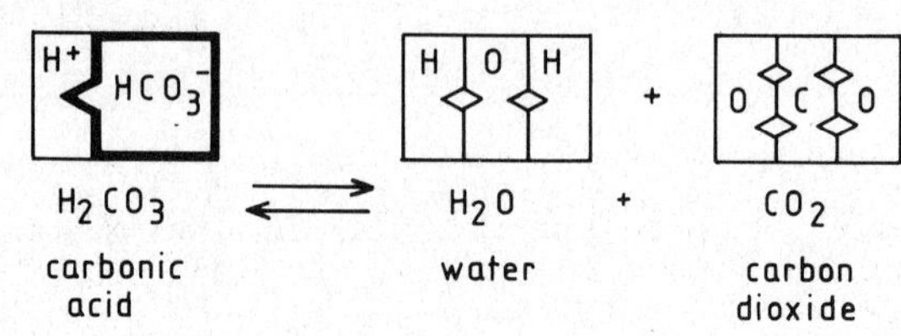

11.2 Bases

WHAT IS A BASE?

Bases are used in the body for neutralizing any increases in acidity. They are a component of the homeostatic mechanism that regulates the acidity of body fluids. Water-soluble bases are called alkalis. The terms "basic solutions" and "alkaline solutions" are generally interchangable.

A base is a substance that donates hydroxyl ions (OH^-) or accepts H+ ions from a solution. A strong base donates relatively large quantities of OH^- or accepts many H+ ions. That is, the strength of a base depends upon the degree of dissociation.

GENERAL PROPERTIES

Taste

Bases have a bitter metallic taste.

Indicator Change

Bases change the colour of substances called indicators. EXAMPLE: Red litmus paper changes to blue in the presence of a base.

React with Acids

Bases neutralize acids by removing protons from solution. This is achieved by combining the proton with an OH^- ion (Eqn 11.6) or another anion (Eqn 11.7).

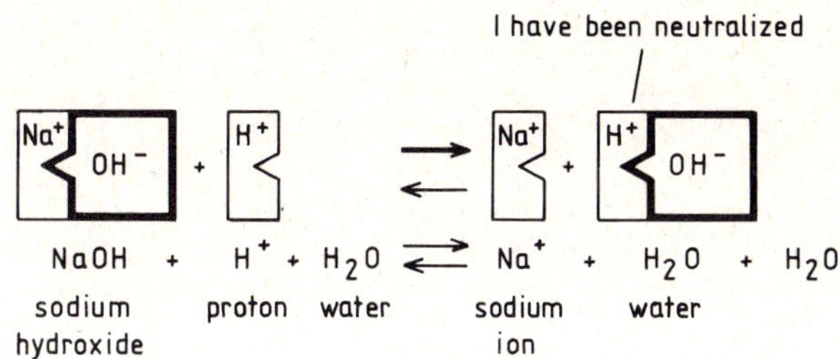

Corrosiveness

Strong bases such as sodium hydroxide (NaOH) are corrosive to body tissue. They can also react with fats and dissolve proteins.

11.3 Measurement of H^+ and OH^-

Whenever the number of H^+ ions and OH^- ions is equal, a neutral solution exists. When the H^+ ion concentration is greater than the concentration of OH^- ions the solution is an acid. Solutions that contain a greater concentration of OH^- ions than H^+ ions are basic solutions.

In one litre of pure water, 10^{-7} moles of water is dissociated into H^+ ions and OH^- ions (Eqn 11.8).

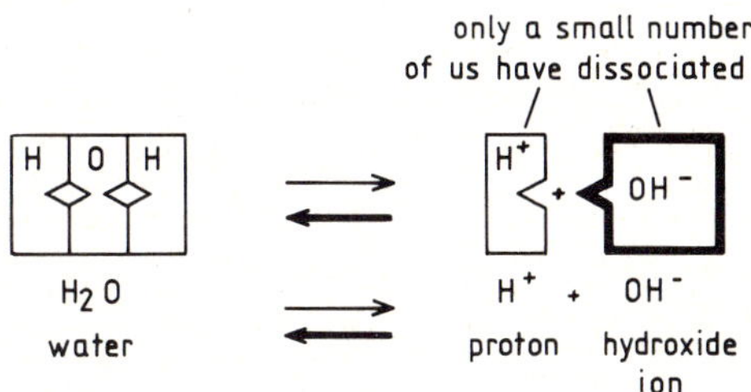

Pure water is a neutral solution since it contains 10^{-7} moles of H^+ ions and 10^{-7} moles of OH^- ions. In water, a balance exists between the proportion of H^+ ions and OH^- ions for any given acid or base. The properties of water are such that the minimum concentration of H^+ ions and OH^- ions is 10^{-14} mol/L. The maximum concentration of H^+ ions and OH^- ions is 1 mole or 10^0, that is, the dissociation properties of water restrict the concentration of H^+ ions and OH^- ions to the range of 10^{-14} to 10^0 mol/L.

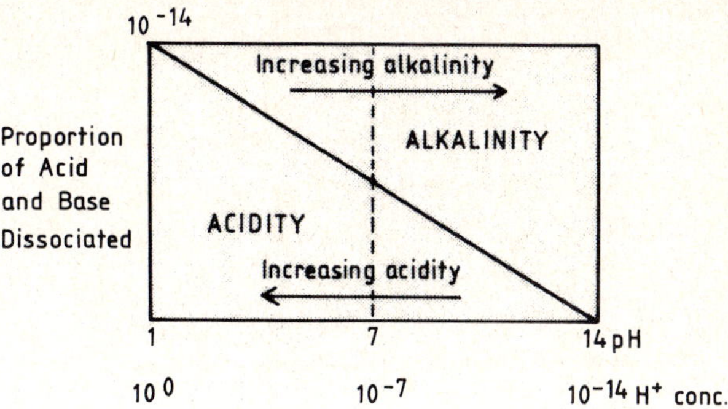

Figure 11.2 Relationship between pH and the concentration of H^+ and OH^- ions. The total number of H^+ and OH^- ions is constant, and at pH 7 the number of H^+ and OH^- ions is equal.

A scale of acidity and alkalinity has been devised that uses the range of 10^{-14} to 10^0 mol/L of H^+ and OH^- ions. The addition of H^+ ions to pure water results in a proportionate decrease in OH^- ions from the initial concentration of 10^{-7} mol/L. A H^+ ion concentration of 10^{-6} mol/L results in a OH^- concentration of 10^{-8} mol/L, whereas a H^+ ion concentration of 10^{-3} mol/L results in a OH^- concentration of 10^{-11} mol/L. Note that in both these examples the combination of the powers of ten totals -14. The combination of the powers of ten for H^+ ions and OH^- ions will always equal -14.

The pH scale represents the powers of ten of H^+ ions from 10^{-14} to 10^0 mol/L. For convenience, this scale uses positive numbers and expresses them as pH 14 to pH 0 (Figure 11.2). At pH 7 an equal number of H^+ ions and OH^- ions occur. Solutions with a pH lower than 7 are acids and those with a pH greater than 7 are bases.

In blood the physiologically normal pH range is 7.35 to 7.45. A blood pH lower than 7.35 is considered acidic whereas a pH greater than 7.45 is alkaline, either of which can have a serious effect on body function.

A change of one pH unit differs in the number of H^+ ions gained or lost when compared with another pH unit.

EXAMPLE:
pH 8 contains 10^{-8} mol/L H^+ or 10 nmol
pH 7 contains 10^{-7} mol/L H^+ or 100 nmol
pH 6 contains 10^{-6} mol/L H^+ or 1000 nmol

A change from pH 7 to pH 8 represents 100 - 10 = 90 nmol, whereas a change from pH 6 to pH 7 is 1000 - 100 = 900 nmol.

Every change in a pH unit represents a tenfold change in H^+ ion concentration. An alteration of pH from pH 7.4 to 7.3 results in a doubling of the H^+ ion concentration. A small change in pH indicates a very significant alteration in H^+ ion concentration. The pH scale actually represents a logarithmic scale and not a linear scale. The H+ ion concentration can also be expressed in nmol/L. This is a linear scale with pH 7 being 100 nmol/L, pH 6 = 1000 nmol/L and pH 8 = 10 nmol/L.

11.4 Buffers

Buffers are chemicals that resist changes in pH on addition of acids or bases. They act by replacing a strong acid or base with a weak one, that is, a stronger acid is replaced by a weaker acid. Since a weak acid releases only a small number of H+ ions then the formation of a weak acid effectively removes most of the H+ ions that were present with a strong acid (Eqn 11.9). EXAMPLE: Lactic acid (strong acid) is replaced by a carbonic acid (weak acid) thereby reducing the number of free H+ ions in solution.

$$(\text{lactate} + H^+) + HCO_3^- \rightleftharpoons H_2CO_3 + \text{lactate}^-$$

lactic acid — bicarbonate ion — carbonic acid — lactate ion

APPLICATION: BICARBONATE/CARBONIC ACID BUFFER

In the bicarbonate/carbonic acid buffer system bicarbonate combines with excess H+ ions whereas carbonic acid combines with excess OH⁻ ions. When excess H+ ions are added to the blood they may combine with the bicarbonate ion to form carbonic acid, which is a weak acid and thus does not release H+ ions readily (Eqn 11.7). Removing the majority of H+ ions results in the blood retaining its original pH. Addition of OH⁻ ions to the blood can cause carbonic acid to combine with it and form bicarbonate ions (Eqn 11.10). The removal of excess hydroxyl ions results in the blood retaining its original pH.

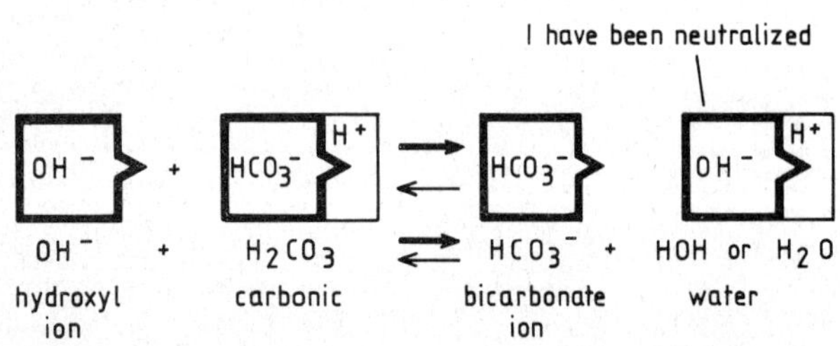

Buffers are used by the body to maintain very narrow pH ranges of body fluids (Table 11.3). These narrow ranges are necessary since biochemical reactions are extremely sensitive to changes in H+ ion concentration. The pH at which biochemical regulators or enzymes function will vary

Table 11.3 The narrow pH range within fluids, and the diversity of pH between different fluids

Body fluid	Locality	pH
Cerebrospinal fluid	Associated with the central nervous system	7.4
Blood	Vascular system	7.35 - 7.45
Bile	Gall bladder and bile duct	7.6 - 8.6
Gastric juice	Stomach	1.2 - 3.0
Intestinal juice	Small intestine	7.6 - 7.8
Pancreatic juice	Pancreas	7.1 - 8.2
Saliva	Mouth	6.5 - 7.5
Semen	Vas deferens	7.35 - 7.50
Urine	Bladder	5.5 - 7.5
Vaginal fluid	Vagina	5.2

for different enzymes. An enzyme that operates in the highly acidic stomach cannot function effectively in the alkaline small intestine. Also, the elimination of carbon dioxide and the transport of oxygen are dependent upon blood pH, small variations result in marked alterations in the amount of carbon dioxide removed by the lungs. Buffers are needed to counteract the non-stop production of acids. These acids mainly result from the breakdown of carbohydrates, fats and proteins. EXAMPLES: Lactic acid and ketone bodies.

11.5 Acid-base balance

BODY-FLUID BUFFERS

Carbonic Acid-Bicarbonate Buffer System

This buffer system is important in the regulation of pH changes that result from alterations in blood carbon dioxide levels. The relationship between carbon dioxide and this buffer system is shown in Eqn 11.11.

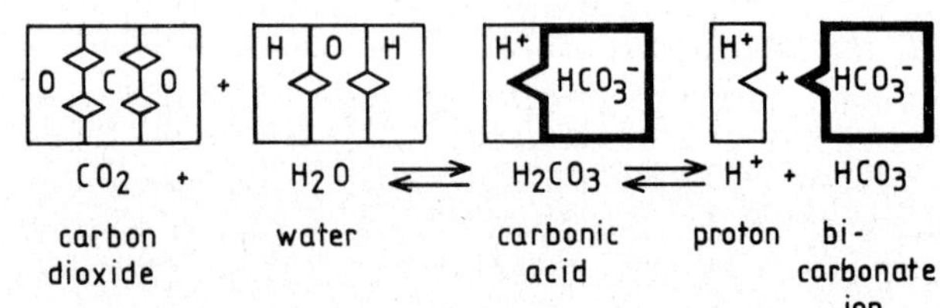

Carbon dioxide combines with water to produce carbonic acid. This is a weak acid and thus a small proportion of the acid is dissociated into H+ ions and HCO_3^- ions.

The bicarbonate and carbonic acid buffer system acts in the following way. Factors that elevate the H+ ion concentration cause a proportionate release of HCO_3^- ions. HCO_3^- combines with excess H+ ions to produce carbonic acid (Eqn 11.7). Excess carbonic acid dissociates into carbon dioxide and water which is removed by the lungs. An occurrence of excess OH^- ions results in carbonic acid releasing a proportionate quantity of H+ ions to neutralize the excess OH^- ions (Eqn 11.10).

The body tends to acidify the blood and thus needs more HCO_3^- than carbonic acid. It is necessary to have a proportion of one carbonic acid molecule to twenty HCO_3^- ions (1.2 mmol/L : 24 mmol/L) if the pH of the blood is to remain in the normal range. Alterations to this ratio will result in a proportionate change in the relative concentration of carbonic acid and HCO_3^- so that the chemical equilibrium ratio is returned. Therefore, any factor that causes an increase in the ratio of carbonic acid to HCO_3^- indicates a greater acidity, since a greater amount of HCO_3^- has been used to remove the excess H+ ions. Thus a greater proportion of carbonic acid has been formed. To overcome an increase in acidity, HCO_3^- is administered to a patient so that the proportion of carbonic acid to HCO_3^- is returned to the ratio of one to twenty.

Whenever a buffering reaction occurs, the concentration of one chemical in the buffer pair is increased while a decreased concentration results in the other. An elevated carbonic acid concentration results from a decreased HCO_3^- concentration.

In summary, HCO_3^- soaks up excess H+ ions and carbonic acid soaks up excess OH^- so that the pH is maintained at normal levels. The excess carbonic acid can be eliminated by the lungs in the form of carbon dioxide and water vapour.

Phosphate Buffer System

This buffer system consists of an acid phosphate, $H_2PO_4^-$ and an alkaline phosphate, HPO_4^{2-}. Addition of H+ converts the alkaline phosphate into the weak acid phosphate. Likewise, addition of OH^- results in alkaline phosphate formation.

The phosphate buffer system is an important control mechanism in red blood cells, kidney tubular fluid and cellular contents. In kidney tubules excess H+ ions combine with HPO_4^{2-} to form the weak acid $H_2PO_4^-$ which is then excreted.

Protein Buffer System

This is the most abundant buffer in cells and plasma. Proteins consist of chains of amino acids (see 24.2). The general structure of an amino acid is shown in Figure 11.3a. These acids consist of a weak acid group called the carboxyl group (Figure 11.3b) and a base structure known as the amine group (Figure 11.3c). The carboxyl group donates its proton to the amine group forming a zwitterion. This is a covalent compound containing positive and negative ionic regions (Figure 11.3d). When amino acids exist as zwitterions they can function as buffers. When excess H+ ions are present the COO^- combines with the H+ (Figure 11.4).

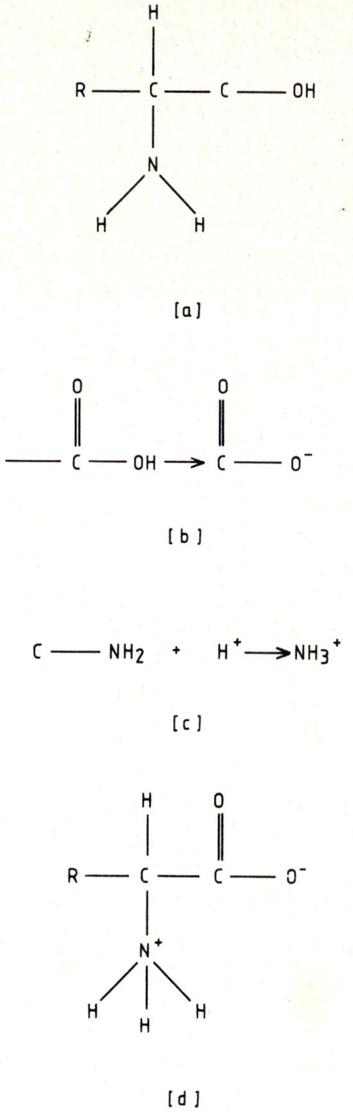

Addition of excess OH^- ions results in the release of H+ from the amine group to combine with the OH^- ions to form water (Figure 11.4).

Figure 11.3 (a) General structure of an amino acid, (b) carboxyl group acts as an acid by releasing H+ ions, (c) the amine group acts as a base by accepting H+ ions. (d) the net result of donation and acceptance of a H+ ion is an internal self-neutralization called a zwitterion.

Ammonia-Ammonium Buffer System

This buffer system is used by the kidneys to eliminate H+ ions from the body. Amino acids such as glutamine are broken down to form the base ammonia (NH_3). This base combines with excess H+ ions to form the weak acid ammonium (NH_4^+) which combines with Cl^- to form ammonium

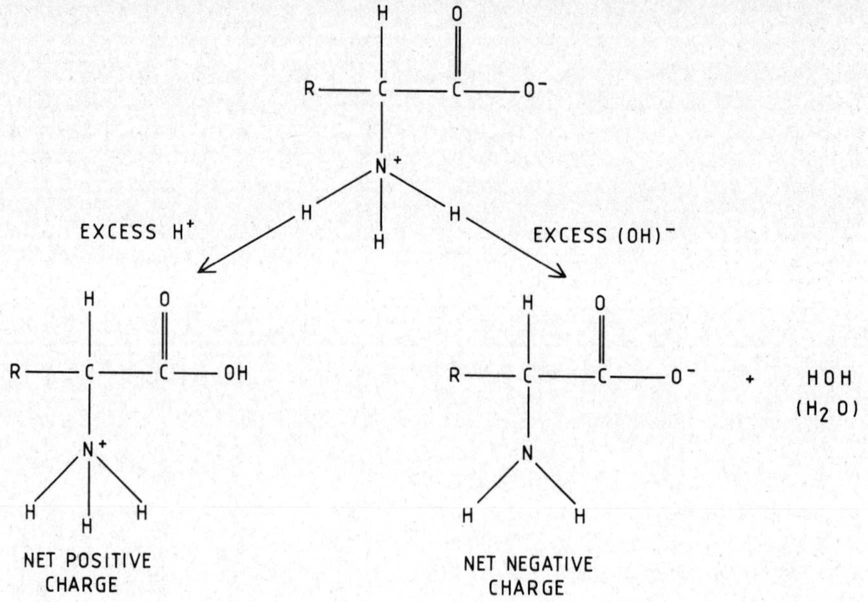

EXCESS H^+

EXCESS $(OH)^-$

NET POSITIVE CHARGE

NET NEGATIVE CHARGE

+ HOH (H_2O)

Figure 11.4 An individual zwitterion acts as either an acid buffer or a base buffer according to the relative concentration of H+ and OH^- ions present in solution.

chloride (NH_4Cl). Removal of ammonium chloride in the urine results in the removal of H+ ions without the need for a high urine acidity.

Haemoglobin Buffer System

In red blood cells carbon dioxide and water are converted to carbonic acid by the action of an enzyme. The H+ ions released by this acid combine with a base known as reduced haemoglobin (Figure 11.5). When haemoglobin loses oxygen it is called reduced haemoglobin. The formation of a weaker acid than carbonic acid results in a lower concentration of H+ ions within the red blood cells. Reduced haemoglobin soaks up most of the H+ ions released by carbonic acid and in turn fewer H+ ions are donated to the intracellular fluid. Hence fewer H+ ions have been added to the intracellular fluid than would have been added without an additional buffer.

The presence of large quantities of oxygen promotes the loss of H+ ions from haemoglobin. In pulmonary capillaries the high oxygen concentrations convert reduced haemoglobin into oxygenated haemoglobin or oxyhaemoglobin. Since H+ ions are released more readily with oxyhaemoglobin than with reduced haemoglobin, the former is therfore a stronger acid than reduced haemoglobin.

The H+ ions released in the presence of high oxygen concentrations are buffered by bicarbonate to form carbonic acid. This acid breaks down to carbon dioxide and water. The carbon dioxide moves into the blood and subsequently into the lungs (Figure 11.6).

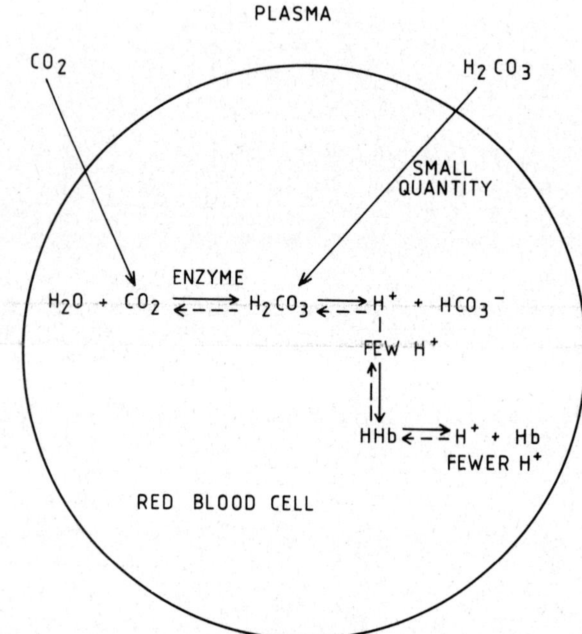

Figure 11.5 The buffering of carbonic acid by reduced haemoglobin (Hb) results in the addition of fewer H+ ions to the red blood cell intracellular fluid. A small quantity of carbonic acid originates from the plasma and the removal of this acid results in a lower H+ ion concentration in the blood.

In summary, the haemoglobin buffer system is important in both the regulation of blood pH as it passes through the tissues and the release of carbon dioxide in the pulmonary capillaries.

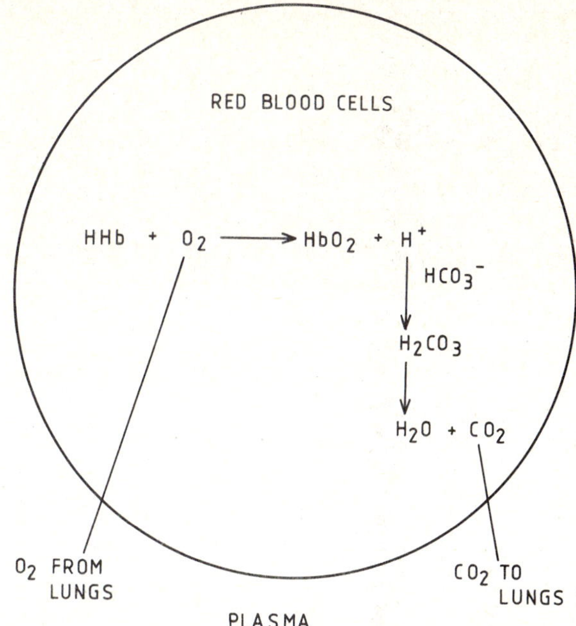

Figure 11.6 The influence of oxygen on reduced haemoglobin (HHb) to form the stronger acid oxyhaemoglobin (HbO_2). The subsequent release of H+ ions is buffered by bicarbonate to form carbonic acid. This acid is broken down to water and carbon dioxide with the carbon dioxide being released to the lungs.

RESPIRATORY REGULATION

An increase in cell respiration effects an elevation in serum carbon dioxide concentration which results in an increase in acidity (Eqn 11.11).

Removal of carbon dioxide by breathing results in an alteration of the chemical balance in Eqn 11.11. The net effect of breathing is that elimination of carbon dioxide results in a proportionate decrease of carbonic acid and H+ as these substances are converted to carbon dioxide and water and carbonic acid respectively.

Alterations in breathing rate can greatly influence the pH of the blood since alterations in carbon dioxide elimination change serum H+ ion concentration. Doubling the breathing rate effects an increase in pH by 0.23 units whereas reducing the ventilation rate by a quarter results in a pH decrease of 0.4. From Eqn 11.11 an increase in serium carbon dioxide causes an elevation of H+ ions whereas a decrease in serum carbon dioxide results in a decrease in H+ ions.

Eqn 11.11 also shows why H+ ion concentration influences ventilation rates. The greater the acidity in blood the greater is the need for the body to remove these excess H+ ions. The more carbon dioxide that is eliminated by the lungs, the greater is

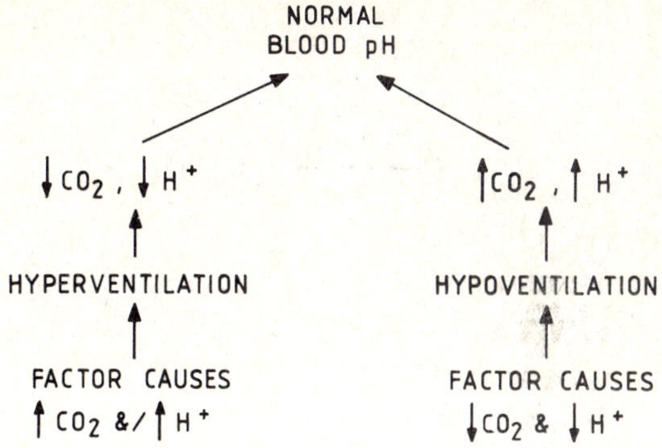

Figure 11.7 The relationship of ventilation to alterations in blood acidity.

the amount of H+ ions and bicarbonate that is converted to carbonic acid. The resulting disproportionate levels of carbonic acid is altered by a proportion being converted to carbon dioxide and water so that the relative concentrations of each substance is returned to their stable proportions. Thus the effect of increased acidity is to increase the levels of carbon dioxide.

Removal of the excess carbon dioxide occurs by an increase in ventilation rates (hyperventilation). High blood acid levels stimulate an increase in ventilation rates. Likewise, a decrease in serum H+ concentration requires the retention of carbon dioxide by slowing the breathing rate (hypoventilation). The buildup of carbon dioxide results in a proportion of it being converted to carbonic acid, H+ and bicarbonate.

The sensitivity of the respiratory rate to changes in pH is such that the ventilation rate is approximately doubled if the pH decreases by 0.1 units, and halved if the pH increases by 0.1 units.

In summary, hyperventilation decreases carbon dioxide and thus H+ ions, whereas hypoventilation increases carbon dioxide and thus H+ ions (Figure 11.7). The rate of respiration stabilizes the carbonic acid to bicarbonate ratio by altering the rate of ventilation so that the proportion of one carbonic acid to twenty bicarbonate ions is maintained.

RENAL REGULATION

The kidney serves two important functions in H+ ion homeostasis:

1. They reabsorb bicarbonate from the urine to the blood. This prevents the loss of bicarbonate and the subsequent alteration of the carbonic acid/bicarbonate buffer system.
2. They counteract a rise in serum H+ concentration by secreting the excess H+ ions in the form of weak acids of the phosphate and ammonium buffer systems. Acids such as sulphuric, uric and keto acids which cannot be removed by the lungs are eliminated.

Kidney cells effect both of these functions by the

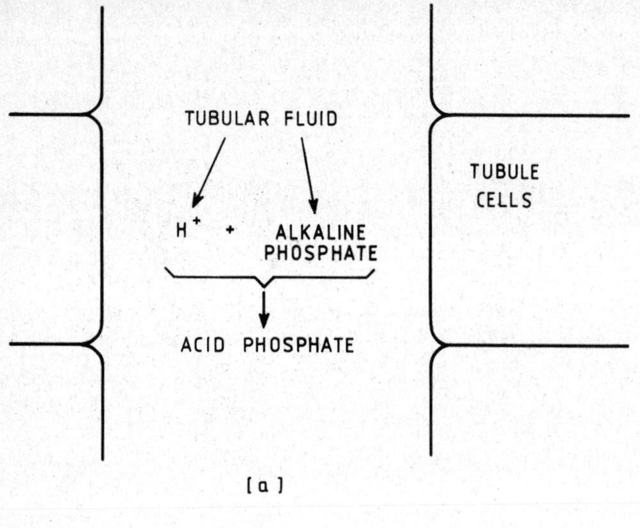

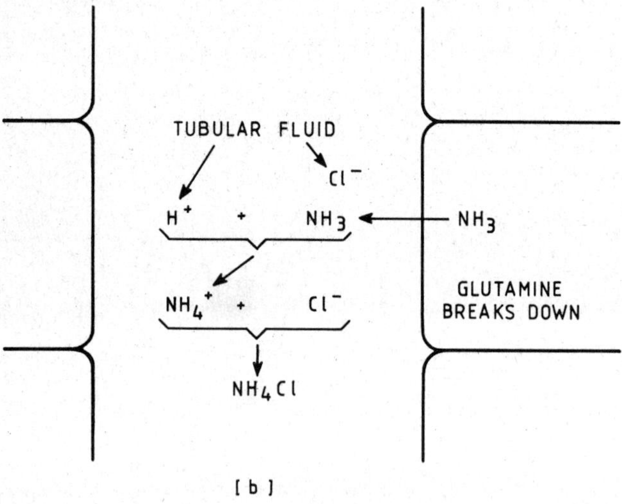

Figure 11.8 Mechanisms involved in regulation of elevated H+ ion concentration in tubular fluid: (a) elevated levels of acidity are buffered by phosphate, whereas (b) a substantial increase in H+ ion concentration requires the ammonia buffer system.

following mechanisms. The kidney tubule cells are sensitive to changes in blood pH. Within these cells a decrease in pH accelerates the conversion of carbon dioxide and water into H+ and bicarbonate. The bicarbonate is released into the blood and assists in returning the blood pH back to normal levels.

H+ ions are secreted into the kidney tubular fluid. This fluid contains alkaline phosphate which combines with the excess H+ ions to form a weak acid called acid phosphate. The effective removal of most of the H+ ions returns the urine pH to normal levels (Figure 11.8a).

When acidity increases substantially in the tubular fluid, the tubule cells form ammonia (NH_3) by breaking down the amino acid glutamine. The ammonia is secreted into the tubular fluid where it combines with H+ ions to form the weak acid NH_4^+. The NH_4^+ combines with Cl^- to form NH_4Cl or ammonium chloride (Figure 11.8b).

Decreased H+ ion concentration results in large quantities of bicarbonate not being reabsorbed. Thus the ratio of serum carbonic acid to bicarbonate is returned to normal since bicarbonate levels are lowered.

11.6 Acid–base imbalance

Acidosis occurs when the serum pH value falls below 7.35. There are two forms of acidosis called respiratory acidosis and metabolic acidosis. Acidosis depresses the central nervous system, which, if the acidosis remains uncorrected, will result in a decreased mental capacity, delirium, coma and death.

Alkalosis exists at serum pH levels greater than 7.45. As with acidosis, there are two forms of alkalosis called respiratory alkalosis and metabolic alkalosis. In this pH imbalance, the nervous system becomes overexcited. Failure to correct this disorder results in muscle spasm, convulsions and eventually death.

RESPIRATORY ACIDOSIS

This results from any condition that decreases the removal of carbon dioxide by the lungs, that is, the removal of carbon dioxide does not keep pace with its production. Elevated arterial carbon dioxide concentration causes a rise in blood H+ ion concentration. Decreased carbon dioxide removal results from chronic respiratory diseases such as pneumonia and emphysema. In addition, drug abuse such as excess morphine therapy and barbiturate poisoning can diminish respiratory function and therefore carbon dioxide removal. Arterial carbon dioxide can also increase as a result of impaired pulmonary circulation.

Elevated arterial carbon dioxide results in an increased carbonic acid and subsequently bicarbonate level (Eq 11.11). The bicarbonate is increased so that the proportion of carbonic acid to bicarbonate is returned to its normal buffering capacity of one to twenty (Figure 11.9). Thus over a period of time the body compensates for the elevated carbonic acid by stimulating the kidneys to release an increased quantity of bicarbonate ion. At the onset of respiratory acidosis the ratio may have changed from one : twenty to two : twenty (Figure 11.10a). The kidneys respond to this acidosis by releasing bicarbonate ions until the ratio is once again one to twenty, that is, over one to two days the bicarbonate ion concentration increases so that the

Figure 11.9 To maintain the normal blood pH range of 7.35 to 7.45 a ratio of 1 H_2CO_3:20 HCO_3^- is required.

ratio is two to forty (Figure 11.10b). Compensation therefore effects an increased blood bicarbonate level.

METABOLIC ACIDOSIS

Metabolic acidosis results from an abnormal increase in acidic metabolic products other than carbon dioxide. The acidosis may be caused by:
1. an inability of the kidneys to eliminate the dietary H^+ ion load due to renal disorders,
2. elevated H^+ ion load due to diseases such as diabetes mellitus, starvation, excess lactic acid production due to insufficient oxygen supply or ingestion of salicylates,
3. excessive bicarbonate loss occurs in renal tubular failure where bicarbonate reabsorption is diminished, and also severe diarrhoea since large amounts of sodium bicarbonate are lost in the gastrointestinal contents.

The accumulated metabolic acids react with bicarbonate in the extracellular fluid to form carbonic acid, that is, the ratio of carbonic acid to bicarbonate is altered such that a greater proportion of carbonic acid is formed. This change in the buffering capacity results in a lower blood pH. The decreased pH stimulates hyperventilation or increased breathing rate which effects a greater removal of carbon dioxide from the lungs. Increased carbon dioxide levels, which according to Eqn 11.11, lowers the concentration of carbonic acid in the blood. If the carbonic acid is reduced sufficiently, the ratio of carbonic acid to bicarbonate will return to normal. Figure 11.11a shows a change in the ratio during metabolic acidosis of one carbonic acid to ten bicarbonate ions.

Increased breathing rates reduce the blood carbonic acid concentration to a level such that the ratio is returned to normal, that is, carbonic acid concentration is decreased so that the ratio of carbonic acid to bicarbonate is 0.5:10, which is effectively 1:20. A respiratory compensation has maintained the carbonic acid/bicarbonate ion buffer system, and thus near normal pH, but at the expense of decreased bicarbonate (Figure 11.11b).

When the production of acid is too great for the compensatory mechanisms, the blood pH falls causing disturbances to the nervous system. In summary, compensated metabolic acidosis has low levels of blood bicarbonate and carbon dioxide along with a normal pH, whereas uncompensated metabolic acidosis has a decreased blood bicarbonate and pH but a normal arterial carbon dioxide level. Metabolic acidosis is treated by correcting the factors causing the imbalance, plus administration of sodium bicarbonate.

RESPIRATORY ALKALOSIS

Respiratory alkalosis results from hyperventilation removing excessive quantities of carbon dioxide. A decrease in blood carbon dioxide and carbonic acid levels results. The carbonic acid to bicarbonate ratio is altered, with the carbonic acid decreasing with respect to the bicarbonate. A ratio of 0.5:20 indicates that a relative excess of bicarbonate ions exist (Figure 11.12a).

Body compensatory mechanisms or treatment may adjust the pH to normal levels by returning the carbonic acid to bicarbonate ratio to 1:20 (Figure 11.12b). This ratio can be achieved by either decreasing the concentration of bicarbonate (0.5:10) or by increasing the carbonic acid concentration (1:20). Some causes of hyperventilation are hysteria, anxiety, high fever, aspirin overdose, hypoxia, brain injury and excessive artificial ventilation.

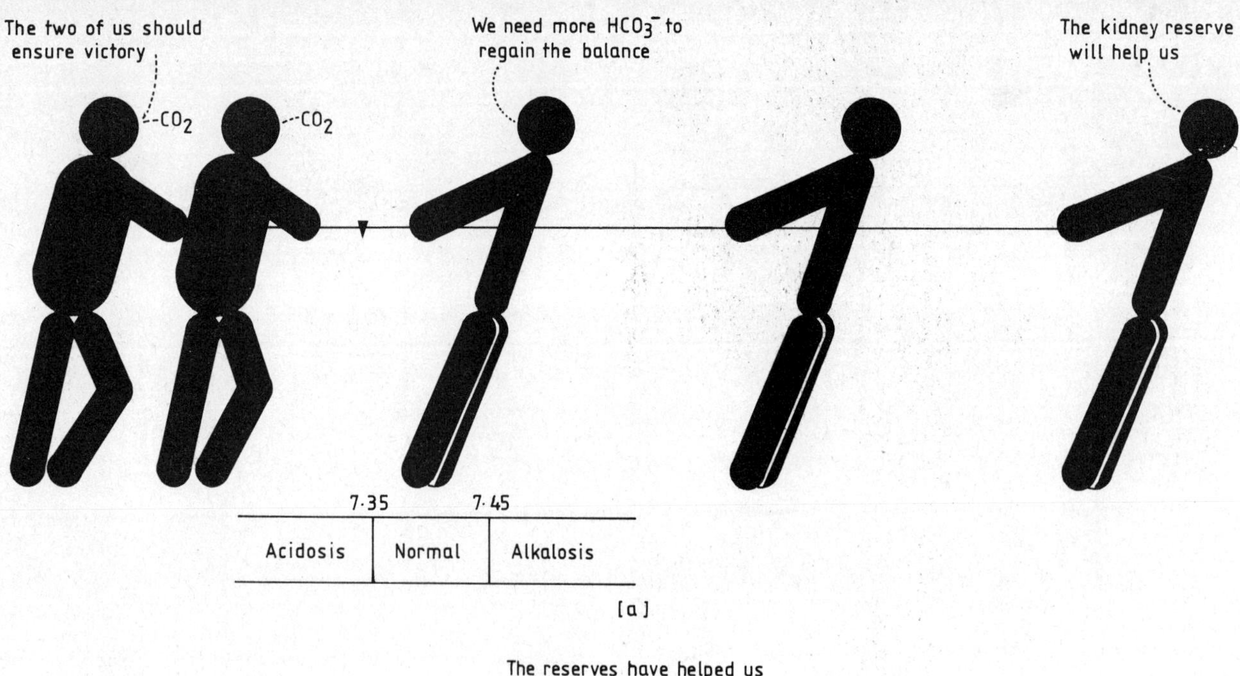

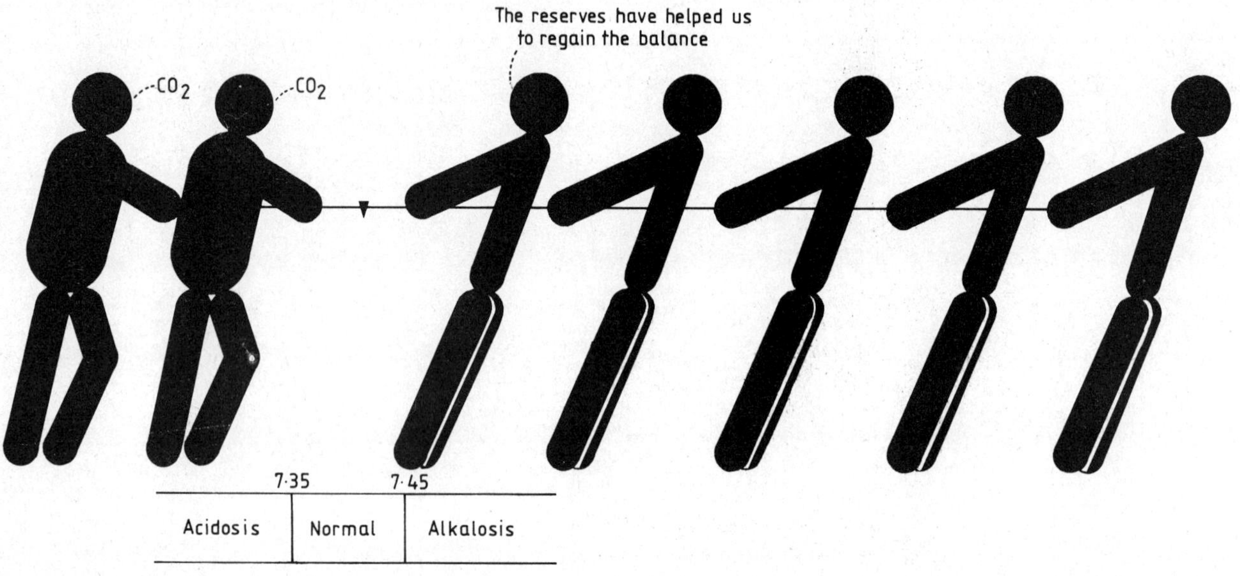

Figure 11.10 Respiratory acidosis: (a) uncompensated respiratory acidosis results in elevated CO_2 and H_2CO_3. The ratio of H_2CO_3:HCO_3^- is now 2:20, (b) Compensated respiratory acidosis results in increased CO_2, H_2CO_3 and HCO_3^-. The ratio of H_2CO_3:HCO_3^- is now 2:40, that is 1:20.

METABOLIC ALKALOSIS

Metabolic alkalosis results from a decreased H^+ ion concentration or an elevated bicarbonate ion concentration. Two causes of this imbalance are the loss of acid due to vomiting of gastric juice and the injestion of large amounts of antacids for indigestion relief.

The relative excess of bicarbonate may be compensated by decreasing carbon dioxide removal by the lungs and thus increasing carbonic acid in the blood. Elevating carbonic acid will alter the alkalotic state by returning the carbonic acid : bicarbonate ratio from an excess of bicarbonate ions (1:40) to the normal ration of 2:40 or 1:20 (Figure 11.13). The kidneys assist this compensation by excreting excess bicarbonate ions.

Metabolic alkalosis can also result from a decrease in blood K^+ concentration. EXAMPLE: A decrease in K^+ concentration can result from the use of diuretics such as frusemide (lasix, frusid) and ethacrynic acid (edecril). In these cases administration of K^+ is prescribed.

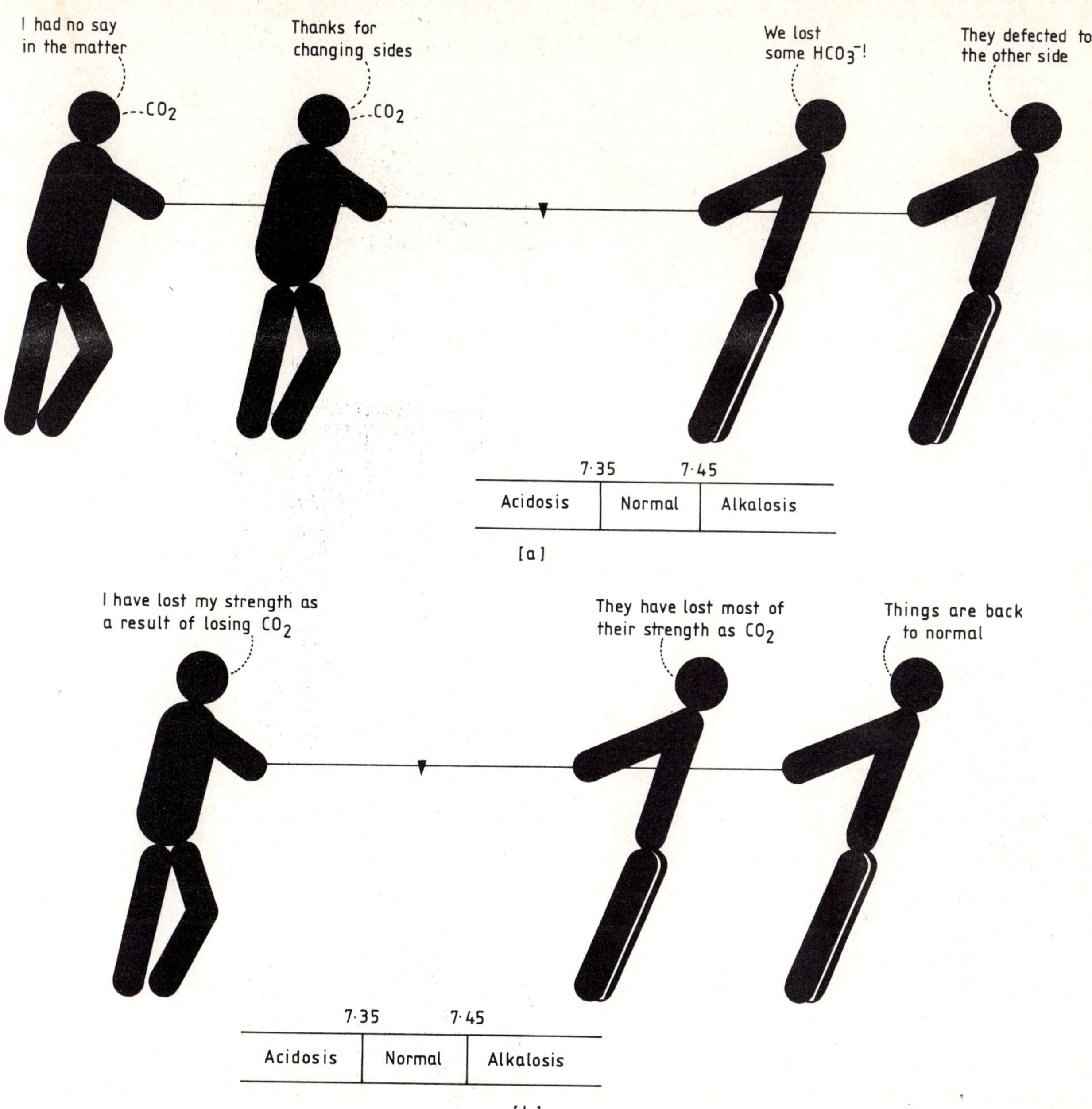

Figure 11.11 (a) Metabolic acidosis results in a decreased HCO_3^- concentration and an elevated H_2CO_3 concentration. The ratio of H_2CO_3:HCO_3^- is now 1:10. (b) Compensation occurs by removing the excess H_2CO_3 as CO_2. The ratio of H_2CO_3:HCO_3^- is 0.5:10, that is 1:20.

11.7 Salts

A salt is an ionic compound, that is, a compound composed of cations and anions other than H^+ or OH^-. The salt is usually formed by a metal combining with a non-metal. A salt can also be formed by combining an acid with a base in the process of neutralization. Salts are important in body function and medical use (Table 11.4).

Generally, salts are soluble in water when the ions in the salt have a greater attraction for water molecules than for oppositely-charged ions within the salt. The greater attraction for water molecules by ions results in the dissociation of the salt into cations and anions in the process of ionization. The process of ionization occurs when the interaction between solute and solvent results in the formation of ions. Unlike acids or bases, the dissociation of salts does not release H^+ or OH^- ions in water.

Not all compounds that are composed of metals and non-metals are salts, as such substances can

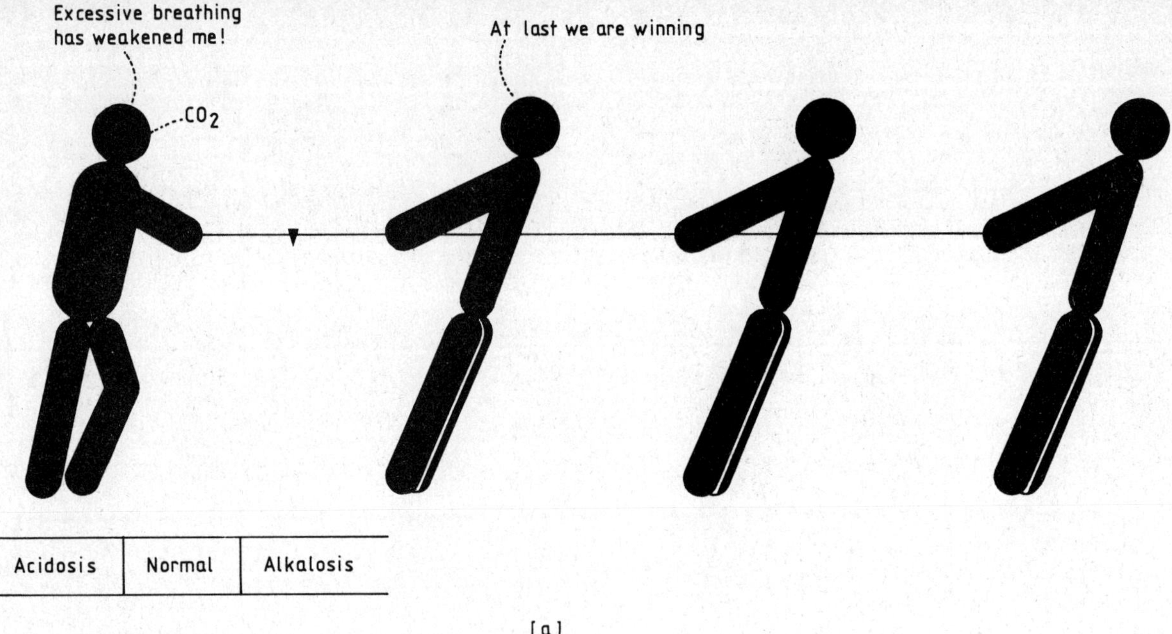

Figure 11.12 Respiratory alkalosis. (a) Hyperventilation causes decreased blood CO_2 and H_2CO_3 level. The H_2O_3:HCO_3^- ratio is 0.5:20. (b) Compensation can occur by either returnign the H_2CO_3 level to normal or by proportionally decreasing the HCO_3^- concentration.

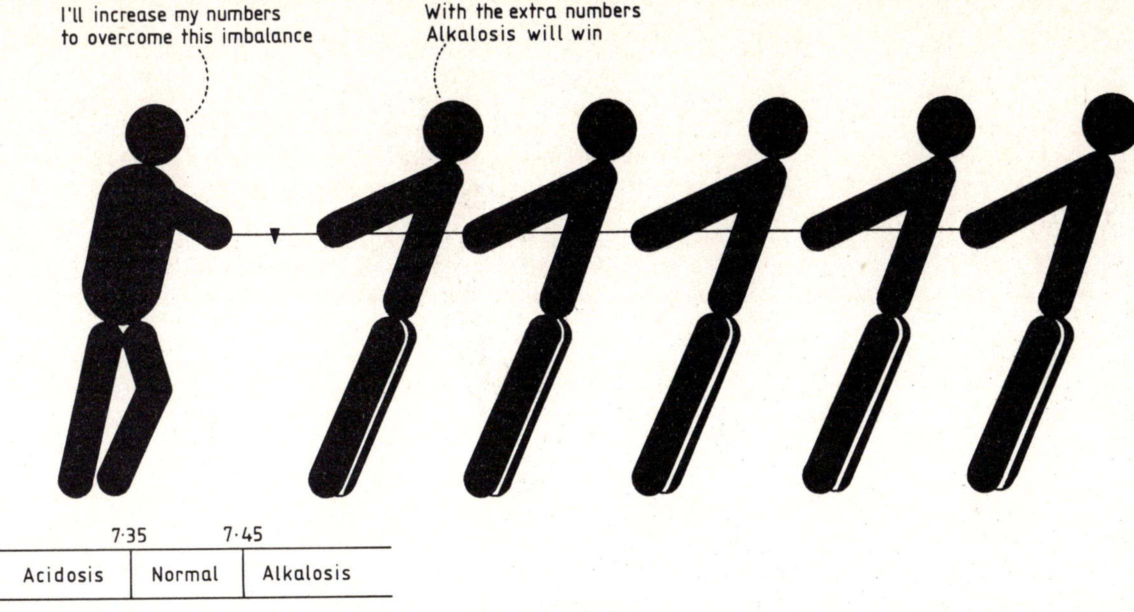

[a]

[b]

Figure 11.13 (a) Metabolic alkalosis can result from HCO_3^- excess. The H_2CO_3:HCO_3^- may alter to 1:40. (b) An increase in the H_2CO_3 results in the H_2CO_3:HCO_3^- ratio returning to 1:20.

consist of covalent bonds. EXAMPLE: Aluminium chloride ($AlCl_3$). These substances are distinguished from salts by their property of not being able to conduct electricity when dissolved in water. These solutes are called non-electrolytes. Salts and their breakdown products or ions can conduct electricity when dissolved in water and are called electrolytes.

Table 11.4 Examples of salt use

Salt	Use
Calcium and phosphorus salts	Formation of teeth and bones
Iodine salts	Thyroid function
Sodium, potassium and chloride salts	Maintenance of cell function
Barium sulphate	X-rays of gastrointestinal tract
Calcium sulphate (plaster of Paris)	Limb casts
Magnesium sulphate (Epsom salts)	Purgative action

11.8 Electrolytes

Electrolytes are solutes that contain a charge in aqueous solutions. Most acids, basis and salts that are soluble in water are electrolytes, whereas most organic compounds are non-electrolytes. EXAMPLE: Na^+ and Cl^- are electrolytes and glu-close is a non-electrolyte. A solution of electrolytes consists of positive and negative ions that are independent of each other. If oppositely-charged regions or poles exist in an electrolyte solution the independent ions move to the oppositely-charged regions, that is, the cations move to the negatively-charged region. The movement of these charges represents an electric current.

The greater the difference in charge (potential difference) between two regions, the greater is the potential for ions to move to their oppositely-charged regions (See 5.3). In the body, cells have a potential difference between the extracellular fluid and the cell's contents. Membrane barriers restrict the flow of ions from one region to another. The combination of the potential difference and the membrane properties results in the maintenance of cell electrolyte concentrations. Nerve and muscle membranes can be altered to allow the flow of ions (current) between the positive extracellular region and the negative intracellular region. This flow of ions results in nerves conducting information and muscles contracting.

Summary

An acid is a substance that releases H^+ into a solution. The greater the number of H^+ in a solution, the greater is the acidity of that solution. A base is a substance that donates OH^- or accepts H^+. Neutralization of an acid solution is achieved by adding a base to the solution.

The pH scale is used as a relative indicator of acidity and alkalinity. Acidic solutions have a pH range of 1 to 7 whereas alkaline solutions have a pH range from 7 to 14. At pH 7 a solution is regarded as neutral. In blood, the normal physiological pH range is 7.35 to 7.45.

Buffers are chemicals that resist changes in pH on the addition of acids or bases. They act by replacing strong acid or base with a weak acid or base. The major buffer systems involved in acid-base balance are:

1. Carbonic acid-bicarbonate ion which is important in the regulation of pH changes that results from altered blood carbon dioxide levels. To maintain normal blood pH, it is necessary to have one carbonic acid molecule to twenty bicarbonate ions. Alterations to this ratio indicate changes in either acidity or alkalinity.
2. Phosphate which is important in the maintenance of pH in red blood cells, kidney tubular fluid and cellular contents.
3. Protein which is the most abundant buffer in cells and plasma. The formation of zwitterions by its constituent amino acids enables proteins to soak up either excess acid or base.
4. Ammonia/ammonium which is used by the kidneys to eliminate H^+ in th form of ammonium.
5. Haemoglobin which is important in the regulation of blood pH as it passes through the tissues and because it releases carbon dioxide in the pulmonary capillaries.

Elevated blood carbon dioxide concentrations result in the formation of greater carbonic acid concentrations. The net effect is an elevated acidity of the blood. Any factor that increases or decreases carbon dioxide in the blood thus influences blood pH.

Hyperventilation removes greater quantities of carbon dioxide which results in a more alkaline blood, whereas hypoventilation increases blood carbon dioxide and thus H^+.

The kidneys serve two important functions in H^+ homeostasis. They reabsorb bicarbonate ions thereby preventing alterations to the carbonic acid to bicarbonate ion ratio of one to twenty. They also counteract elevated blood H^+ concentration by removing excess H^+ in the form of the weak acids in the phosphate and ammonium buffer systems.

Acidosis occurs when the serum pH value falls below 7.35. This fall can be due to a decreased removal of carbon dioxide from the lungs (respiratory acidosis). This may be compensated by the kidneys over one to two days by releasing additional quantities of bicarbonate to return the ratio of carbonic acid to bicarbonate to normal. Metabolic acidosis results from an abnormal increase in acidic metabolic products other than carbon dioxide. A respiratory compensation may maintain the normal pH by hyperventilation causing increased removal of carbon dioxide. Compensated metabolic acidosis has a decreased carbon dioxide and bicarbonate ion whereas uncompensated metabolic acidosis has a decreased blood bicarbonate ion and pH but a normal blood carbon dioxide level.

Alkalosis occurs when the serum pH levels are greater than 7.45. Respiratory alkalosis results from hyperventilation removing excessive carbon dioxide. Metabolic alkalosis can result from excessive loss of acid from the stomach, the ingestion of large amounts of antacids or from a decrease in blood K^+ concentration. The body can usually compensate by decreasing respiratory rate and increasing bicarbonate secretion by the kidneys.

A salt is composed of cations and anions. Salts usually dissociate in water which results in the presence of free ions which are known as electrolytes. An electrolyte has the ability to move to oppositely-charged regions in a solution. Non-electrolytes are not attracted to charged regions.

Chapter 12

Composition and properties of body fluids

Objectives

At the completion of this chapter the student should be able to:
1. relate the properties of fluids to body fluid composition,
2. describe the differences in fluid composition between the different body fluid compartments,
3. relate the properties of fluids to kidney function and urine composition,
4. describe the process of dialysis.

12.1 Composition of body fluids

Body fluids consist of a combination of solutes, suspended particles and colloidal particles. Body fluid characteristics depend upon the properties of solutions, suspensions and colloids. Substances that can penetrate membranes (solutes) are dispersed throughout all body fluids. Substances that are unable to pass through membranes (suspended particles and colloidal particles) are restricted in their distribution.

The widespread distribution of solutes results in these substances being capable of performing a diverse range of functions (Table 12.1). Suspended particles and colloidal particles are restricted in their distribution since cell membranes act as a barrier to the movement of these particles. The relative distribution of substances in the various body fluid compartments is shown in Figure 12.1. The properties of solutes, suspended particles and colloidal particles along with the properties of each fluid compartment can be used to explain the composition of body fluid compartments. Body fluids can be separated into intracellular fluid and extracellular fluid, with the latter being subdivided into various compartments.

INTRACELLULAR FLUID

Each cell contains an individual mixture of different substances, but the concentration of these substances is similar for all cells. Thus the fluid composition of cells is considered as one large fluid compartment known as the intracellular fluid. This fluid is formed from substances passing through cell membranes and by the manufacture of substances within cells.

Intracellular fluid contains mainly electrolytes, glucose, large quantities of protein (four times plasma concentration) and small quantities of lipids.

Table 12.1 A list of common body solutes. The importance of solutes to the body is shown by their diversity of function

Solute	Function
Amino acids	Chemical building blocks of protein
Ammonium	Involved in the regulation of kidney tubular fluid pH
Bicarbonate	Involved in the regulation of blood and kidney tubular fluid pH
Bilirubin	Breakdown product of haemoglobin
Carbon dioxide	Waste product of tissue metabolism
Calcium	Necessary constituent of bone and teeth; essential for hormone manufacture, blood clotting, nerve and muscle function
Chloride	Important in fluid balance
Creatinine	Produced by muscles; the rate of elimination from the kidneys is used to evaluate kidney function
Glucose	Chemical building block of stored carbohydrate (glycogen); it is the main carbohydrate used in energy production
Hydrogen ion	Released from body acids and metabolism
Iron	Essential constituent of haemoglobin and respiratory enzymes
Lactic acid (lactate)	Waste product of metabolism formed during oxygen deficiency
Phosphate	Important regulator of pH in the kidney tubular fluid, red blood cells and other cells
Potassium	Important in nerve and muscle function
Sodium	Essential in blood volume control, nerve and muscle function
Urea	Chief nitrogenous end product of protein metabolism

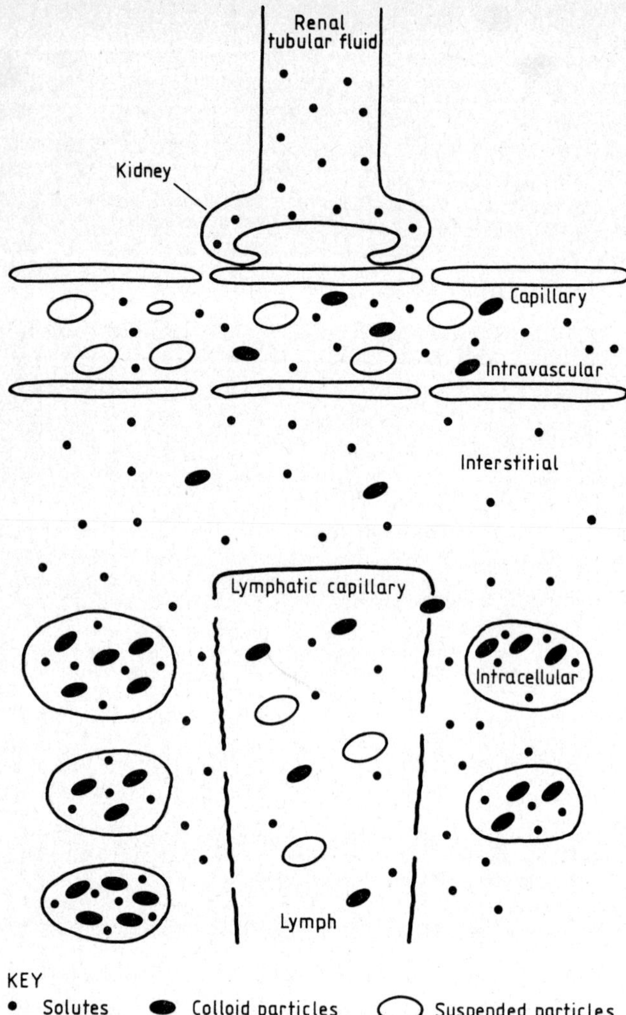

Figure 12.1 Body fluid compartments showing the relative distribution of solutes, colloid particles, and suspended particles.

(Note that most of this protein and lipid has been manufactured within the cell from solute particles.) Electrolytes such as K^+, Mg^{2+}, PO_4^{2-}, and SO_4^{2-} are found in cells in relatively large quantities compared with extracellular fluid (Figure 12.2). The large quantities of protein within cells act as an important intracellular buffer. Phosphate also is involved in the regulation of cell pH.

Cells contain a predominate quantity of negatively charged proteins and other colloid particles. These particles confer a net negative charge within a cell when compared with the extracellular fluid. This potential difference is an important factor in the functioning of nerves and muscles (see 5.3).

EXTRACELLULAR FLUID

Interstitial Fluid

Interstitial fluid (intercellular fluid) is formed from substances passing through either capillary walls or the cell membranes of nearby cells. Interstitial fluid contains only trace quantities of proteins and blood cells, thus indicating that these particles have difficulty in passing through membranes. The small quantity of plasma proteins in interstitial fluid is due primarily to some proteins that leak through the gaps between the cells that form the capillary walls. Damage or alteration to these cells results in a large quantity of proteins and blood cells moving into the interstitial fluid.

Large quantities of solutes, especially electrolytes, are found in this fluid since these substances can readily penetrate membranes and capillary wall gaps. Interstitial fluid contains relatively large concentrations of Na^+, Cl^- and HCO_3^- compared with intracellular fluid (Figure 12.2). The large concentrations of HCO_3^- and the low concentrations of protein and phosphate indicate that the main interstitial buffer system is the HCO_3^-/H_2CO_3.

Lymph

Interstitial fluid contains a small concentration of proteins and suspended particles. These particles have difficulty in passing through the capillary wall and thus they require an alternative route through which they can return to the circulatory system. A drainage system known as the lymphatic system returns interstitial fluid containing colloids and suspended particles to the blood. The lymphatic system is composed of channels that link the interstitial fluid with the circulatory system. Fluid enters this system via porous endings at the beginning of a drainage channel. As soon as the fluid enters the lymphatic channels it is called lymph.

The lymph vessels drain fluid, cellular secretions of proteins such as hormones and enzymes, suspended particles such as blood cells and cell debris. In addition to these particles, cells (lymphocytes) and proteins (antibodies) involved in the body defence mechanisms are released into the lymph from lymphoid tissue such as the lymph nodes.

Intravascular Fluid (Blood)

Blood consists of suspended particles or cells and a liquid referred to as plasma. Serum is another term used in describing the liquid component of blood, but serum differs from plasma in that serum does not contain the colloid factors involved in blood clot formation. Plasma is the naturally-occurring intravascular liquid whereas serum is the liquid formed when blood is allowed to clot in containers. The two terms serum and plasma are often (incorrectly) interchanged when referring to the concentration of substances in the blood.

Plasma is almost identical in solute composition to interstitial fluid (Figure 12.2). However, the concentration of protein is about three and a half times greater in plasma than in interstitial fluid. Proteins and a large quantity of cells found in the

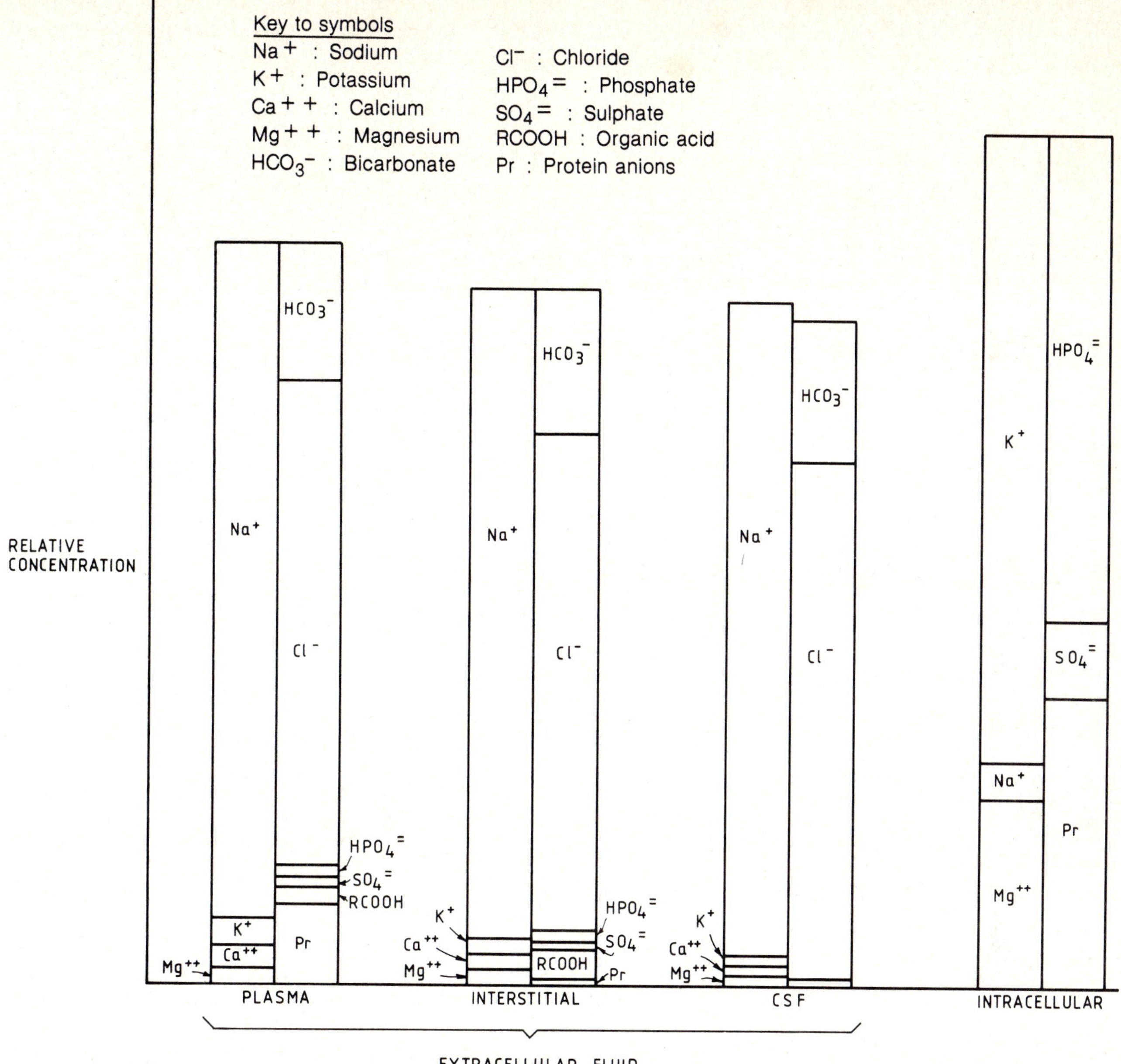

Figure 12.2 Comparison of electrolyte concentrations in plasma, interstitial fluid, cerebrospinal fluid and intracellular fluid.

blood do not enter the vascular system by passing across membranes but pass directly into the blood stream from areas such as the bone marrow.

The high concentration of proteins and HCO_3^- indicates that these substances are the main intravascular pH regulators. The presence of red blood cells and therefore haemoglobin results in the blood also being buffered by the haemoglobin buffer system.

Cerebrospinal Fluid (CSF)

This fluid originates from blood being passed through a network of capillary membranes in the brain. This network is known as the choroid plexi. The CSF functions as a protective cushion for the brain. It also acts as a carrier of nutrient and waste materials between the blood, brain and spinal cord. Fluid leaves the CSF compartment via ducts that connect with the venous system.

The composition of CSF is restricted to substances that can pass through the capillary walls of the choroid plexi. These wall cells appear to selectively secrete solutes and not colloid or suspended particles into the CSF. The composition of electrolytes in CSF is similar to interstitial fluid (Figure 12.2).

Generally any substance that has a molecular weight greater than 500 is excluded from CSF. This restriction is important in the effectiveness of drugs acting on the brain. The CSF differs from interstitial fluid in that it contains lower concentrations of phosphate ions. The CSF contains an excess of cations whereas the interstitial fluid and blood consists of a balance of cations and anions. Thus CSF has a more positive charge, and therefore a potential difference exists between CSF and these fluids such that the CSF is more positive.

The pH of CSF is 7.32 and it is maintained by the bicarbonate buffer system, since only small quantities of proteins and phosphate exists in CSF.

APPLICATION: LUMBAR PUNCTURE

This process involves the insertion of a needle into the CSF in the lumbar region of the spinal cord and taking CSF samples or administering substances to the CSF. Due to the choroid plexi barrier, normally only solutes exist in the CSF. Analysis of CSF for the presence of white blood cells and blood is used for diagnostic purposes. Culturing CSF samples assists in determining the presence of bacteria. In addition, anaesthetics that are unable to penetrate the choroid plexi can be administered via a lumbar puncture.

Urine

Urine is formed from substances that can normally pass through capillary membranes. These substances pass through a membrane filter and then into kidney tubules. Some of these substances are taken up by the kidney cells and returned to the blood. EXAMPLE: Na^+ and glucose. Other substances such as urea pass through the kidney tubules and form urine. The main constituents of urine are waste products of metabolism such as urea and creatinine (Table 12.2). Note that the constituents of urine are solutes. Proteins, cells and hardened moulds of cells known as casts only normally occur in trace amounts.

The inability of proteins to penetrate a membrane results in the kidneys using solutes for buffers. The main solute buffers are the phosphate and ammonium buffer systems. These buffers are used to remove acids from the body in the form of weak acids. These acids can result in a slightly acidic urine.

Damage or alteration of the kidney filtering system results in the presence of an elevated concentration of proteins and suspended particles. Calcium precipitates known as calculi may also be present. When the kidneys are unable to return a substance such as glucose to the blood the excess is eliminated in the urine. The presence of glucose in urine can be due to either high blood glucose concentrations or a decreased capacity of the kidney to return substances to the blood.

In summary, the kidney separates suspended particles and colloidal particles from solutes. The solutes are in turn separated by kidney cells so that toxic waste products are eliminated in urine and essential solutes are returned to the blood. Artificial methods have been developed that mimic the kidney filtration system. These methods are known as dialysis.

Table 12.2 Normal constituents of urine

Constituent	Comment
ORGANIC	
Urea	Usually sixty to ninety per cent of all nitrogen solutes; varies with dietary protein intake
Uric acid	Product of nucleic acid breakdown; low solubility results in kidney stones
Creatinine	Produced during muscle contraction from creatine
Ketone bodies	Usually found in small amounts, the result of protein and fat breakdown; high levels are found in individuals receiving low carbohydrate diets, pregnant women, untreated diabetes mellitus and acute starvation
Hippuric acid	Results from benzoic acid, a toxic substance in fruits and vegetables; diets containing a high vegetable content result in increased levels
INORGANIC	
Sodium chloride	Most abundant inorganic salt
Potassium, magnesium and calcium ions	Combine with anions, chloride, sulphate and phosphate to form salts
Sulphate, phosphate	Combines with sodium to form buffers; used to remove hydrogen ions in the urine
Ammonium	Used as a compensatory mechanism to remove high levels of acids; combines with chloride to form ammonium salt

12.2 Dialysis

RENAL DIALYSIS

Dialysis is the separation of suspended particles and colloids from solutes by a membrane. Renal problems can be overcome by the use of a kidney machine (artificial kidney) in the process of haemodialysis. This is the removal of soluble toxic substances from the blood stream by an artificial membrane. In haemodialysis, a patients blood is pumped through a tube of artificial membrane. This tube has both ends connected to the patient's vascular system. The plastic membrane tube is

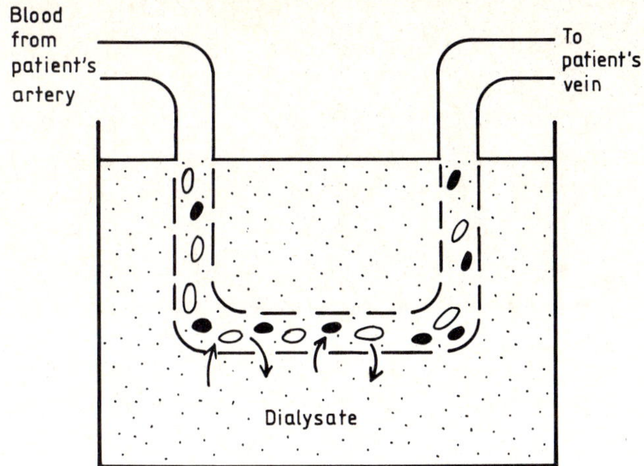

Figure 12.3 Principle of haemodialysis whereby solutes are separated from colloids and blood cells. Toxic solutes flow from a relatively high concentration in the blood to a low dialysate concentration.

immersed in a bathing solution called the dialysate. It contains the required concentration of soluble substances for the patient. Soluble particles pass through a dialysing membrane from a high concentration to a lower concentration (Figure 12.3). Initially, no waste products occur in the dialysate so waste substances move from the blood to the dialysate. No net flow of other subtances will occur since the concentration of these substances is the same as in the blood. Proteins and suspended particles are unable to penetrate the membrane and thus are retained by the dialysing tube.

If a patient requires substances such as glucose, the dialysate concentration is increased. The greater glucose dialysate concentration results in a flow of this substance into the blood. Also, to lower a patient's blood solute concentration, simply decrease the concentration of that substance in the dialysate to the required level.

As the dialysate concentration of toxic substances increases, the efficiency of removal of these waste products from the blood decreases. Therefore it is important to change the dialysate at regular intervals.

PERITONEAL DIALYSIS

In this procedure the dialysate flows via a special catheter into the sealed peritoneal cavity (Figure 12.4). The dense capillary network of the peritoneal cavity is used as the dialysing membrane. This large surface area of membrane enables efficient movement of solutes from the peritoneal capillaries

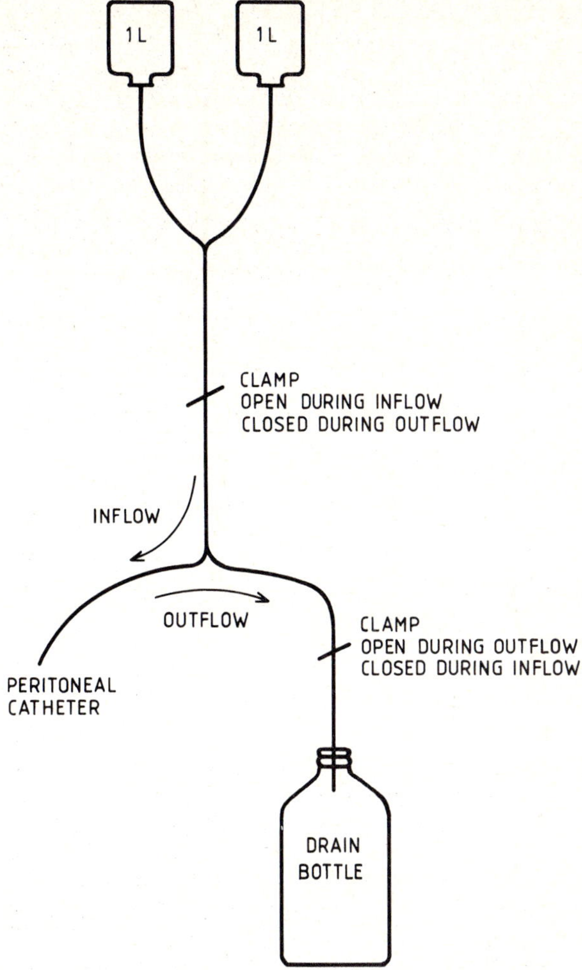

Figure 12.4 Principle of peritoneal dialysis showing the infusion of dialysate into the patient via a peritoneal catheter. Clamps control the inflow and outflow of the dialysate.

to the dialysate and dialysate solutes into the blood. In adults, two litres of sterile body temperature dialysate is infused over a fifteen minute period. The process of dialysis is allowed to occur for thirty to forty-five minutes. This period is called the dwell time. After dialysis the bathing solution is removed by siphonage over approximately twenty minutes. The contents of the outflow and its volume are then measured and recorded.

The infusion, dwell time and outflow are varied according to the patient's symptoms. EXAMPLE: Overhydration and oedema usually require a short dwell time. The procedure is repeated any number of times until the desired plasma concentration of solutes have been attained. Peritoneal dialysis is usually a short-term therapy since the risk of peritonitis is great.

Summary

Body fluid compartments consist of varying mixtures of solutes, suspended particles and colloidal particles.

The distribution of colloids and suspended particles is restricted by their difficulty in penetrating membranes, whereas solutes are widely distributed throughout the body.

Intracellular fluid, lymph and blood contain relatively large quantities of colloids compared with interstitial fluid, CSF and urine.

Lymph and blood contain large quantities of cells.

Body fluid buffers are determined by the distribution of proteins. Blood and intracellular fluid rely on proteins as a major source of buffers, whereas interstitial fluid, CSF and urine rely entirely on solutes such as bicarbonate.

The kidneys separate solutes from suspended particles and colloidal particles. Kidney cells separate the toxic waste products of the body from the essential solutes and eliminate the waste solutes in urine.

Dialysis is a method by which an artificial membrane separates suspended particles and colloidal particles from solutes. The solutes contain toxic waste susbstances that flow into the dialysate. Removal of the dialysate at intervals results in the effective elimination of toxic solutes from the patient.

Unit Six
Pressure

This unit integrates the properties of pressure and applies these properties to normal and abnormal body function. You will learn how pressure influences the supply, movement and removal of nutrients and waste products in the body.

Chapter 13

Properties of pressure

Objectives

At the completion of this chapter the student should be able to:
1. define pressure and apply this definition to the laws relating to pressure,
2. explain the laws of pressure using either physiological, nursing or clinical examples,
3. explain how respirators can artificially control or assist external respiration,
4. describe the principle of hydrostatic pressure,
5. describe how hydrostatic pressure may be altered by either normal or abnormal body function,
6. explain the term blood pressure.

13.1 Principle of pressure

Pressure is the force exerted by particles on an area of matter. Pressure can be developed by all forms of matter. The properties of pressure have many applications in body function and patient care. Throughout the body, pressure is used to move liquids and maintain a fluid balance. In the process of breathing (ventilation), pressure is required to move oxygen and carbon dioxide into and out of the lungs.

With liquids and gases, the degree of pressure on an object is determined by the number of molecules colliding with that object. To determine the amount of pressure imposed on an object the following equation is used:

$$\text{pressure} = \frac{\text{force applied}}{\text{area of application}}$$

This equation implies that the greater the area upon which a given force is applied, the less the pressure.

This principle is shown when pressure sores develop in long-term, bed-ridden patients. EXAMPLE: Spinal injuries. When lying on a firm mattress a large proportion of body weight is born by projections of the skeleton such as the scapula, coccyx and elbows. These small areas of contact bear the body weight and thus a large amount of pressure is exerted on these protrusions, resulting in bed sores (Figure 13.1a). It is therefore advisable to spread the pressure of the body protrusions over a wider area. This is achieved by using materials that will mould with the body contours such as water beds or an individually moulded cast (or bed). Decreased pressure results from increasing the area over which a force is applied (Figure 13.1b).

The force applied for a given pressure is always opposed by an opposite or resistance force. The effect of this opposite force is to resist the further movement of particles (see 7.2). When the pressure exerted on the walls of a container or the surface of

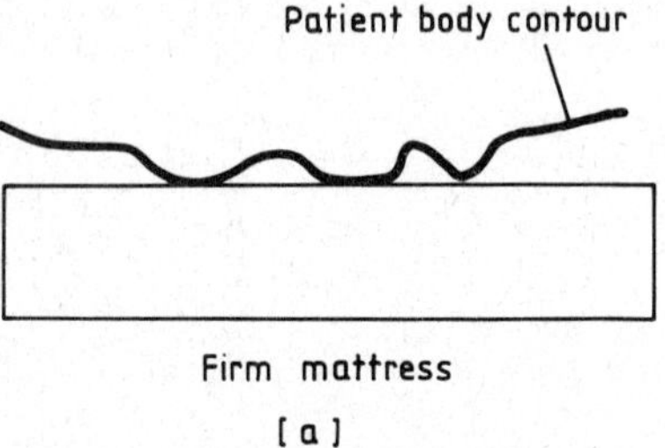

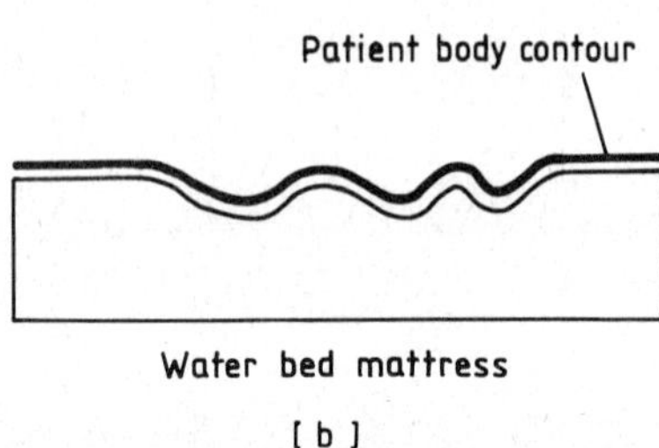

Figure 13.1 Pressure between patient and bed surface. (a) With a firm bed a small area of contact occurs between the patient and the bed resulting in the weight of the patient (force) being born by several small body projections. The flesh between the bones and the bed therefore develops sores. (b) With a flexible bed, a large area of contact occurs between the patient and the bed. Low pressure between the patient and the bed minimizes the development of bed sores.

solids increases, one of the following effects will occur:

1. If the surface on which the pressure is exerted is elastic it will stretch.
2. If the surface is incapable of withstanding this increased pressure it will break. EXAMPLES: In fluids the rupture of blood vessels and subsequent haemorrhaging, in solids the fracturing of bones due to excessive pressure upon the bone.
3. Material that can resist changes in pressure will increase the potential of the matter within the container to exert pressure. EXAMPLE: Compressed gas in a cylinder.
4. Inelastic material that is capable of resisting an increase in pressure exerted by a solid will effect an increase in resistance, i.e. friction between the solid and the surface that is in contact with it. EXAMPLE: Bed sores develop on areas where an increase in pressure occurs between the patient's body and regions of contact with a bed.

Pressure can be increased by accelerating the number of particles that collide with the walls of a container at any given moment, i.e. exerting an increased force on a given area. Thus any body function that increases the number of collisions of particles will increase pressure. EXAMPLES: Expiration, changes in blood pressure and osmotic pressure.

13.2 The gaseous state

Several physical laws assist in the understanding of pressure and how pressure may be altered in gases.

CHARLES' LAW

Charles' law states that if a pressure is constant, then the volume of gas is directly proportional to its absolute temperature. That is, an increase in temperature results in an increase in the movement of particles resulting in an increase in the area that these particles can occupy. When the container walls are inelastic, raised temperature will cause an increased pressure due to a greater number of particles colliding with the wall.

APPLICATION: COMPRESSED GAS CYLINDERS

Compressed gas containers (aerosol cans, gas cylinders) will explode if the temperature inside the container is increased to a level where the gas pressure becomes too great for the container walls to withstand this pressure.

BOYLE'S LAW

Boyle's law. The volume of a gas varies inversely with pressure when the temperature is constant:

$$\text{volume} \propto \frac{1}{\text{pressure}} \quad \text{or pressure} \propto \frac{1}{\text{volume}}$$

This means that if the volume is doubled the pressure is halved since a given number of gas molecules now occupy a greater volume. This results in less molecules colliding with the wall of a container at any particular moment. Conversely if the volume is halved the pressure is doubled. In effect the density of gas has been altered. Thus the number of particles colliding with the walls of the container has been altered.

APPLICATION: VENTILATION

See Figure 13.2

POSITIVE AND NEGATIVE PRESSURE

Pressure that is greater than standard atmospheric pressure (101 kPa or 760 mm Hg) is often referred to as positive pressure, and pressure that is less than 101 kPa is referred to as negative pressure. Positive and negative pressure are used in respiration. Both of these pressures are also used by machines to mechanically assist respiration (respirators). The elimination of air and blood following thoracotomy also uses differences in positive and negative pressure.

APPLICATION I: VENTILATION

Inspiration occurs when a difference in pressure exists between the atmosphere and the lungs, with the pressure in the lungs being negative. Expiration occurs when the pressure in the lungs is positive. In the latter the greater pressure within the lungs forces air to move to the atmosphere. To understand the process of breathing it is necessary to know how pressure changes occur within the lungs, since prior to inspiration pressure in the lungs is approximately equal to atmospheric pressure.

The lungs are surrounded by a sealed thoracic cavity (Figure 13.2). According to Boyle's law, an expansion of this sealed area (volume increase) will result in a decrease in the intra-thoracic pressure. This event occurs when the diaphragm and intercostal muscles contract. This contraction effects an increase in the size of the thoracic cavity (Figure 13.2b). A decreased intra-thoracic (intra-pleural) pressure results in less particles colliding with the walls of the lung that line the intra-thoracic cavity. A relative difference in the number of particles colliding with the two surfaces of the lung (intra-thoracic and intra-alveoli) develops with a greater number of particles colliding with the intra-alveoli wall. Since lung tissue exhibits elastic properties an expansion of the lung occurs until the forces are once again balanced. This expansion of the lung results in a decreased density of air and air pressure within the lung (Figure 13.2c). This

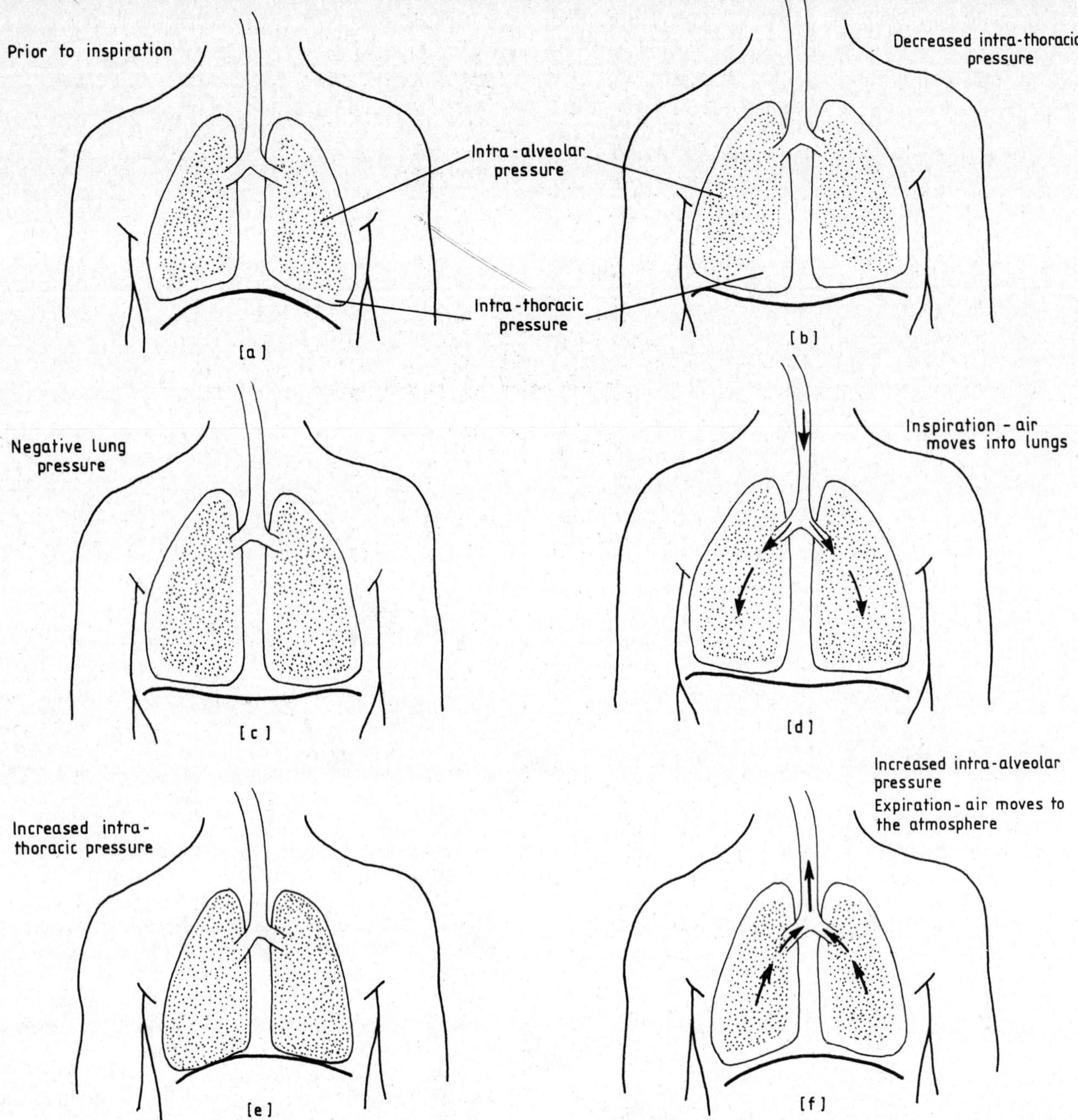

Figure 13.2 Relationship of the lungs to the sealed thoracic cage during a breathing cycle.

(a) Lungs and thoracic cage prior to inspiration, pressure in the lungs equals atmospheric pressure.

(b) Contraction of the diaphragm and intercostal muscles effects an increased volume within the thoracic cage, expansion of this sealed cage decreases the intra-thoracic pressure compared with the intra-alveolar pressure.

(c) The lungs expand until the intra-alveolar pressure has decreased sufficiently to regain the original balance in pressure with the intra-thoracic region. There is now a lower pressure in the lungs compared with the atmospheric pressure.

(d) Air moves from the atmosphere to the lungs until the intra-alveolar pressure equals the atmospheric pressure, that is, inspiration occurs.

(e) Relaxation of the intercostal muscles and diaphragm results in a decreased intra-thoracic volume, and an increased intra-thoracic pressure.

(f) Lungs recoil as a result of this alteration between the intra-alveolar and intra-thoracic pressures, the decreased lung volume develops a positive pressure. Air moves from the lungs in the process of expiration.

decreased pressure (negative) allows particles to flow from the atmosphere into the lungs in an attempt to equalize the lung pressure with that of the atmosphere (Figure 13.2d).

At a certain point in the expansion of the intra-thoracic cavity, stretch receptors are activated, resulting in the inhibition of diaphragm and intercostal-muscular contraction. Both these muscles relax, effecting a decrease in intra-thoracic volume. This allows the stretched elastic tissue in the thoracic cage to recoil to its original unstretched length. Now a greater density has developed within the intra-thoracic cavity, since a decreased volume for a given number of particles exists. This increased density results in a greater number of particles from the intra-thoracic cavity colliding with wall of the lung (Figure 13.2e). The balance of forces between the intra-thoracic and intra-alveolar walls of the lung has been altered, with the elastic tissue of the lungs being forced to recoil. A decrease in lung volume results, causing an increased density and thus pressure of the intra-alveolar air. The pressure within the lungs becomes positive, thus forcing air to move to the atmosphere (Figure 13.2f).

It should be noted that expiration is regarded as a passive process since little muscular contraction occurs. Contraction of the internal intercostal muscles assists expiration to a limited extent.

APPLICATION 2: RESPIRATORS

Breathing can be assisted, and if necessary controlled, by the use of respirators. These machines mechanically alter the pressure difference between the atmosphere and the lungs. Various forms of respirators are available, each designed to overcome specific breathing problems: EXAMPLES: Iron lungs, intermittent positive-pressure respirators (IPPR), positive expiratory end pressure (PEEP) and continuous positive airways pressure (CPAP) respirators. All of these respirators effect a negative or positive pressure in the lungs.

Negative Pressure Respirators

These machines indirectly develop a negative pressure within the lungs to initiate inspiration. The respirators are called iron lungs. They are large, with the machine enclosing the entire body apart from the neck and head (Figure 13.3).

A pump creates a negative pressure between the patient and the machine's wall. This negative pressure effects an expansion of the thoracic cavity. This expansion results in air flowing from the atmosphere to the lower pressure that has developed within the lungs. This expansion is similar to the normal process of inspiration. Expiration occurs when the thoracic cavity recoils to its original size. This step is aided by the use of a positive pressure developed within the machine. Iron lungs can be used with patients suffering from respiratory muscle paralysis and are continually used with patients who have poliomyelitis.

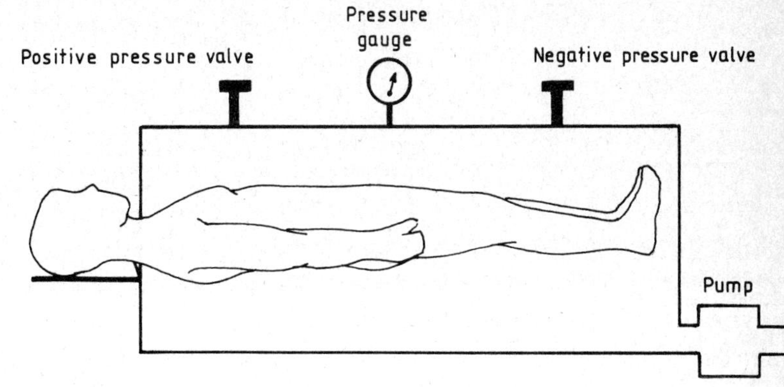

Figure 13.3 Patient in an iron lung.

Positive Pressure Respirators

Positive pressure respirators force gas into the lungs at greater than atmospheric pressure. The intermittent positive pressure respirators (IPPR) force air into the lungs. This forced intake of air results in an expansion of the lungs and inspiration. The elastic recoil of the stretched tissue effects a passive expiration. These machines can be used to regulate breathing rates in unconscious patients or to improve lung expansion in patients suffering from obstructive or restrictive lung disease. Gas mixtures can be altered to regulate blood oxygen levels and pH of the blood.

Often it is necessary to retain a small positive pressure within the lungs throughout breathing. In premature babies immature alveoli may collapse preventing gaseous exchange. Therefore continuous positive pressure must be maintained. This is usually done by continuous positive airways pressure (CPAP) in which the neonate relies on its own respiratory mechanisms but a small constant pressure is retained within the lungs at the end of expiration. This positive pressure prevents the collapse of the alveoli. With IPPR respirators a small positive pressure can be retained at the end of expiration. This pressure allows a longer period for gaseous exchange and is referred to as positive end expiratory pressure (PEEP).

APPLICATION 3: UNDERWATER DRAINAGE APPARATUS

Underwater drainage is used to remove plasma, blood and air from the thorax following thoracotomy. The underwater drainage apparatus is shown in Figure 13.4. The tube is inserted at the time of operation and it is essential that it has air-tight fittings. The leakage of air from the atmosphere into the tube will cause collapse of the lung, therefore care must be taken at all times to ensure that the apparatus is not accidently knocked.

A positive pressure is created in the thoracic cavity by coughing or sneezing. This positive pressure forces air and blood to flow via the tube to the water. The air then passes through the water to the

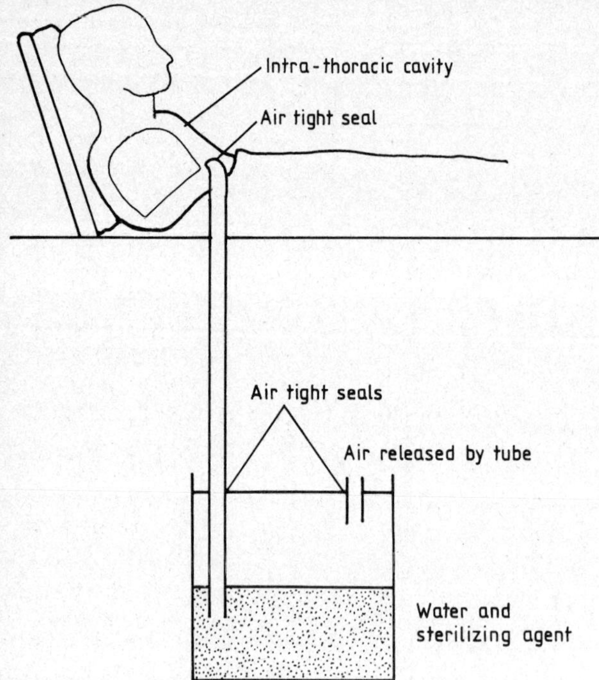

Figure 13.4 Underwater drainage apparatus consisting of a sealed tube connecting the intra-thoracic cavity to a solution of water and sterilizing agent about one metre below the patient's chest.

atmosphere. On inspiration, expansion of the intra-thoracic cavity creates a greater negative pressure in the thoracic cavity. Atmospheric pressure forces water into the tube. Deep inspiration results in water rising approximately forty centimetres in the tube, thus to prevent fluid entering the thoracic cavity the underwater drainage bottle is placed one metre below the patient's chest.

DALTON'S LAW

Each gas in a mixture of gases exerts its own pressure as if all other gases were not present. That is, each gas exerts a partial pressure. The total pressure exerted by a mixture of gases is the sum of the partial pressures of the constituent gases. This is not the case with gases that combine (react) within the mixture. When gas molecules combine a decrease in the total number of random moving molecules occurs.

In one atmosphere of pressure (101 kPa or 760 mm Hg), the pressure consists of a summation of individual gas components within the atmosphere:
101 kPa or 760 mm Hg = PO_2 + PCO_2 + PN_2 + PH_2O (water vapour).

To determine the partial pressure of a gas, multiply the percentage of the component gas present within the total amount of gas by the total gas pressure of the mixture. Oxygen represents approximately twenty-one per cent of total air content, and thus:
atmospheric PO_2 = 21/100 × 101 kPa or 760 mm Hg
= 21.2 kPa or 160 mm Hg

Partial pressure is important in determing the relative pressure differences of individual gases, such as oxygen and carbon dioxide, between the lungs and blood and the blood and body cells. Pressure differences indicate the likelihood of gas movement between two regions and thus our ability to exchange gases such as oxygen and carbon dioxide.

APPLICATION: EXCHANGE OF ALVEOLAR AND BLOOD GASES

Oxygen has an alveolar PO_2 of 13.3 kPa (100 mm Hg) and a venous PO_2 of 5.3 kPa (40 mm Hg). The wall separating these two regions allows oxygen to pass through it and thus a net movement of oxygen molecules will move from the region of higher pressure (concentration) to the region of lower pressure in the blood. A decrease in the difference of partial pressure between these two regions would result in less oxygen moving to the blood thus resulting in dizziness. For example, at high altitudes the PO_2 of the atmosphere, and therefore the alveoli, decreases. The lower amount of oxygen moving from the alveoli to the blood results in altitude sickness.

Carbon dioxide has a greater partial pressure in the blood (PCO_2 6.0 kPa or 45 mm Hg) returning to the lungs than in the alveoli (PCO_2 5.3 kPa or 40 mm Hg). Thus, carbon dioxide moves from the blood to the alveoli, as the lung wall allows carbon dioxide to pass through it.

HENRY'S LAW

The amount of gas dissolved in a liquid varies proportionately with the partial pressure of the gas, when the temperature remains constant. Thus the greater the PO_2 in the alveoli the greater is the amount of oxygen that can dissolve in blood.

APPLICATION: HYPERBARIC CHAMBERS

Hyperbaric chambers are used to increase the PO_2 of the alveoli. A hyperbaric chamber contains oxygen at approximately two to three times the atmospheric pressure. Thus the level of dissolved oxygen in the blood is increased proportionately. The increased PO_2 of blood results in a greater ability for oxygen to move to the tissues. This principle is used in the treatment of anaerobic infections, spinal injuries etc. where the previous oxygen levels were too low for the effective movement of oxygen to the damaged area of tissue.

GAS FLOW IN TUBES

The flow of gas molecules through a tube is influenced by the diameter of the tube. The greater the diameter the less chance a molecule has of colliding frequently with the tube wall and losing kinetic energy. Excessive collisions of molecules with the wall result in the slowing down of gas molecules. This slowing down occurs when the tube diameter has been decreased or an obstruction blocks part of the tube, that is, an increase in resistance to gas flow has occurred.

APPLICATION: OBSTRUCTIVE AIRWAYS DISEASE

A decrease in the effective diameter of the airways occurs by obstruction or constriction of vessels. Resistance to air flow increases, resulting in a need to overcome this resistance. Use of respirators can assist patients by forcing air into the lungs with a positive pressure. This greater than atmospheric pressure overcomes the airway resistance, enabling the patient to receive the correct volume of gas into their lungs. In asthma patients, bronchodilators such as salbutamol dilate the air passage thus reducing the resistance to air flow.

13.3 The liquid state

The properties of liquids under pressure are important for body function. EXAMPLES: Movement of blood and lymph, fluid movement across membranes, bladder function.

PASCAL'S LAW

Pressure exerted on a confined liquid is transmitted equally in all directions. A change in pressure at any point will result in the pressure change being transmitted equally throughout the liquid.

HYDROSTATIC PRESSURE

Hydrostatic pressure refers to the force that pushes liquids against the walls of its container. This pressure is developed by the heart and bladder. The pressure developed will depend upon the ability of the container wall to compress the fluid within it. Since liquids consist of molecules that are loosely held together (see 6.3), the distance that molecules can be pushed towards each other (compressed) by a given force is minimal, when compared with the freely-moving gas molecules. Thus a relatively large force is required to compress liquids. Small changes in the volume of a confined liquid result in a large increase in hydrostatic pressure. This increased pressure will be applied, according to Pascal's law to all regions of the container. If a region of the wall is unable to withstand this pressure increase, then that region will collapse and fluid will be pushed through it. This principal is used by valves in the heart and sphincters in the bladder wall to release fluid into vessels. The amount of fluid moving from the bladder or heart into vessels will depend upon the resistance of the vessels to fluid.

APPLICATION 1: MICTURITION

Micturition, also known as urination or voiding, refers to the process of expelling urine from the bladder. As the bladder fills, it expands due to the relaxation of muscles. This expansion, with the resulting increase in volume, enables fluid to be stored with only slight increases in pressure. When full, the bladder exerts an equal pressure on the surrounding tissue which results in pain if the pressure is not relieved.

When the volume reaches 300 to 400 ml, stretch receptors are activated causing a contraction of muscles in the bladder wall and a relaxation of the internal sphincter located at the entrance to the urethra. Contraction of the muscles effects an increase in pressure on the fluid. Since the internal sphincter is the region of lowest resistance and is not strong enough to oppose the increase in pressure, fluid from the bladder will flow through it in order to reduce the pressure within the bladder. Urine then flows to the external sphincter (consciously controlled) and urethra resulting in urination.

APPLICATION 2: BLOOD PRESSURE

The pressure developed by the heart is dependent upon the size of the chambers and the volume of blood within a chamber. When blood flowing into the atria creates a hydrostatic pressure greater than in the ventricles, the resistance to blood flow through the atrio-ventricular valves is diminished. This region of low resistance allows blood to flow through it from the atria to the ventricles. At the point where intra-ventricular pressure is greater than the intra-artrial pressure, a backflow of blood closes the atrio-ventricular valves. The ventricles contract and develop a greater hydrostatic pressure. The semilunar valves (between ventricles and arteries) are forced open since the pressure in the aorta and pulmonary arteries is lower than the intra-ventricular pressure. That is, the resistance to the flow of blood from the ventricle has been overcome by an increased ventricular pressure, resulting in blood entering the arteries. The blood (hydrostatic) pressure in the arteries increases as the ventricle contracts. This increased pressure is referred to as systolic blood pressure.

The loss of blood from the ventricle decreases intra-ventricular pressure to a point where a greater arterial hydrostatic pressure exists. The semilunar valves are forced to close and the arterial pressure is now called the diastolic pressure. This pressure decreases until

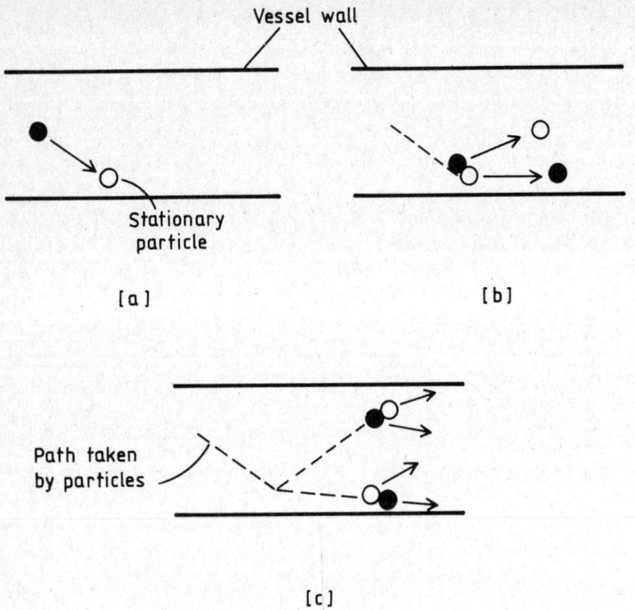

Figure 13.5 The collision of particles in a liquid results in a flow of the liquid. (a) particle 1 is pushed towards particle 2, (b) particle 1 collides with particle 2, (c) particle 1 continues moving forward but in a different direction, particle 2 moves forward as a result of the collision and collides with other particles in its path forcing these particles to move forward.

the intra-ventricular pressure is again greater than the arterial pressure, at which point the above cycle is repeated. Throughout the cardiac cycle, blood has moved from an area of high pressure to an area of low pressure by the assistance of valves.

FLOW OF LIQUIDS THROUGH VESSELS

A force that increases the movement of particles in a specific direction will cause these particles to collide with any other particles in the former's path. These collisions result in the forward movement of other particles in the liquid (Figure 13.5).

The greater the force imposed on these particles the greater is the kinetic energy gained by them, and this results in more frequent collisions. This increased frequency of collisions results in an increased flow of particles and thus a greater number of particles collide with the vessel wall per unit time. The pressure on the walls due to these collisions is called hydrostatic pressure.

The net movement of liquids through a vessel can be influenced by several factors:

Viscosity

The greater the density of a liquid, the greater is the force required to move that liquid. Viscosity of a fluid increases with an increase in the tendency to resist flow and bind to each other. The unit of measurement is the poise. A thick fluid with a high viscosity such as grease pours more slowly than a less dense liquid such as water.

APPLICATION: BLOOD VISCOSITY

Normal human blood is from two to five times more viscous than water, due to plasma proteins and red blood cells within the blood. A condition that increases the number of red blood cells results in the heart having to use more force to overcome this liquid resistance. The increased work load placed upon the heart may result in cardiac failure if the viscosity remains elevated for a long period of time. An elevated viscosity results in an increased blood pressure.

Vessel Diameter

The smaller the diameter of a vessel the greater is the resistance to the movement of particles. An increased resistance results from a greater probability of particles colliding with the vessel wall (Figure 13.6). When a particle collides with the wall some of the particles kinetic energy is lost on impact, resulting in a slowing of the particle. In a smaller diameter vessel, excessive collisions of particles with the vessel wall will result in a marked decrease in the speed of particles moving through the vessel. This results in a decrease in hydrostatic pressure at regions further away from the constriction.

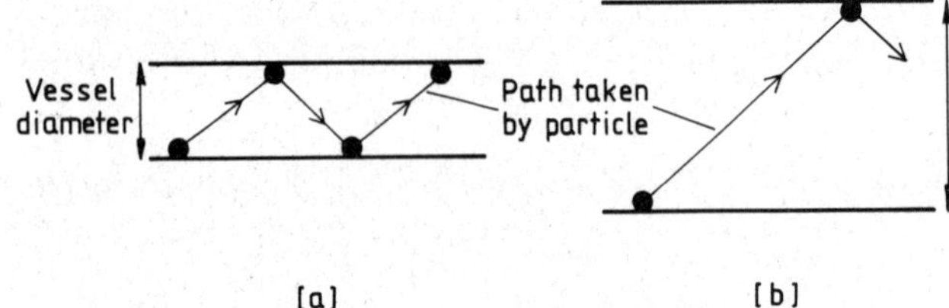

Figure 13.6 Effect of vessel diameter on the flow of a liquid. Particles collide more frequently with the wall of vessel (a) compared with (b). Each collision results in the loss of some kinetic energy, and so the greater number of collisions in vessel (a) results in a slower movement of the particle than in vessel (b). That is, the resistance to particle movement is greater in vessel (a) than with vessel (b).

APPLICATION I: PERIPHERAL VASCULAR RESISTANCE

Peripheral resistance refers to the resistance of small diameter blood vessels (arterioles, capillaries) to blood flow. The small diameter arteriole blood vessels can vary their diameter by vasoconstriction and vasodilation. These alterations in diameter are a result of muscular contraction and relaxation. The resistance to blood flow developed by the arterioles accounts for approximately half of the total systemic circulation resistance.

The greater the general resistance of the peripheral vessels to blood flow the greater is the amount of blood left in the arteries. A sustained increase in arterial pressure is referred to as hypertension. The greater the peripheral resistance, the greater is the force required by the heart to move blood through the body. This increased work load by the heart may lead to cardiac failure.

APPLICATION 2: ATHEROSCLEROSIS

In this disease an accumulation of fat occurs in an arterial wall. This mass of fat is called an atheroma. As atheromas grow the build-up of fat decreases the vessel diameter resulting in the flow of blood being impeded. Damage to tissues supplied by the blood vessel will depend upon the location and degree of atheroma development.

Vessel Length

The longer a vessel, the greater is the resistance to the flow of liquid through it. A longer vessel will require a grêater pressure to force a given volume of liquid through it than a shorter vessel. An increased pressure is required to overcome the resistance developed from the greater number of collisions of particles with the wall of the longer vessel.

Difference in Pressure Between Two Regions

The net flow of a liquid depends upon the difference in pressure existing between two regions. A flow of liquid will occur from a region of high pressure to a region of low pressure (Figure 13.7a). If no pressure difference occurs between the two regions no net flow of liquid occurs (Figure 13.7b). An increased pressure difference results in an increase in the flow of liquid (Figure 13.7c). A decreased pressure difference between the regions reduces the flow of liquid (Figure 13.7d).

APPLICATION I: INFLUENCE OF BLOOD PRESSURE ON BLOOD FLOW

Blood flows throughout the body from regions of high pressure to regions of low pressure. Increasing the pressure difference between the large arteries and veins results in a greater flow of blood to the veins. That is, an increased flow of blood, containing oxygen and nutrients, occurs to the tissues. In exercise, muscles require a large increase in oxygen and glucose. These chemicals are supplied by increasing the arterial pressure and decreasing the peripheral pressure. That is, increasing the difference in pressure between the heart and tissues.

An increase in the pressure that opposes blood flow results from an increase in the resistance to blood flow. Increases in peripheral resistance cause a decrease in the pressure difference between the heart and tissues. This decrease results in a diminished flow of blood. An increase in the arterial pressure is required to maintain normal blood flow levels (Figure 13.7e). A sustained elevation of resistance, due to constriction of vessels by atherosclerosis and arteriole constriction results in chronic hypertension. As a result, an increased workload is placed on the heart. This increase may lead to cardiac failure if hypertension is not corrected.

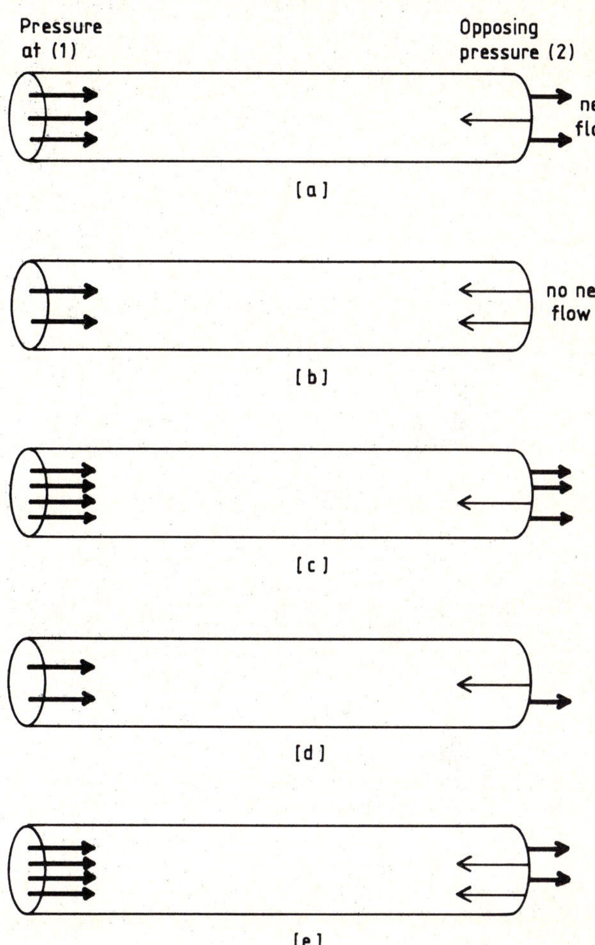

Figure 13.7 Effect of pressure difference on the flow of liquids. (a) Pressure at (1) is greater than an opposing pressure at (2) which results in a flow of liquid. (b) There is no net movement of liquid, since equal but opposite pressures exist. (c) An increased difference in pressure between (1) and (2) results in an elevated flow of liquid. (d) A decreased pressure difference reduces the net flow of a liquid. (e) To overcome an increased opposing pressure that reduces net flow, an increased pressure is required at (1). Regions (1) and (2) can be compared with the heart (1) and peripheral vessels (2).

BLOOD PRESSURE

Blood pressure consists of two separate pressures due to the pulsatile pumping action of the heart. The greater pressure is developed by the left ventricle which pumps blood into the aorta. This is called the systolic pressure. The lesser pressure develops when the left ventricle is closed and being filled. This is called the diastolic pressure. The later pressure is created by the recoiling of large elastic arteries.

These two pressures are recorded as follows:

16.0 kPa (120 mm Hg)	systolic pressure
10.6 kPa (80 mm Hg)	diastolic pressure

These values are for an average person at rest. Factors such as exercise, age, weight and stress will alter these values. Pressures that are considered normal are: systolic—12.0 to 18.6 kPa (90 to 140 mm Hg) and diastolic—6.6 to 12.0 kPa (50 to 90 mm Hg).

Blood pressure is usually measured by a sphygmomanometer. Compression of the upper arm by the cuff of the sphygmomanometer prevents blood flowing through the brachial artery. A gradual release of pressure allows the higher blood pressure (systolic) to flow through the arm, and is detected by a stethoscope placed over the artery. When the first sounds are heard on releasing the cuff pressure, the pressure on the sphygmomanometer should be noted. At the same time the mercury in the sphygmomanometer moves up and down in pulses. This is the systolic pressure. Further release of the cuff pressure allows the diastolic pressure to move blood through the arm. Since this pressure creates a continuous flow and not a pulse, the sounds due to the systolic pressure become muffled and will eventually disappear. Diastolic pressure occurs at the point where the regular beating sound becomes muffled. The pulsatile movement in the mercury column also ceases near the diastolic pressure.

The pulsatile property of systolic pressure can be used to determine heart beat, since every systolic pulse indicates another ventricular contraction. Systolic blood pressure and thus heart beats, can be felt wherever an artery occurs near the skin's surface. A firm background is required so that the artery can be compressed against this background. The commonest site for pulse rate determination is the radial artery in the lower arm. Other sites are also used (Figure 13.8) when it is difficult to use the radial artery. For example, if a patient is suffering from burns or fractures in this region.

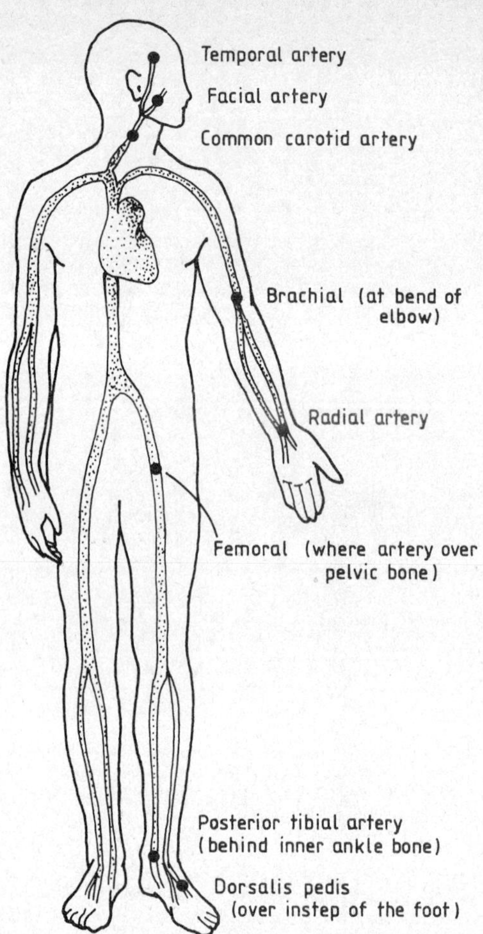

Figure 13.8 Location of alternative body sites for determining pulse rate.

Summary

Pressure is the force applied by particles over a given area: pressure = force applied/area of application.

With solids, particles exert pressure mainly through an area in contact with another surface. The particles of gases and liquids exert a pressure by colliding with the walls of their container. Several factors may alter the rate of collision with a containers wall: temperature—Charles' law, volume—Boyle's law.

Any pressure greater than atmospheric pressure (101 kPa, 760 mm Hg) is regarded as positive pressure, and any that is less will be a negative pressure.

In a mixture of gases, each gas exerts its own pressure as if all the other gases were not present—Dalton's law. This is known as the partial pressure of a gas, e.g. PO_2.

The ability of a gas to dissolve in a liquid is proportional to the partial pressure of that gas—Henry's law.

The flow of gases and liquids through a tube is altered by the tubes diameter, length, and the difference in pressure between two regions of the tube. Fluid movement is also affected by its viscosity.

An increase in the kinetic energy of particles elevates pressure. With liquids this kinetic energy increase is called hydrostatic pressure. In blood, the heart develops a hydrostatic pressure referred to as systolic blood pressure. The recoil of large elastic arteries produces a continuous diastolic blood pressure that is lower (6.6 to 12.0 kPa or 50 to 90 mm Hg) than the pulsatile systolic pressure (12.0 to 18.6 kPa or 90 to 140 mm Hg).

Chapter 14

Diffusion

Objectives

At the completion of this chapter the student should be able to:
1. describe the principle of diffusion,
2. explain how rates of diffusion may be altered within the body,
3. show how chemical gradients relate to rates of diffusion,
4. describe the effect of different substances diffusing across the same membrane,
5. describe how diffusion rates across a membrane are altered in body function.

14.1 Principle of diffusion

Diffusion is the net movement or flow of particles from a region of high concentration to a region of low concentration. The process of diffusion has many applications in body function. EXAMPLES: Absorption of nutrients, exchange of oxygen and carbon dioxide, kidney function and dialysis.

Diffusion results from the energy of motion or kinetic energy of particles constantly colliding with each other. This constant motion of particles in random directions is called Brownian movement. The greater the concentration of particles in a region the greater is the chance of collision between particles. A region with a higher number of collisions will cause particles to be forced to a region where less collisions occur (Figure 14.1).

When the rate of collisions between particles in two regions is approximately equal (concentrations equal), no net movement of particles occurs. An equal movement of particles occurs between the two regions but the concentrations of particles in the two regions will remain the same.

Diffusion is rapid over short distances but slows down as the distance between the regions increases. The ability of a particle to diffuse over a certain distance will be influenced by the following factors:

1. The number of particles that are between the given particle and its destination. The greater the number of particles, the greater will be the resistance to diffusion. Gas molecules diffuse a greater distance per unit time when they diffuse through a gaseous region than when they diffuse through a liquid. This difference between gases and liquids is due to the more dense liquid resisting the gas molecules movement. Generally, particles diffusing within gas will move a greater distance than particles diffusing in liquids. In the body's liquid environment, forces such as the pressure generated by the heart are necessary to move the particles further than by simple diffusion.

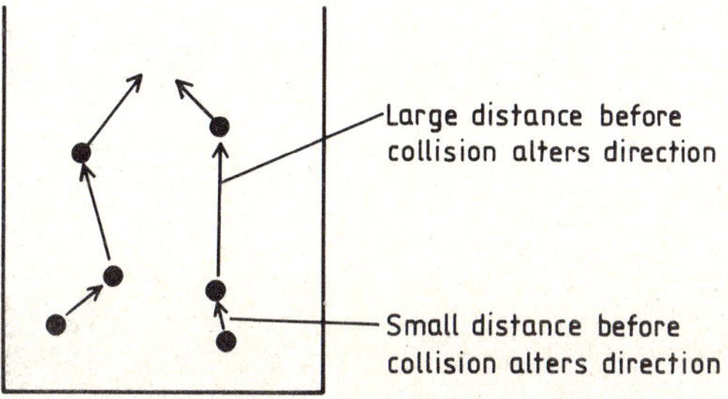

Figure 14.1 Diffusion of a substance from a concentrated region to a less concentrated region.

APPLICATION: DIFFUSION OF OXYGEN AND CARBON DIOXIDE

During inspiration a concentration difference develops between the lungs and the atmosphere. This causes oxygen and carbon dioxide to move the relatively large distance from the atmosphere to the alveoli. Differences in concentrations of both oxygen and carbon dioxide result in the diffusion of dissolved oxygen from the alveoli to the blood and dissolved carbon dioxide from the blood to the alveoli. The limited distance that substances can diffuse in liquids is shown by the body requiring a heart to pump blood around the body to the tissue site. Since diffusion in liquids occurs only through short distances, a network of capillaries ensures that this distance is not too large to prevent the effective diffusion of oxygen from the lungs to the blood and from the blood to the tissue (Figure 14.2).

An increase in the capillary/alveoli and capillary/tissue distance due to oedema results in a greater resistance to diffusion. A decreased ability of oxygen and carbon dioxide to diffuse to the appropriate region occurs. The greater the degree of oedema the less oxygen and carbon dioxide that can diffuse. This lack of oxygen diffusion to the blood and tissue can be altered by increasing the difference in concentration between the alveoli/blood and blood/tissue regions. Hyperbaric chambers increase the concentration of oxygen in the lungs which result in more oxygen being able to diffuse the greater distance from the alveoli to the blood or from the blood to the cells.

2. The larger the size and weight of a particle the greater is its inertia. Slower movement occurs which results in a lower rate of diffusion. To enable a comparison of the diffusion rates of different particles a term called the diffusion coefficient is used. The greater the weight of a particle the lower the diffusion coefficient.

3. An irregular-shaped molecule may diffuse at a slower rate than a compact regular-shaped molecule that has a larger weight. A large globular plasma protein can diffuse faster than a long thread of a small molecule such as collagen.

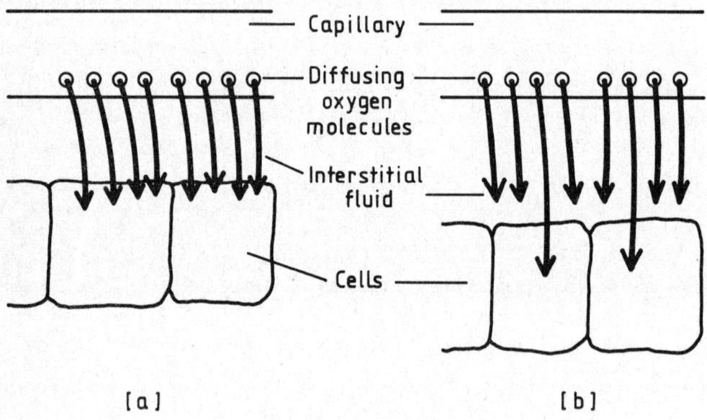

Figure 14.2 Diffusion of oxygen between a capillary and cells. (a) A normal amount of interstitial fluid separating the capillary and cells results in sufficient oxygen diffusing to the cells. (b) An increased amount of interstitial fluid between the capillary and cells results in an elevated resistance of oxygen diffusion to the cells.

14.2 Diffusion across membranes

If a membrane is permeable to a solute (i.e. solute can diffuse across it) and a difference in concentration of the solute exists between the two sides of the membrane, the solute will diffuse across the membrane from the region of high concentration to the region of low concentration (Figure 14.3).

In solution 1, the greater number of oxygen molecules results in twice the chance that a given molecule will collide with the membrane than does a

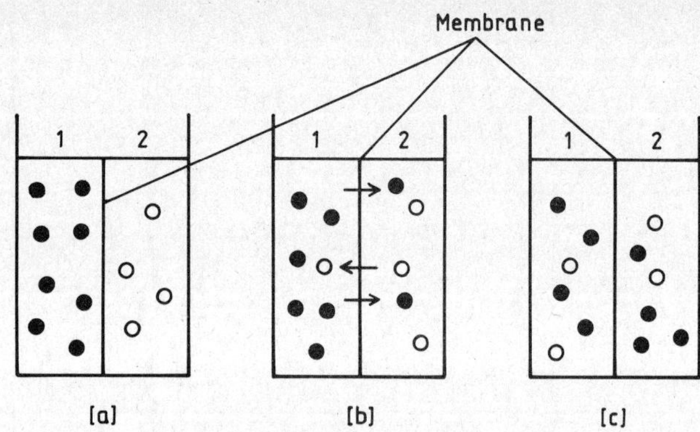

Figure 14.3 Diffusion of one substance across a membrane. (a) Solution 1 contains twice the number of oxygen molecules as solution 2, and they are separated by a membrane that allows oxygen to pass through it. (b) Twice as many oxygen molecules diffuse across the membrane from solution 1 to solution 2 than from solution 2 to solution 1; net diffusion of oxygen occurs from solution 1 to solution 2. (c) Chemical equilibrium results from the net movement of oxygen, i.e. the number of oxygen molecules in each solution becomes equal.

molecule from solution 2 (Figure 14.3a). If a particular oxygen molecule collides at the site of a hole (pore) in the membrane, the oxygen molecule will pass through the membrane pore as long as the pore is larger than the molecule.

This difference in the rate of collisions with the membrane results in a net movement of oxygen from solution 1 to solution 2 (Figure 14.3b). Molecules will also move from a low concentration to a high concentration but twice as many oxygen molecules move from the higher concentration causing a net flow of molecules until the concentration of oxygen in solutions 1 and 2 is equal (Figure 14.3c), that is, the chance of an oxygen molecule hitting a membrane pore is equal on both sides of the membrane (chemical equilibrium). The greater the difference in concentration across a membrane, the greater is the rate of diffusion.

If the membrane separates two different solutes of neutral charge such as oxygen and carbon dioxide, these solutes will diffuse independently of each other (Figure 14.4). The oxygen molecules will diffuse down their concentration gradient by colliding with the membrane pores whilst the carbon dioxide will collide independently with the membrane pores. Diffusion of oxygen and carbon dioxide across the alveolar membrane relies on this principle.

Several factors may alter the diffusion rate of a solute across a membrane:

1. If the particles are charged, this charge will

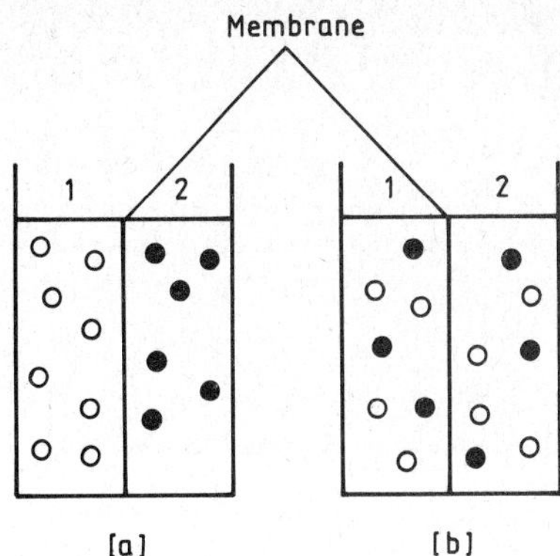

Figure 14.4 Diffusion of two substances across a membrane. (a) Solution 1 contains oxygen whereas solution 2 contains carbon dioxide. These two solutions are separated by a membrane that allows both oxygen and carbon dioxide to pass through it. (b) Both oxygen and carbon dioxide diffuse across the membrane *independently* of each other. Independent diffusion results in a chemical equilibrium between solution 1 and solution 2 for oxygen and a separate chemical equilibrium for carbon dioxide.

influence diffusion, as like charges repel. With charged particles the rate of diffusion depends upon the chemical and electrical gradient (difference in potential). This balance of chemical and electrical gradients is important in the functioning of nerves and muscles (see 5.3).

2. If the surface area of a membrane separating solutions is decreased the rate of diffusion of particles across the membrane decreases. Fewer membrane pores results in less particles being able to pass through the membrane at any moment.

APPLICATION 1: EMPHYSEMA

In emphysema the total surface area of the alveolar respiratory membrane decreases due to the loss of alveolar walls. The diffusion of gases across the alveolar membrane is significantly impeded, even at rest, if the loss of total membrane surface area is greater than one-third of normal.

APPLICATION 2: SMALL INTESTINE STRUCTURE

The principle of an increased surface area creating an elevated rate of diffusion is shown by the extensive folding in the small intestine. This folding occurs both macroscopically and microscopically as is shown in Figure 14.5. The absorptive area is increased by about 600-fold to provide a membrane surface area of 250 square metres. This large surface area enables a maximum absorption of nutrients across the small intestine and provides a large reserve for small intestine function. Removal of up to seventy-five per cent of the small intestine can occur without the patient suffering serious malnutrition.

3. An increase in membrane thickness results in a greater resistance to particles diffusing across it. The particles are also required to move a greater distance. Both of these effects can results in decreased efficiency of respiratory gas exchange. EXAMPLE: Pulmonary fibrosis.

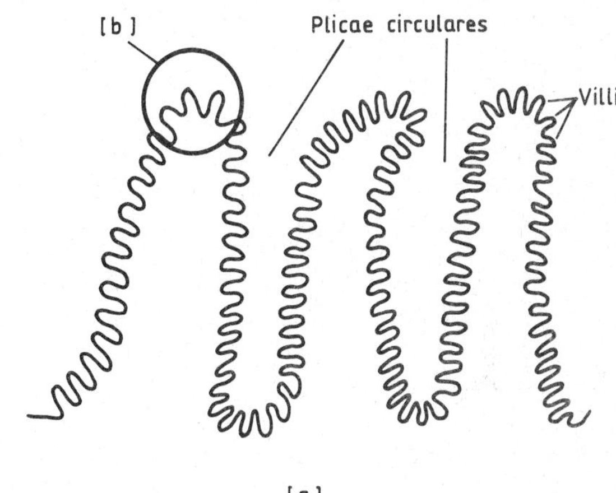

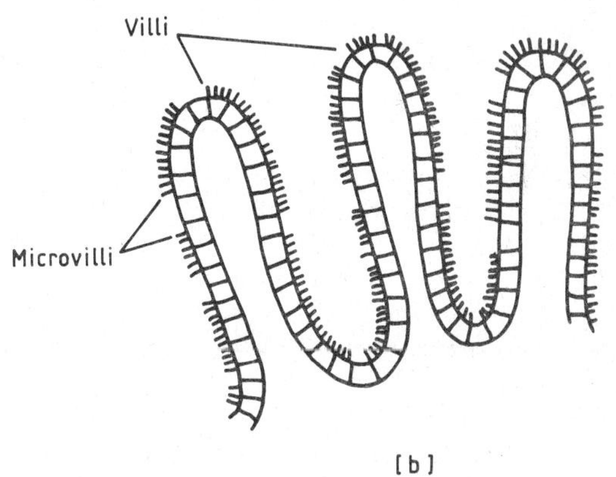

Figure 14.5 Cross-section of a small intestine wall showing folds (plicae circulares) and projections (villi and microvilli) which vastly increase the surface area. (a) Macroscopic view of the small intestine showing the deep folds of the plicae circulares which contain small projections (0.5 to 1.0 mm) called villi. (b) Microscopic view of villi shows a large number of finger-like projections from the plasma membranes of cells involved with nutrient absorption.

Summary

Diffusion is the net movement or flow of particles from a region of high concentration to a region of low concentration. This net movement is due to the Brownian motion of particles. The ability of a particle to diffuse a certain distance will be influenced by the:
1. number of particles that it must diffuse through,
2. size and weight of the diffusing particle,
3. shape of the particle,
4. charge of the particle.

A net diffusion of particles will occur across a membrane if the membrane separates two solutions of different concentrations. That is, more particles collide with the membrane on one side and then pass through it. When different non-electrolytes are in a solution, each non-electrolyte will diffuse across the membrane independently of other non-electrolyte solutes.

The rate of diffusion across a membrane is altered by a change in the surface area of membrane and the membranes thickness.

Chapter 15

Osmosis and body fluid balance

Objectives

At the completion of this chapter the student should be able to:
1. describe the principle of osmosis,
2. define osmotic pressure,
3. compare and contrast the effects of isotonic, hypotonic and hypertonic solutions on cells,
4. define the term osmol,
5. explain how osomotic equilibrium exists between cells and the extracellular fluid,
6. describe how a balance exists between hydrostatic pressure and osmotic pressure at the capillary,
7. explain why this balance of pressure is important in the maintenance of body fluid,
8. explain how an imbalance in hydrostatic or osmotic pressure alters body fluid volume and nutrient supply.

15.1 Principle of osmosis

A large proportion of a human body consists of water. The diffusion of water is an important component of normal cell function and this diffusion is referred to as osmosis.

Osmosis is the net movement of water through a semi-permeable membrane from an area of high water concentration to an area of low water concentration. A semi-permeable membrane is a membrane which allows the movement of water but not other substances, through it. When a semi-permeable membrane separates solutions of different concentrations, water will diffuse from the region of high water concentration to the region of low water concentration.

In Figure 15.1a two solutions of equal volume but different concentration are separated by a semi-permeable membrane. Solution 1 contains a larger number of water molecules than solution 2, that is,

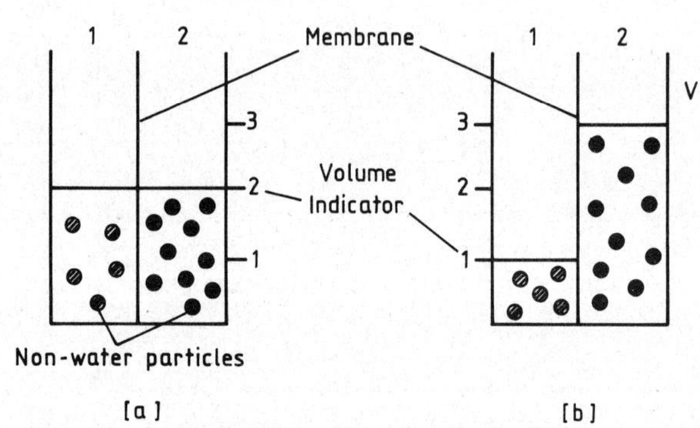

Figure 15.1 Movement of water by the process of osmosis across a semi-permeable membrane.
(a) Two solutions that contain the same quantity of particles (equal volumes) are separated by a semi-permeable membrane.
(b) Solution 1 contains a greater proportion of water molecules than

does solution 2. The greater number of water molecules in solution 1 results in a greater chance of water molecules colliding with the membrane, than occurs with solution 2. A net movement of water develops from the region of high water concentration (solution 1) to the region of low water concentration (solution 2) until the water concentrations in either solution are equal.

in solution 1 a greater proportion of water molecules make up the total volume. A net movement of water molecules occurs from 1 to 2 since there is a greater chance that a water molecule from solution 1 will collide with the membrane than there is from solution 2. The kinetic energy of water molecules results in a movement of water molecules in both directions, but a net flow will occur until the water concentrations are equal (Figure 15.1b). In balancing the water concentrations, a change of volume has occurred. This change of volume will occur whenever a change in concentration of a solution develops.

APPLICATION: PURGATIVE AGENTS

Purgative (cathortic) agents such as epsom salts (magnesium sulphate) act by increasing the water content of the faeces (see Figure 15.2).

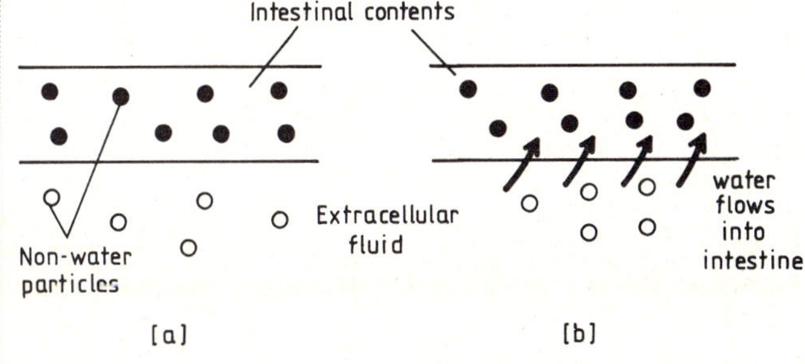

Figure 15.2 Action of purgative agents such as epsom salts in the intestine.
(a) Purgative agents are retained within the intestine since they cannot penetrate the intestinal membrane. Retention of epsom salts results in a lower water concentration within the intestine than in the surrounding body fluid.
(b) Water moves by osmosis from the high water concentration of the body fluid to the intestinal content until the water concentrations are approximately equal. This movement results in an increased fluidity of the intestinal contents which results in easier evacuation of the bowels.

15.2 Terms relating to osmosis

OSMOTIC PRESSURE

This is the pressure required to exactly oppose the net movement of water by osmosis, that is the resistance that is required to just stop the net flow of water. Osmosis results from the pressure difference developed on opposite sides of a membrane by a greater number of water particles colliding with one side of the membrane than the other side. The greater the difference in water concentration the greater is the osmotic pressure that is required to prevent water molecules moving via osmosis.

The osmotic pressure of a solution depends upon the proportion of non-electrolytes, electrolytes and colloids present compared with water molecules. Compounds that separate in solution (dissociate) into ions, create a greater osmotic pressure than compounds that remain intact in solution.

APPLICATION: ELECTROLYTE AND NON-ELECTROLYTE INFLUENCE ON OSMOTIC PRESSURE

Glucose contains the same number of particles whether it exists as a solid or is dissolved in a liquid. Sodium chloride consists of twice as many particles in solution than when it exists as a solid. This increase in particles is a result of sodium ions and chloride ions separating (dissociating) when placed in solution. Thus 100 mmol of sodium chloride in solution contains double the number of particles and develops twice the osmotic pressure as 100 mmol of glucose (see Figure 15.3).

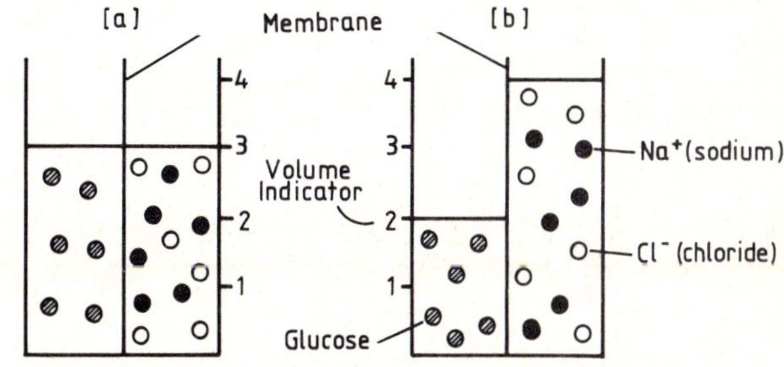

Figure 15.3 A comparison of the effect of glucose (left chamber) and sodium chloride (right chamber) on osmosis.
(a) 100 mmol of glucose and a 100 mmol of sodium chloride are dissolved in water.
(b) The sodium chloride solution contains twice as many particles in water as the glucose solution and the proportion of water molecules is less in the sodium chloride solution than the glucose solution. This difference in water concentration results in water diffusing from the glucose to the sodium chloride solution until both water concentrations are equal.

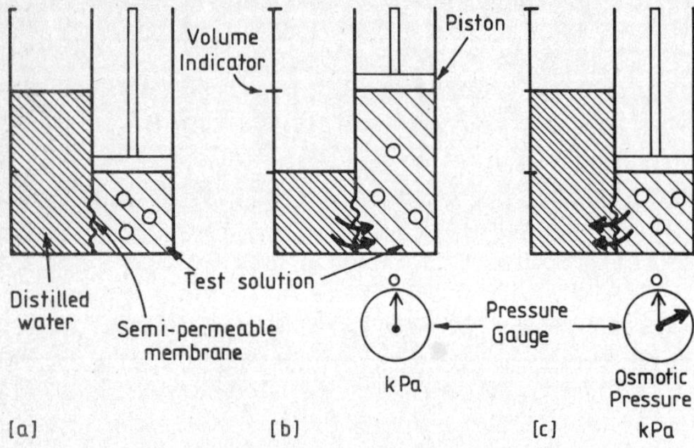

Figure 15.4 The measurement of osmotic pressure.
(a) A solution of unknown osmotic pressure is placed in a cylinder containing a piston. This cylinder is separated from a measured volume of distilled water by a semi-permeable membrane.
(b) Water will flow from the high water concentration to the test solution if the piston is not exerting any pressure.
(c) The osmotic pressure or resistance to the flow of water is determined by applying pressure on the piston. The osmotic pressure is that pressure exerted by the piston which returns the distilled water to the original water level. That is, the required pressure that will prevent the net flow of water.

Classically osmotic pressure was measured by the use of a hydraulic cylinder connected with a pressure gauge (Figure 15.4a). To measure the potential degree of water movement it was easier to determine the pressure that was required to prevent the flow of water. Thus, osmotic pressure was always greatest on the side with a lower water concentration. In Figure 15.4b, a net flow of water develops from the distilled water when no external forces are present. This flow is due to the greater probability of water molecules colliding with the membrane surface that adjoins the distilled water than the surface that adjoins the test solution. The pressure developed by this difference in probability can be opposed by exerting an equal but opposite pressure. This opposing pressure will increase the number of water molecules in the test solution colliding with the membrane surface. This opposing pressure is developed by compressing the piston against solution B which forces more water molecules to collide with the membrane at any given moment (Figure 15.4c). The osmotic pressure developed by solution B can be determined by measuring the pressure that is required to maintain the volume of solution A at its original level (Figure 15.4c).

Osmotic pressure is always greatest on the side with the lowest water concentration and in real terms is a measure of the pressure required to oppose osmosis.

OSMOL

Osmol is the measurement of a solute's ability to induce osmosis and osmotic pressure and is a measure of the total number of particles present in a solution. One osmol equals one mole of particles present in a solution. Note that when one mole of electrolyte (e.g. sodium chloride) is placed in solution, separation into sodium and chloride ions results in a greater number of particles than occurs with a substance that remains intact in a solution. The number of osmols in a electrolyte solution is determined by multiplying the molar concentration by the number of ions present in one molecule. EXAMPLE: Calcium chloride ($CaCl_2$) contains one calcium ion and two chloride ions. A one molar concentration will therefore contain three osmols.

In physiological solutions the low concentration of particles requires the use of a term one-thousandth of an osmol, called the milliosmol (mosmol).

OSMOLARITY AND OSMOLALITY

The osmol concentration of a solution can be described in two ways. These are osmolality when expressed per kilogram of water and osmolarity when expressed per litre of solution. These terms are interchangeable when body fluid concentrations are used. The osmolal concentration of isotonic body fluid is 280 to 290 mosmol/L. An approximate indication of correct body fluid osmolality can be found by determining if the serum sodium concentration is in the range 137 to 149 mmol/L. An elevated serum sodium would indicate a loss of water and an increased osmolality.

ISOTONIC OR ISOSMOTIC SOLUTION

This solution will cause no shrinking or swelling of normal body cells when the latter are placed in it. The osmotic pressure in the fluid and cells is equal. No net movement of water will occur between the fluid and cell (Figure 15.5a). EXAMPLE: Five per cent glucose solution, 0.9 per cent sodium chloride solution.

HYPOTONIC SOLUTION

This is solution that will cause normal cells to swell. These solutions exert a lower osmotic pressure than cells. Water flows from the bathing solution into cells, which results in a swelling of the cells (Figure 15.5b). When a low osmotic pressure stretches the cell membrane to its limit, the cell will rupture, i.e. cells lyse. EXAMPLE: Erythrocytes (red blood cells) will haemolyse when placed in a hypotonic solution such as distilled water or 0.45 per cent sodium chloride.

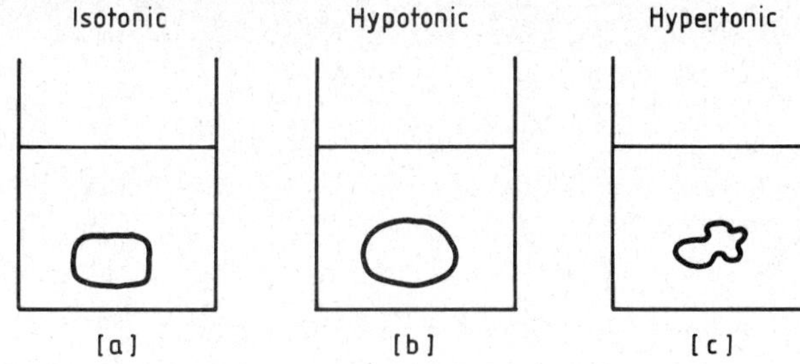

Figure 15.5 Effect of varying solution concentrations on red blood cells. (a) In an isotonic solution no net movement of water occurs. Cells retain their original size and shape. (b) Hypotonic solution creates a net movement of water into the cells, the cells swell and may rupture (haemolysis). (c) Hypertonic solution causes cells to shrink (crenation).

HYPERTONIC SOLUTION

This is solution that develops a greater osmotic pressure than cells. Hypertonic solutions have a lower water concentration than cells, thus resulting in normal cells losing water and shrinking (Figure 15.5c). EXAMPLE: 1.8 per cent sodium chloride.

15.3 Osmotic equilibrium between cells and their surrounding fluid

The concentration of particles in the extracellular fluid compartment equals the concentration of particles in the intracellular compartment (Figure 15.6). Throughout the body there is an osmotic equilibrium between these compartments. A change of concentration in one compartment also changes the concentration in the other compartment (Figure 15.7).

APPLICATION: DEHYDRATION

Diarrhoea, vomiting, fever and burns causes net water loss which results in dehydration. In dehydration, the extracellular fluid compartment loses water. This results in a hypertonic extracellular fluid when compared with the intracellular fluid compartment. Water flows via osmosis from the cell to the extracellular fluid, causing the cell to shrink. Receptor cells in the thirst centre of the brain are stimulated by shrinking. These cells initiate a homeostatic mechanism that returns the osmolality of the extracellular and intracellular fluids to normal (Figure 15.7).

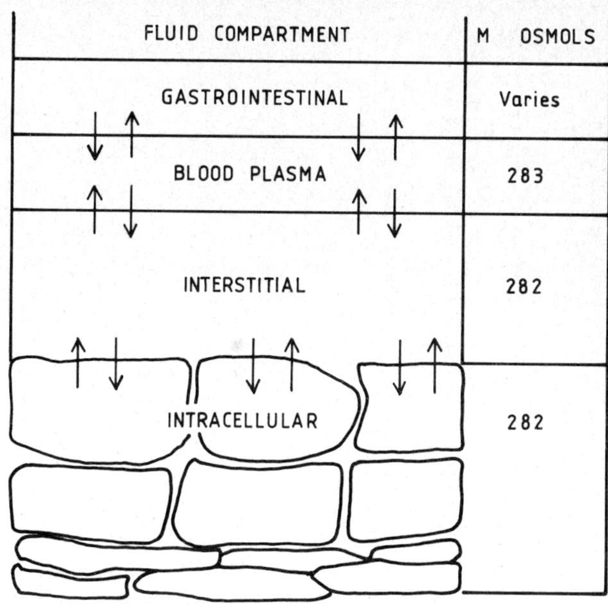

Figure 15.6 Osmotic balance between different body fluid compartments.

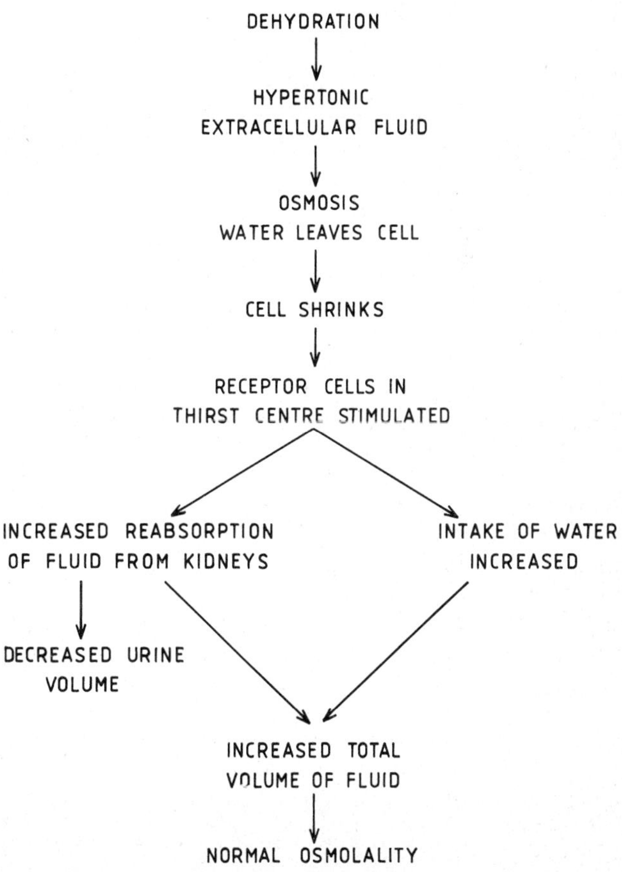

Figure 15.7 Effect of dehydration on the homeostatic mechanisms for the control of osmotic balance.

15.4 Capillary hydrostatic pressure and osmotic pressure influence on interstitial fluid concentration

In capillaries, a balance exists between differences in osmotic pressure and hydrostatic pressure. This balance enables particles to be moved to the cells and cell products being returned to the vascular system (Figure 15.8). An imbalance in this system can result in alterations to body fluid distribution, such as oedema.

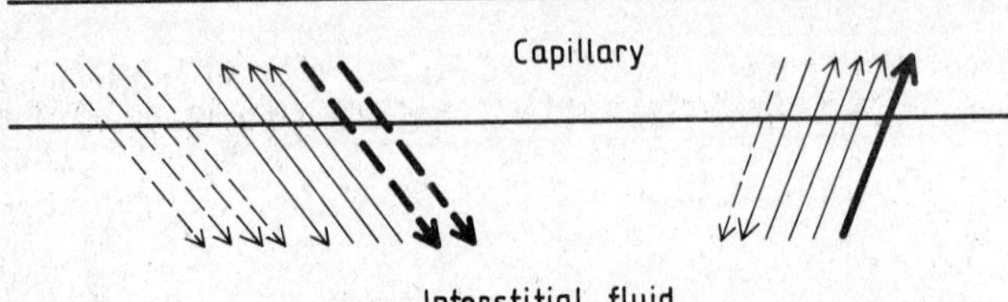

Figure 15.8 Exchange of fluid and nutrients at the capillary according to Starling's law of the capillary. The thick arrows indicate the net movement of fluid. The --- arrows show the relative hydrostatic pressure and the —— arrows the osmotic pressure.

In the capillary, the hydrostatic pressure is sufficient to force solutes and water through gaps in the capillary walls. As blood flows along the capillary, an increase in resistance occurs due to both the small diameter and length of the vessel. This resistance results in a decreased hydrostatic pressure at the venule end of the capillary. The hydrostatic pressure decreases from 4.0 kPa to 1.0 kPa.

The osmotic pressure in the capillary is mainly due to plasma proteins exerting a pressure of 3.0 kPa in the opposite direction to the hydrostatic pressure. The interstitial fluid has an osmotic pressure of 1.0 kPa and no hydrostatic pressure. This results in a greater number of particles colliding with the vascular wall of the capillary than the interstitial side of the capillary. A net flow of particles into the interstitial fluid occurs. Large molecules such as the plasma proteins normally cannot pass through the capillary gaps. The retention of proteins in the blood maintains the osmotic pressure along the length of the capillary. Towards the venule end, the hydrostatic pressure decreases below the osmotic pressure. Fluid returns to the capillary by the influence of osmosis as a result of the greater number of water and solute particles colliding with the interstitial wall of the capillary.

APPLICATION: OEDEMA

Hypertension can result in a decreased osmotic influence. Hydrostatic pressure is greater than normal and pushes excess fluids into the interstitial space. Osmotic pressure is insufficient to overcome all of the hydrostatic influence, and thus fluid is retained in the interstitial region.

Summary

Osmosis is the net diffusion of water through a semi-permeable membrane from a region of high water concentration to a region of low water concentration, that is, from a region of low solute concentration to a region of high solute concentration.

Osmotic pressure is that pressure which will oppose osmosis. The region with the lowest water concentration contains the greatest osmotic pressure.

Electrolytes can dissociate in water, creating a greater number of particles and thus a greater osmotic pressure than substances that remain intact.

The osmotic pressure of an isotonic solution is the same as intracellular fluid, whereas with hypertonic solutions the pressure is greater, and in hypotonic solutions it is less.

The ability of a substance to induce osmosis and osmotic pressure is measured in osmols. An osmol is the total number of particles present in a solution and is thus related to the molarity of a solution. In body fluids, osmolarity and osmolality are interchangeable terms describing the osmol concentration.

An osmotic equilibrium exists between the blood, interstitial and intracellular body fluid compartments. An alteration in the concentration of one compartment has corresponding affects in the other compartments.

In capillaries, a balance exists between differences in osmotic pressure and hydrostatic pressure—Starling's law of the capillary. This balance enables an exchange of respiratory gases and nutrients between the blood and interstitial fluid.

Unit Seven
Intracellular structure and function

This unit outlines the structure and function of cells, with particular reference to plasma membrane and nucleus function.

Chapter 16

Cell structure

Objectives

At the completion of this chapter the student should be able to:
1. describe the structure of a typical animal cell and relate this structure to cell function,
2. describe the differences between eucaryotic and procaryotic cells,
3. differentiate between cilia and flagella,
4. define cell inclusion and give examples of beneficial and harmful cell inclusions,
5. relate the use of drugs to bacterial structure,
6. relate capsules and endospores to bacteria function.

16.1 Cell theory

All organisms, from bacteria to humans, consist of matter that is arranged in structural and functional units called cells. A cell is generally the smallest and simplest unit of matter that has the potential to duplicate itself. The only forms of matter that do not consist of cells but which are capable of reproducing themselves are the viruses. Viruses are termed acellular, that is, non-cellular.

All cells consist of a complex arrangement of molecules. These molecules are almost entirely composed of the elements carbon, hydrogen, oxygen and nitrogen. Cells function by altering the structure of intracellular molecules. These molecules are either broken down to provide energy or are altered into new molecules that are required by the cell or other cells in the body. Cellular homeostasis enables cells to maintain their chemical function in a changing extracellular environment.

The study of cells is known as cytology.

16.2 Types of cells

Cells are divided into two types according to their internal structure. The procaryotic cells are cells that consist of a simple internal structure. Eucaryotic cells have a highly organized internal structure.

16.3 Eucaryotic cells

A generalized eucaryotic cell consists of the following structures:
1. Plasma (cell) membrane—this separates the cell's internal contents from the extracellular environment. The term protoplasm is used to describe a cell's contents along with the plasma membrane. Protoplasm is usually divided into the nucleus and the cytoplasm.
2. Nucleus—a relatively large membrane-bound structure that controls and regulates cell multiplication and maintains cell function.
3. Cytoplasm—this is a general name for all the contents of a cell apart from the nucleus. The cytoplasm can be divided into three parts: organelles or small functional structures surrounded by membranes, inclusions which are secretions and storage areas and the cytoplasmic matrix which is the fluid in which the organelles and inclusions lie.

Both animals and plants are composed of eucaryotic cells. These two forms of eucaryotic cells contain many common organelles. Each also have some unique organelles. EXAMPLE: Chloroplasts are present in plant cells but not animal cells. The general structure of an animal cell is shown in Figure 16.1.

A cell is a microcosm of life in a city. This analogy will be used to assist in understanding and remembering the function of organelles. An introductory account of the main functions of each animal cell component is described below. Note that there are many variations in the form that a cell can take. The difference cell forms are described in 19.1.

CELL MEMBRANES

The plasma membrane consists of one layer of membrane. The structure of a membrane is described in 17.1. Most membranes that surround organelles also consist of one layer. The nucleus and mitochondria are surrounded by two layers of membrane. The function of the membranes of a cell is discussed with the respective organelles, apart from the plasma membrane which is described in Chapter 17. ANALOGY: The membranes of a cell act as protective wrappers similar to the plastic wrappers used to seal food goods.

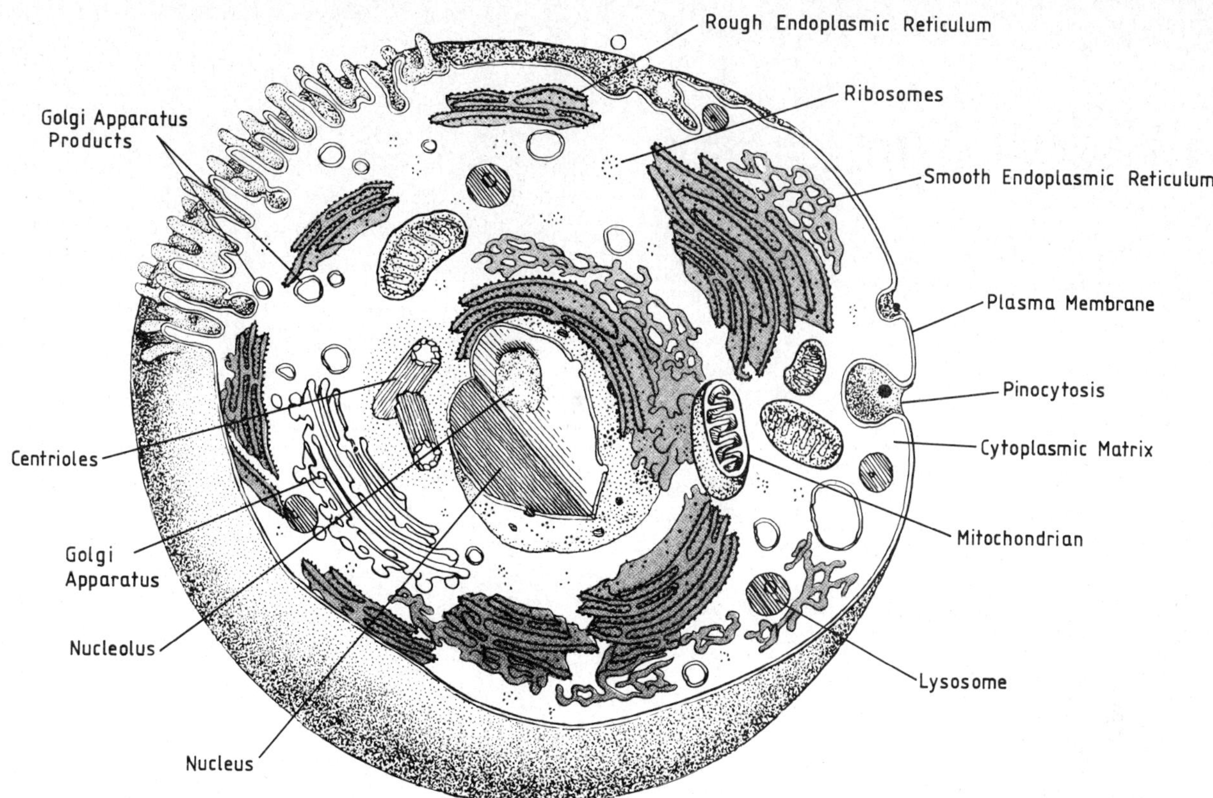

Figure 16.1 The general structure of an animal cell. (After figure 3.1 in *Principles of Anatomy and Physiology*, 3rd edition, Copyright © 1981 Gerard J. Tortora and Nicholas P. Anagnostakos, reproduced by permission of Harper and Row Publishers Inc.)

NUCLEUS

The nucleus is the largest structure within a cell. It is surrounded by two membranes that contain openings or pores. These pores probably allow the movement of large molecules that carry genetic information between the nucleus and the cytoplasm. It is possible that these pores can be closed to prevent the leakage of substances through the membrane.

Within the nuclear membrane there exists a gel-like material called nucleoplasm, one or more spherical bodies called nucleoli (nucleolus singular) and genetic material in the shape of rods. These rods are the basis of hereditary information and are termed chromosomes. The chromosomes control cell reproduction whereas the nucleoli are involved in the transfer of hereditary information into a form that regulates cell chemical reactions. ANALOGY: The nucleus acts as an information resource centre from which material can be organized for use by the cytoplasmic organelles.

CYTOPLASMIC MATRIX

The cytoplasmic matrix consists of a mixture of solutions, sols and gels. The solution component contains mainly dissolved proteins, electrolytes, fatty acids, glucose and cholesterol. The sol consists of large particles such as fat globules, glycogen granules, inclusions, secretory packages and cytoplasmic organelles dispersed in a liquid. A gel often exists near the plasma membrane.

CYTOPLASMIC ORGANELLES

Endoplasmic Reticulum

The endoplasmic reticulum or ER consists of a membraneous network of canals that runs through the entire cytoplasm. The ER system of canals can be linked directly to both the nuclear membrane and the plasma membrane (Figure 16.1). The ER can be in the form of either canals or flattened canals which terminate in sac-like structures.

This canal system conveys information from the nucleus to the cytoplasm and allows intercommunication between the outside and the inside of a cell. EXAMPLE: In muscles, the ER plays an important role in the coupling of muscle depolarization at the plasma membrane with the initiation of the contractile state deep inside a muscle cell. The presence of a canal system enables an efficient movement of substances throughout a cell which bypasses the relatively inefficient, slow movement of chemicals via diffusion through the cytoplasmic matrix.

The endoplasmic reticulum exists in two forms, either rough ER or smooth ER.

The rough endoplasmic reticulum bears a covering of tiny organelles called ribosomes. The ribosomes are the site of protein manufacture. Instructions are carried from the nucleus via the ER to the areas of rough ER where the appropriate proteins are manufactured. The proteins can then be transported via the ER to storage sites or released out of the cell.

Endoplasmic reticulum that lacks a coating of ribosomes is referred to as smooth endoplasmic reticulum. The smooth ER is involved in the production of fat and fat-soluble hormones such as steroids.

Both forms of ER in liver cells are involved in the breakdown and modification of drugs. EXAMPLE: The sedative pentobarbitone is degraded into biologically inactive compounds. ANALOGY: Instructions are sent from the head office or nucleus via a freeway network or ER to the manufacturing plants which are either areas of rough ER or smooth ER. This system is far more efficient than information being carried by couriers via tedious, time-consuming side streets or cytoplasmic matrix.

Ribosomes

Ribosomes are small organelles (fifteen to twenty nanometres) that are located on the ER. They are also spread throughout the cytoplasm, usually in the form of clusters called polyribosomes. As previously mentioned, ribosomes are the site of protein manufacture. ANALOGY: The sites of chemical manufacture are located on the freeway network or in industrial centres known as polyribosomes.

Mitochondria

Mitochondria are double-membraned organelles (Figure 16.2). The inner membrane is arranged in folds or cristae which protrude into the inner mitochondrial cavity called the mitochondrial matrix. The presence of cristae result in increased surface area of the internal membrane. This relatively large surface area is important for cell function since almost all of a cell's energy is produced on the surface of the cristae. This energy is only formed in the presence of oxygen. The cristae are the end site for oxygen that has been transported through the body. Mitochondria are often referred to as the powerhouse of a cell and thus cells that require large amounts of energy contain numerous mitochondria. EXAMPLE: Cardiac muscle cells. ANALOGY: Factories can only manufacture goods when energy is present. This energy is mainly produced at powerhouses which are located in the cell at the mitochondrial cristae.

Golgi Apparatus

The golgi apparatus is probably an extension of the endoplasmic reticulum. It consists of a network of flattened tubes that are stacked on top of each other in a similar manner to a group of stacked plates (Figure 16.3).

It appears that the golgi apparatus acts as a storage and distribution centre for the proteins and fats that were formed at the ER. These substances travel from the ER either directly or indirectly to the golgi apparatus. The golgi apparatus stores the manufactured chemicals and packages them for release. The membrane-bound packages are released into the cytoplasmic matrix. They then are either used by the cell or are transported to the plasma membrane where they are released from the cell, that is, secreted from the cell. The golgi apparatus is very prominent in secretory cells. EXAMPLE: Secretory cells of the gastrointestinal tract contain a prominent golgi apparatus in the side of the cell that lines the tract. The packages that are utilized by the cell are usually released in the form of membrane-bound organelles called lysosomes which are described under the next heading.

The golgi apparatus is also involved in the manufacture of mucus which is later incorporated into the cell's membrane or secreted. Other substances that are formed and secreted, form the basis of the interstitial fluid.

In summary, the golgi apparatus acts as a storage

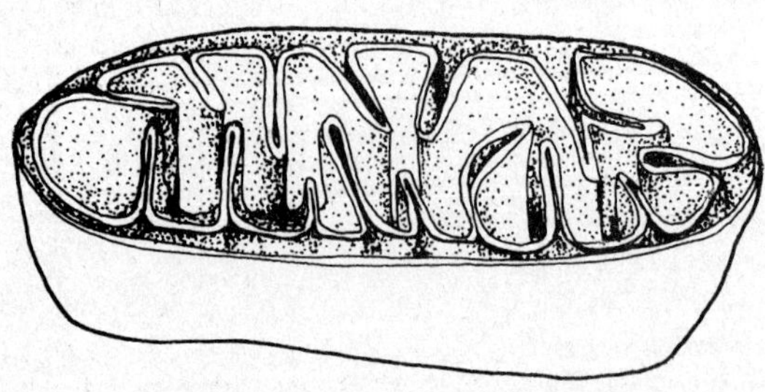

Figure 16.2 Representative mitochondria showing the folds or cristae of the inner membrane.

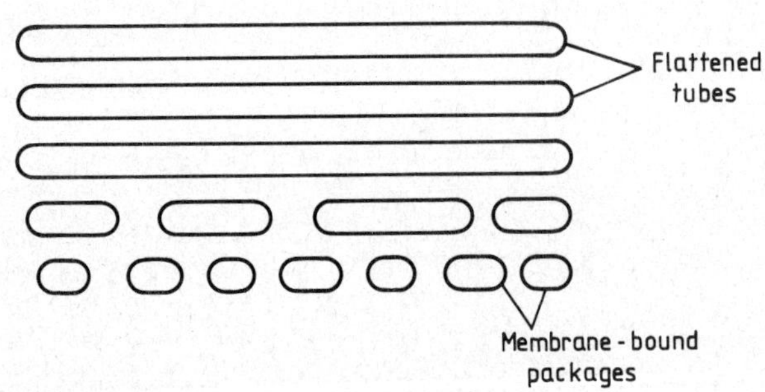

Figure 16.3 Golgi apparatus showing a network of flattened tubes.

and distribution centre. In addition, it is also the manufacturing site of substances such as mucus. ANALOGY: When goods are manufactured they are then stored in warehouses from which they are distributed according to demand.

Lysosomes

Lysosomes are small organelles formed by the golgi apparatus. They contain powerful digestive substances that are used to break down unwanted molecules and particles such as fragments of organelles and bacteria. The lysosomes thereby remove unwanted substances and structures from the cytoplasm. These substances and structures are broken down and then are either released into the cytoplasm for recycling or are excreted from the cell. ANALOGY: Manufactured goods usually end up as waste products which have to be disposed of. Waste disposal units break down unwanted refuse and recycle reusable goods as do lysosomes.

The rupture of lysosomal membranes releases powerful digestive substances into the cytoplasmic matrix. If sufficient lysosomes are ruptured the whole cell is broken down in the process of autolysis. EXAMPLE: Exposure to sufficient sunlight causes sunburn and the rupture of lysosomes in the skin cells. The release of the lysosomal contents kills cells near the skin's surface and leads to blistering and later to the "peeling" of a layer of skin.

Centrioles

Centrioles are small cylindrical structures that play an important role in cell division and are discussed in chapter 18.

Cilia and Flagella

Both cilia and flagella are structural projections of the plasma membrane which have a motile property. Cilia are short hair-like motile projections that occur in numerous quantities on the plasma membrane. Flagella (flagellum singular) are long motile projections that occur in small numbers.

Cilia usually propel material from one cell to the next by moving in coordinated waves. EXAMPLES: The respiratory tract is lined with cilia that propel mucus from the lungs. The fallopian or uterine tubes are lined with ciliated cells that move the ovum from the ovaries to the uterus. Flagella move like a whip. They are used for propelling cells. EXAMPLE: Sperm are propelled by the movement of their long tail or flagellum.

Cell Inclusions

Cell inclusions are a diverse group of structures that usually contain materials retained by a cell for purposes of cellular function. Cell inclusions are used for a variety of functions. EXAMPLES: Haemoglobin for carrying oxygen in red blood cells, glycogen and fats for cell chemical reserves, melanin for cell pigmentation and mucus as a cellular secretion.

A virus is a cell inclusion that is harmful to the cell. Viruses are small acellular structures that consist of hereditary information surrounded by protein. Viruses use the cell contents for multiplication. They can remain dormant in their intracellular position until a satisfactory change affects the cell or the virus. When in the dormant or latent state they probably do not interfere with the function of their host cell.

APPLICATION: HERPES

Herpes simplex is the virus responsible for the development of cold sores. This virus remains in a latent state in cells until some stress such as fever, emotional upset, sun exposure, being run down or ill with another disease activates it. The virus causes the formation of vesicles, especially in the skin of the lips. Rupture of these vesicles results in the release of viruses from them. Herpes can thus be transmitted by direct personal contact with the ruptured vesicles. Body defence mechanisms develop to overcome the outbreak and return the skin to normal. However, some of the virus remains in the latent state. This ensures that the virus will remain in a person for life.

IMPORTANT PLANT CELL STRUCTURES

Plant cells contain the above organelles plus several other structures that are necessary for plant cell function. Plants contain a cell wall which confers rigidity to the shape of a cell. Many plant cells contain chloroplast organelles. These organelles play an important role in life since they use solar energy to produce carbohydrate from carbon dioxide and water. Large cell inclusions occur in the form of membrane-bound fluid filled spaces called vacuoles. This fluid contains salts, sugars, organic acids and other soluble substances necessary for cell function.

16.4 Procaryotic cells

Procaryotic cells differ substantially from eucaryotic cells by lacking a distinct nucleus and cytoplasmic organelles (Figure 16.4). Procaryotic cells are found only in bacteria and blue-green algae. These cells exist and function as single cells which are referred to as unicellular organisms. The important structural features of bacteria and their application to health are described below.

CELL STRUCTURES

Cell Membrane

The plasma membrane differs from that found in eucaryotic cells in that is probably performs the function of the mitochondria of animal and plant cells.

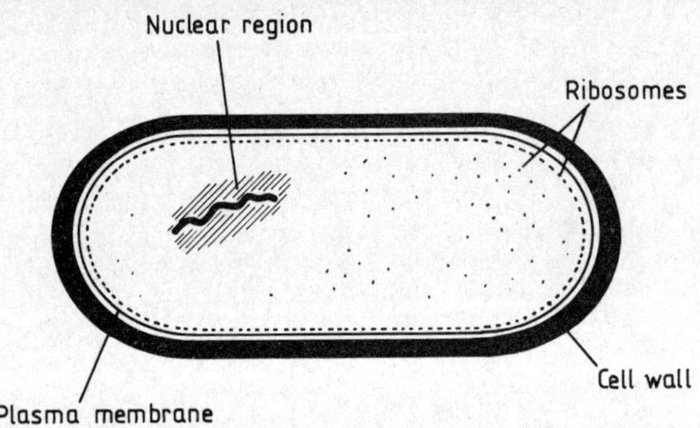

Figure 16.4 General structure of a procaryotic cell.

Ribosomes

Ribosomes are found either in the cytoplasmic matrix or are attached to the plasma membrane. The ribosomes manufacture bacterial proteins.

Nuclear Region

This region is not a distinct nucleus as no membrane is present. It contains one strand of hereditary information which is responsible for both reproduction and protein synthesis. Drugs such as tetracycline interfere with the growth of bacteria by interfering with the synthesis of proteins.

Flagellum

These can be found on mobile bacteria, that is, rod shaped (bacilli) and coiled (spirillia) bacteria. Non-mobile bacteria such as spherical bacteria (cocci) lack a flagellum.

Cell Wall

Most procaryotic cells contain a cell wall which has a protective function. The cell wall maintains the cell's shape during harmful environmental changes such as osmotic imbalances. Substances such as penicillin, vancomycin and bacitracin all act on cell wall manufacture, whereas enzymes such as lysozyme (found in tears, saliva) and gastric juices can rupture the cell wall.

APPLICATION: GRAM POSITIVE AND NEGATIVE BACTERIA

Bacteria are often identified according to their reaction to the stain, crystal violet. Gram stains are widely used in medicine and are often the first step in making a bacteriologic diagnosis of an infectious disease. Bacteria that form a blue or purple colour in the presence of crystal violet are termed gram positive bacteria. This colour response results from a thick cell wall which is composed of a different chemical to the cell walls of other bacteria. These latter bacteria stain red and are referred to as gram negative bacteria.

The gram positive bacteria are sensitive to penicillin and lysozyme whereas gram negative bacteria are sensitive to drugs such as ampicillin.

Capsules

Some bacteria have a slimy mucous-like coat surrounding the cell wall which is called a capsule. It can act as a protective structural covering, a nutrient reservoir, a site for the disposal of waste products and as a mechanism by which the bacteria can adhere to a host. EXAMPLE: The bacteria *Streptoccoccus mutans* causes tooth decay by being able to attach to teeth using a capsule. Capsules increase the virulence of a bacteria, that is, the ability to cause disease, because they resist ingestion by the host defence cells.

Endospores

Certain bacteria produce dormant resting cells within the cytoplasm of the bacteria when the environment becomes unfavourable for normal growth and reproduction. These dormant cells are termed endospores or spores (Figure 16.5). Endospores resist harsh physical and chemical treatments and will remain in a state of suspended animation until the environment becomes favourable. When satisfactory environmental conditions exist the endospore actively grows and at this stage is referred to as a vegetative form.

Normal sterilization procedures, that is measures which destroy or remove all living organisms, can be used on the vegetative form, whereas endospores must be treated with sporicidal sterilization. EXAMPLE: Vegetative bacteria are readily killed when exposed to boiling water for a few minutes whereas spores are resistant to this measure. Spores can be killed by using an autoclave which is an apparatus that produces temperatures in the order of 134°C.

The main spore-bearing bacteria include some of the most harmful bacteria. EXAMPLES: Anthrax, tetanus and gas gangrene.

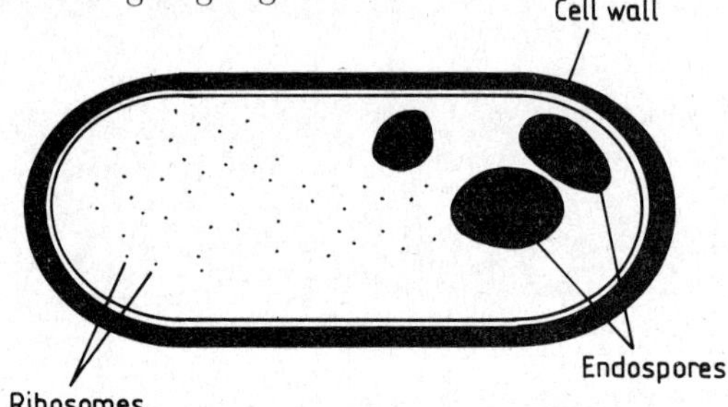

Figure 16.5 Bacterium containing endospores.

CELL INCLUSIONS

Viruses

These acellular structures can invade a bacteria where they can live and multiply by using the cell's chemicals. A virus may lie dormant for long periods in its intracellular position interfering little, if at all, with the cell's activity. The virus remains in this latent condition until some change affects the cell or the virus, after which the virus multiplies.

Summary

A cell is the structural and functional unit of all organisms. Each cell has the potential to duplicate itself. Viruses are acellular structures that are capable of duplicating themselves. This duplication can only occur when they are within cells.

Animals and plants are composed of eucaryotic cells whereas bacteria and blue green algae consist of procaryotic cells. The contents of a cell and its surrounding membrane is often referred to as protoplasm.

The protoplasm of eucaryotic cells is subdivided into the nucleus and cytoplasm. The nucleus controls and regulates cell division and maintains cell function. The cytoplasm is the site of the manufacture, storage and breakdown of substances. It is composed of organelles and inclusions embedded in a fluid known as the cytoplasmic matrix.

A generalized animal cell contains the following structures:

1. membranes that act as protective barriers for cell organelles and for the entire cell, the latter is called the plasma membrane,
2. endoplasmic reticulum is a canal system that enables an efficient movement of substances throughout the cell, rough endoplasmic reticulum consists of ribosomes attached to the ER,
3. ribosomes are the sites of protein manufacture,
4. mitochondria are the sites of oxygen dependent energy production,
5. golgi apparatus is a network of flattened tubes where substances are packaged into membrane bound packages,
6. lysosomes are membrane-bound packages that digest unwanted molecules and particles,
7. centrioles are structures involved in cell division,
8. cilia and flagella are structural projections that are found in certain cells.

An animal cell may also contain cell inclusions that can be either beneficial or harmful to the cell.

Plant cells contain the above features plus chloroplasts for converting solar energy into carbohydrates and large fluid filled spaces called vacuoles.

Procaryotic cells differ substantially from eucaryotic cells because they lack a distinct nucleus and cytoplasmic organelles. These cells contain a nuclear region, cell membrane and ribosomes. They usually contain a cell wall which has a protective function. The composition of this wall varies between different bacteria and may be used to divide them into gram positive and gram negative bacteria. Capsules and endospores assist various bacteria in resisting harmful environments. Bacteria can also contain viruses as a cell inclusion.

Chapter 17

Plasma membrane dynamics

Objectives

At the completion of this chapter the student should be able to:
1. describe the functions of the plasma membrane,
2. relate the structure of a plasma membrane to its functions,
3. list the factors that influence the diffusion of a substance through a membrane,
4. explain the difference between the diffusion of electrolytes and non-electrolytes,
5. describe the mechanisms of facilitated diffusion, active transport, pinocytosis and phagocytosis and give examples of how they are involved in body function,
6. differentiate between diffusion, facilitated diffusion, active transport, pinocytosis and phagocytosis.

The plasma membrane plays an important role in the functioning of cells. It acts as a protective barrier for the cell by restricting the entry of unwanted substances. Cells can contain slightly different membrane structures as a result of differences in cell requirements. A second function of plasma membranes is their regulation of the entry of substances required by the cell.

17.1 General structure of plasma membranes

The plasma membrane consists mainly of protein and lipid molecules. A knowledge of how these molecules constitute the plasma membrane will assist in understanding the mechanisms of how substances move through membranes. Currently, most membrane scientists believe that plasma membranes have a structure similar to the fluid-mosaic model (Figure 17.1). This model suggests that the membrane consists of two layers of lipids that contain phosphates attached to their outer surface. Protein molecules are embedded within the lipid layers. Some of these protein molecules span the lipid layers while others are interspersed within the layers. It appears that the proteins are not rigid but are able to move, and this movement is thought to be associated with the movement of substances through membranes. The relative proportion of proteins and phosphates appears to be different at the outer membrane surface when compared with the cytoplasmic membrane surface. This difference suggests that the ability of a substance to pass through a membrane may depend upon the direction of the substance's movement. A substance may be able to move into, but not out of, a cell.

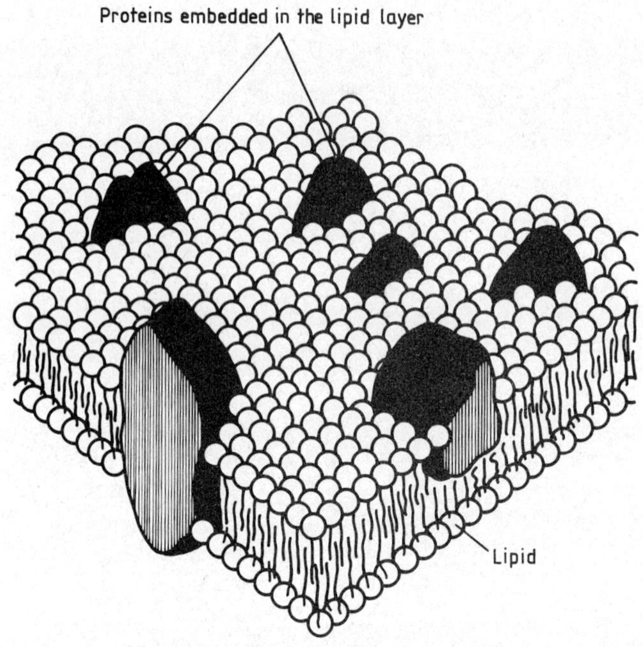

Figure 17.1 The fluid—mosaic model of cell membranes proposes a double layer of lipid interspersed with mobile protein molecules.

Another feature of the plasma membrane is the probable existence of holes of diameter 0.8 nm. These holes which are known as pores, act as channels for small substances such as electrolytes.

17.2 Diffusion through plasma membranes

PERMEABILITY

The term permeability is used to describe the ease with which a substance can diffuse across a membrane. When a substance is unable to pass through a membrane, the membrane is said to be impermeable for that substance.

Plasma membranes are differentially or selectively permeable, that is, they are permeable to a selected number of substances. There are several mechanisms by which substances cross the membrane, and these mechanisms determine its selective properties. The mechanisms include:

1. diffusion through lipid,
2. facilitated diffusion,
3. diffusion through pores.

DIFFUSION THROUGH LIPID

The plasma membrane consists mainly of lipids and this acts as a boundary between intracellular and extracellular fluid by only allowing lipid-soluble substances to pass through it. Lipids are non-polar substances and thus only non-polar substances are capable of dissolving in the lipid. EXAMPLE: Oxygen.

The rate of flow of a lipid-soluble substance across a membrane is determined by the difference in concentration that exists between the extracellular and intracellular fluids. The greater the concentration gradient, the greater is the rate of diffusion across a membrane. Remember that the net diffusion of a substance occurs from a region of high concentration to a region of low concentration (see 14.2).

FACILITATED DIFFUSION

Facilitated diffusion is a process by which a lipid-insoluble substance such as glucose is rendered soluble by combining it with a molecule from within the membrane. This complex shuttles across the membrane and then breaks up, resulting in the release of the substance into the nearby fluid. Since the membrane molecules, in effect, carry the substance from one side of a membrane to the other, they are referred to as carriers.

A carrier only facilitates the diffusion of a substance from one side of the membrane to the other. The normal requirements such as a difference in concentration are necessary for facilitated diffusion to occur (see Figure 17.2).

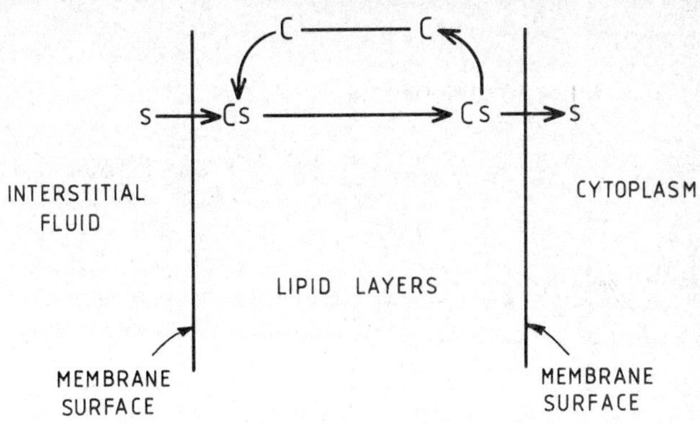

Figure 17.2 Mechanism of facilitated diffusion. A lipid insoluble substance (S) combines with a protein carrier (C) near the surface of the membrane. This carrier—substance complex (CS) is soluble in lipids and diffuses across the membrane where the complex breaks up liberating the substance into the cytoplasm. The carrier then returns to combine with another molecule.

Membranes contain many different shaped carriers. Each type of carrier is probably a protein with a unique chemical structure which enables the carrier to only combine with a complementary shaped substance. Carriers exhibit the property of specificity, that is, they can only carry one specific substance at maximum efficiency. Other similar shaped substances may be carried less efficiently and all other substances are not able to be carried.

Facilitated diffusion differs from simple diffusion by the dependence on carriers for movement. In simple diffusion, any increase in the concentration difference results in a proportionate increase in the diffusion of a substance and there is no upper limit to the flow of a substance (Figure 17.3). With facilitated diffusion, an elevated concentration difference does not necessarily result in a proportionate increase in the flow of a substance. The rate of movement of a substance using facilitated diffusion depends upon the following factors:

1. difference in concentration between both sides of the membrane,
2. amount of carrier available,
3. time taken for the carrier complex to be formed and broken down,
4. type of carrier present, as each carrier is relatively specific for similar structured substances.

There is a limit or saturation point reached in facilitated diffusion which occurs when all of the carriers are occupied. Any further increase in the concentration gradient will result in no further increase in the flow of the substance (Figure 17.3). The amount of substance that is carried when the carriers are saturated is referred to as the threshold or transport maximum (T_m) of that substance.

The transport of a substance can be altered by the presence of similar structured substances, as both

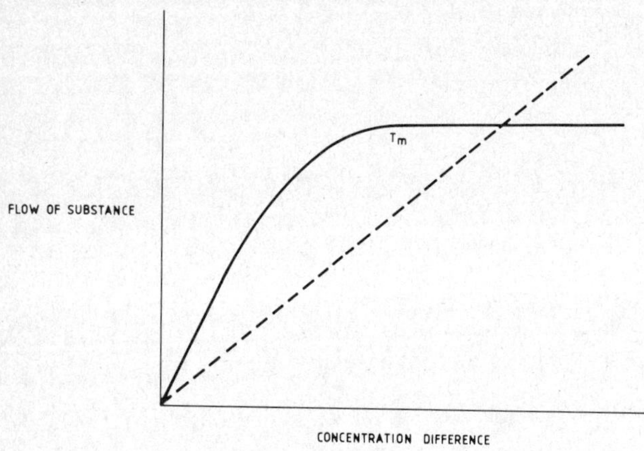

Figure 17.3 In simple diffusion (---) the greater the concentration difference the greater is the flow of substance across the membrane. The flow of substance in facilitated diffusion (——) is controlled by protein carriers. These carriers transport substances more rapidly than simple diffusion at low concentration differences, but reach a saturation point (T_m) when all the carriers are occupied. Any further increase in concentration difference results in no change in the flow of a substance.

substances will compete for the available carriers. When two substances with the same concentration exist in a solution, the probability of a molecule of either substance colliding with the membrane and combining with a carrier is equal.

Facilitated diffusion only transports a net flow of substance from a high concentration to a low concentration, that is, it operates on a downhill gradient. The transport of substances by combining with protein carriers does not require cell energy. Many substances that are unable to readily diffuse through the lipid barrier are carried by facilitated diffusion. EXAMPLES: Sugars, amino acids and certain ions such as sodium.

DIFFUSION THROUGH PORES

Effect of Pore Size

Substances considerably smaller than a membrane pore can move at high rates through a membrane, that is, the membrane exhibits a high permeability to them. EXAMPLES: Water, chloride ions, urea, lactic acid. These substances have a small hydrated diameter. Remember that the hydrated diameter of a substance is the combined diameter of the solute and the water molecules that are loosely attached to it (see Figure 10.1). Any substance that has a hydrated diameter greater than the pore diameter of 0.8 nm is unable to move through the pore, and can therefore only penetrate a membrane if it is able to move through the lipid region. EXAMPLE: Glucose is too large to pass through the pores and so it has to move through the lipid region. The diameter of a pore therefore assists in selecting which substances can pass through a membrane.

Effect of Electrical Charges on the Flow of Ions

Small-sized positive ions such as K^+ and Na^+ pass with greater difficulty through a membrane pore compared with anions of a similar size EXAMPLE: Cl^- passes through membrane pores approximately 500,000 times faster than K^+. This observation suggests that the pores have positive charges surrounding them. Figure 17.4 shows how positive charges lining a pore restrict the flow of positive ions.

The field of influence of the positive charges lining a pore decreases the effective pore diameter for positive charges. Any positive charge that approaches the field of influence is repelled because like charges repel. Anions are able to move into this field of influence since opposite charges attract. This results in a pore having a greater effective diameter for anions than similar-sized cations (Figures 17.4a, 17.4b). The larger effective pore diameter for negative ions results in a faster flow of these ions through a pore.

The anions are prevented from adhering to the lining of a pore by the following factors:
1. The flow of water due to osmosis, pushes negative ions through the membrane before they can hook onto a positive charge. ANALOGY: If a person is being pushed along a rapidly flowing river, the speed of the river's current may be too fast to enable the person to grasp objects on the river bank.
2. The hydrated diameter of anions results in the anion being separated from the pore's lining by

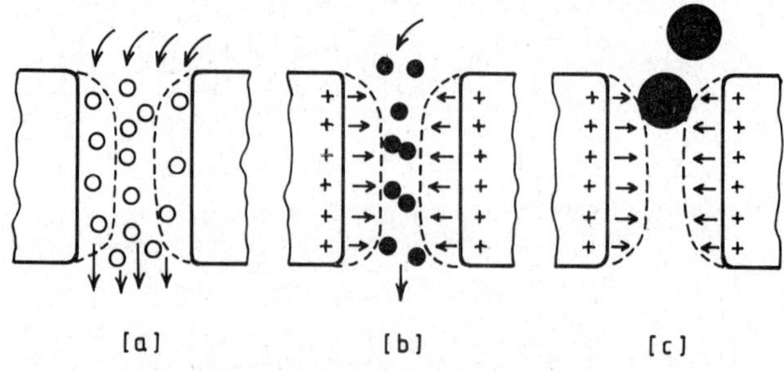

Figure 17.4 Diffusion of ions through a membrane pore: (a) Cl^- is able to flow through the entire pore diameter, (b) K^+ can only flow through the region outside the influence of the positively-charged lining, (c) the hydrated diameter of Na^+ is too large to flow through the central region of the pore.

water molecules. The presence of water molecules between the opposite charges restricts the ability of these charges to form a strong bond.

Plasma membranes have a high permeability to K^+ and a low permeability to Na^+ due to the larger hydrated diameter of Na^+ ions. This difference in permeability results in a high extracellular fluid concentration of Na^+ ions and a high intracellular concentration of K^+ ions.

Effect of Calcium on Pore Permeability

Calcium ions are thought to be present at the pore edges and involved in the positive charge of the pore's lining. An excess of extracellular Ca^{2+} results in an increased number of positive charges lining the pore, which results in a decreased pore diameter to cations. A decreased extracellular Ca^{2+} concentration results in less positive charges lining the pore and thus an increased pore diameter to cations.

APPLICATION: HYPOCALCAEMIA

Hypocalcaemia is a deficiency of calcium which may occur as a result of dietary deficiency, decreased intestinal absorption, hypoparathyroidism or impaired kidney function. The extra demands for Ca^{2+} during pregnancy can also result in a relative deficiency of calcium, even though the calcium intake of a person may be the same as that prior to pregnancy. A deficiency of Ca^{2+} results in plasma membranes increasing their permeability to Na^+ ions (Figure 17.5). The influx of Na^+ into nerve and muscle cells can initiate nerve impulses and muscle contraction, e.g. twitching of facial muscles, "pins and needles" and numbness in the extremities. With more severe Ca^{2+} deficiencies, spasms of skeletal muscles in the hands occur which may be followed by convulsions.

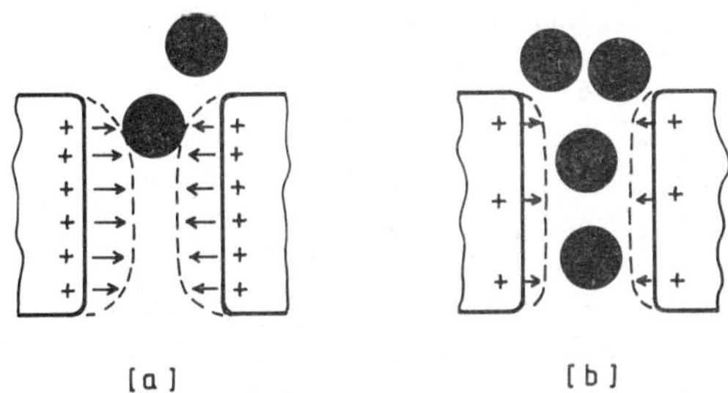

Figure 17.5 The effect of a deficiency of Ca^{2+} on Na^+ flow: (a) normal extracellular concentrations of Ca^{2+} results in the development of a positively-charged field that reduces the diameter of the pore and prevents Na^+ flow, (b) a deficiency of extracellular Ca^{2+} reduces the positive charges lining the pore and results in an increased pore diameter for Na^+.

Effect of Anti-Diuretic Hormone on Pore Permeability

Anti-diuretic hormone increases the physical diameter of membrane pores in the kidney. The larger diameter results in an increased reabsorption of water from the kidney. Inhibition of anti-diuretic hormone by substances such as alcohol and caffeine results in less water being able to pass through the membrane pores of the distal parts of the kidney tubules and ducts. This decrease in water permeability results in excess water being retained in the tubules and ducts. The excess water flows out of the ducts causing diuresis, that is, an increased excretion of urine.

FACTORS INFLUENCING NET DIFFUSION

Concentration Difference

The rate of diffusion through a membrane is dependent upon the difference in concentration between the two sides of the membrane. The greater the concentration difference, the greater is the rate of diffusion, apart from facilitated diffusion where a maximum diffusion rate exists.

Electrical Potential Difference

Since opposite charges attract, an electrical force will exist across a membrane whenever a difference in charge exists between the solutes on either side of the membrane. In human cells the cytoplasm is negative with respect to the extracellular fluid. This difference in charge is due to the cytoplasm containing negatively-charged substances that are unable to penetrate the plasma membrane. These "fixed" charges create an excess of negative charges in the cytoplasm.

The negatively-charged cytoplasm attracts positive ions across the membrane. Potassum ions are able to move across the membrane which results in a high intracellular K^+ concentration. Note that the force of attraction is of sufficient strength so that K^+ ions are pulled across the membrane from a low concentration to a high concentration. Even with this high intracellular concentration of K^+ ions, the net intracellular charge remains negative. The extracellular fluid is thus positive with respect to the cytoplasm. This positive charge repels the diffusion of K^+ from a high concentration in the cytoplasm to the low extracellular concentration.

The low permeability of the membrane to Na^+ ions prevents them from entering a cell. The Na^+ ions have the potential to flow but are prevented from doing so by a membrane barrier. Any increase in the membrane permeability of Na^+ will result in the flow of these positive ions into a cell.

In nerve cells an alteration in membrane permeability results in the flow of Na^+ ions into the nerve. This flow of positive ions or current results in the initiation of a nerve impulse (see 5.3).

Electrochemical Gradient

Substances that lack a charge (non-electrolytes) diffuse down a difference in concentration or concentration gradient. A potential difference across a membrane has no influence on the movement of these substances.

Charged particles diffuse under the influence of two factors:

1. the chemical gradient,
2. the potential difference or electrical gradient.

When the influence of the electrical gradient is greater than the chemical gradient, a net flow of ions can occur from a low to a high concentration. EXAMPLE: K^+ ions flow into cells. To determine the direction of net diffusion, the influence of the electrical gradient is compared with the chemical gradient. The combined influence of these two gradients is termed the electrochemical gradient. Charged substances are unable to diffuse against the direction of an electrochemical gradient, that is, against a downhill electrochemical gradient which drives ions in a specific direction.

To overcome a net force that drives substances in a specific direction requires the use of energy. Some substances have been found to move against an electrochemical gradient. EXAMPLE: Na^+ from the cytoplasm to the extracellular fluid. This suggests that an alternative mechanism to diffusion exists for the movement of substances across membranes.

17.3 Movement of substances by active processes

An active process occurs when the movement of substances across a membrane requires energy from the cell. The three main forms of active process are active transport, pinocytosis and phagocytosis.

ACTIVE TRANSPORT

Mechanism of Active Transport

Active transport occurs when the movement of substances by carriers is independent of the electrochemical gradient. This independence results from the use of cell energy. The carrier mechanism is similar to that of facilitated diffusion. The difference between the two modes of transport is that active transport requires energy to enable the carrier mechanism to function.

Most plasma membranes exhibit active transport. EXAMPLES: Na^+ and K^+ in nerve transmission, absorption of nutrients from the small intestine, kidney reabsorption and electrolyte movements in muscle function.

The movement of a substance will only occur if a specific carrier for that substance is present. Similar structured substances can be used by the same carrier but these substances will be carried at different rates.

A carrier often transports a second substance in the opposite direction to the first substance. This is known as coupled transport (Figure 17.6). Coupled transport utilizes a carrier in both directions instead of one. The most common coupled system is that of Na^+ and K^+. Usually Na^+ that has "leaked" into a cell is carried out, whereas K^+ is transported from the extracellular fluid into the cell by the same carrier. The following sequence summarizes the mechanism of active transport (Figure 17.6).

A substance in the surrounding fluid (S) combines with a molecule that has enzyme-like properties on, or inside, the membrane's surface. This enzyme-like molecule (C) has a specific preference for that substance. Other carriers have a preference for different substances. Energy may be required to form the complex (CS). The carrier-substance complex (CS) is soluble in the lipid and moves across the membrane to the opposite surface.

Enzymes may effect a splitting of the complex which requires the use of energy. Carrier (C) and substance (S) are separated and the impermeable substance is now trapped on the opposite side of the membrane.

The carrier either combines with another substance in a similar way as with the first substance and takes it across the membrane (coupled transport) or the carrier returns to the initial surface unoccupied.

The movement of substances by active transport is limited by the ability of that substance to combine with, and be transported by, a carrier.

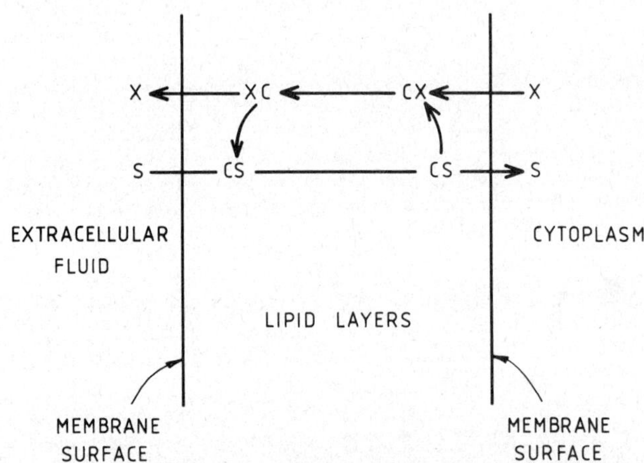

Figure 17.6 Coupled active transport involves the formation of a carrier-substance complex (CS) which moves to the opposite side of the membrane. On arrival the CS splits and the carrier is free to combine with a similar substance (X) and transport it in the opposite direction.

APPLICATION: ACTIVE TRANSPORT OF GLUCOSE

The rate of formation and breakdown of a glucose-carrier complex restricts the quantity of glucose that can enter a cell to below the amount that is necessary for cell function. The hormone insulin increases the rate of formation and breakdown of the glucose-carrier complex. This results in an increase in glucose transport by as much as ten times the rate of transport as when no insulin is secreted.

The number of carriers available for the transport of glucose also has an influence on glucose transport. In kidney cells, the capacity of carriers to transport quantities of glucose from the tubular fluid to the blood far exceeds the normal quantities of glucose that are filtered by the kidneys. Elevated blood sugar levels above 10 mmol/L result in a greater quantity of glucose being filtered than can be reabsorbed by the active transport carriers.

The appearance of glucose in the urine indicates the saturation of glucose carriers and is referred to as the renal threshold for glucose (Figure 17.7). The maximum amount of glucose that can be transported by carriers per minute is called the transfer maximum or transport maximum (T_m). The renal threshold varies with individuals, and in some patients the transport mechanism is defective so that glucose is found in the urine at normal blood glucose levels.

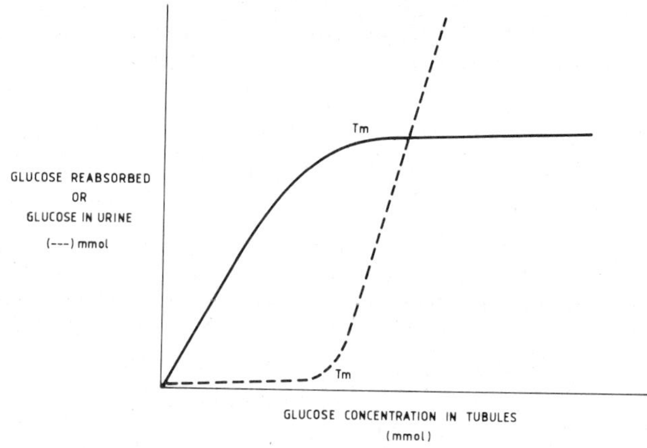

Figure 17.7 The relationship between glucose in urine and the active transport mechanism of glucose reabsorption.

Influence of Active Transport on Moving Other Substances

If a charged ion is carried across a membrane, an imbalance of potential difference develops between the fluids on either side of the membrane. This alteration in potential difference changes the electrochemical gradient for other ions via diffusion down the electrochemical gradient. EXAMPLE: If Na+ is actively transported across a membrane as in Figure 17.8 the receiving fluid will become more positive than the donating fluid. To maintain the original balance in potential difference between the two fluids, an anion such as Cl^- is driven across the membrane. Note the movement of Cl^- is passive, that is, independent of cell energy and can move through membrane pores instead of moving through the lipid combined with carriers (Figure 17.8b).

The movement of these ions has also resulted in an osmotic imbalance. This imbalance results in flow of water in the direction of the Na+ and Cl^- ions (Figure 17.8c).

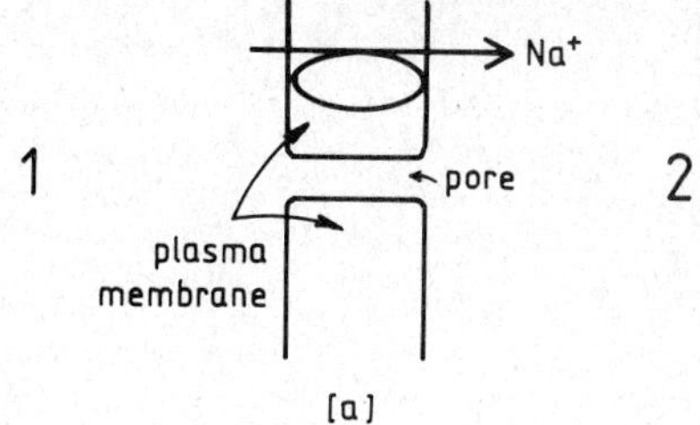

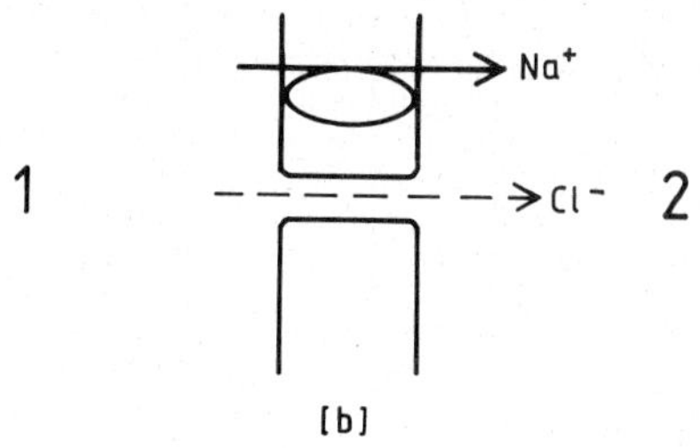

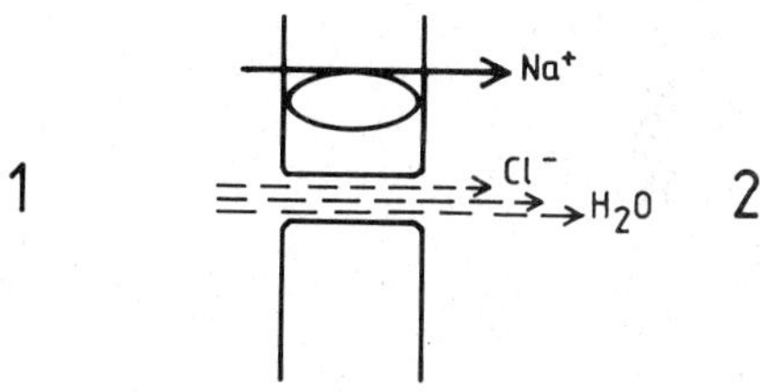

Figure 17.8 The effect of active transport on the movement of other substances: (a) Na + is actively transported from fluid 1 to 2, resulting in an increased positive charge in fluid 2, (b) Cl^- diffuses from 1 to 2 to neutralize the positive charges, (c) water flows from 1 to 2 as a result of the osmotic imbalance.

APPLICATION: ANION AND WATER REABSORPTION IN KIDNEYS

In the kidneys, Na+ is actively transported from the lumen of the tubules into kidney cells from which it returns to the blood. This results in the blood becoming momentarily more positive than the tubular fluid. Anions such as Cl^- are passively reabsorbed down the new electrochemical gradient.

The movement of Na^+ and anions into the blood results in the development of a higher osmotic pressure in the blood. This osmotic imbalance drives water from the tubules into the blood in order to establish osmotic equilibrium. About eighty per cent of water is reabsorbed from the kidneys by this process which is dependent on the active transport of Na^+. The inhibition of the active transport of Na^+ results in the indirect inhibition of Cl^- and water reabsorption. This is the principal behind the function of most diuretics, that is, drugs that cause increased fluid loss in urine.

PINOCYTOSIS

Pinocytosis is a general term used to describe the active movement of substances via membrane-bound packages from the extracellular fluid to the cytoplasm or vice versa. These packages are called vesicles. The movement of substances into cells by this mechanism is used by most body cells. EXAMPLES: Absorption of macromolecules such as proteins and lipids across the small intestine. Secretion of hormones and enzymes by secretory cells.

Pinoctyosis can be separated into endocytosis and exocytosis. By the process of endocytosis substances are taken into the cell, and by exocytosis they are removed from the cell.

Endocytosis is an active process where extracellular fluid containing macromolecules such as proteins is engulfed by the plasma membrane and taken into the cell. This process requires energy. The stages of endocytosis are shown in Figure 17.9. The process of pinocytosis is initiated by the plasma membrane invaginating in response to positive charges on the macromolecules. These molecules become attached to the membrane's surface by adsorption, that is, the molecules are held to the surface (see 10.4). The presence of these molecules causes the membrane to invaginate and eventually encircle the molecules (Figure 17.9d). The two ends of the invagination fuse forming a small sac or vesicle (Figure 17.9c). The pinocytotic vesicle then moves towards the centre of the cell (Figure 17.9f). Note that the vesicle still contains extracellular fluid with its macromolecules. Lysosomes attach to the pinocytotic vesicle (Figure 17.9g) and release their digestive agents into the vesicle causing a breakdown of the macromolecules (Figure 17.9h) after which the vesicular membrane breaks and releases its contents into the cytoplasm.

Substances can also be released or secreted by a cell using the mechanism of pinocytosis. This form of pinocytosis is called exocytosis. Vesicles that are formed by the golgi apparatus may contain macromolecules such as hormones and enzymes that need to be secreted by the cell. The size of such molecules prevents them from passing through the plasma membrane. The secretory vesicles move from the golgi apparatus to the plasma membrane. They then attach to the plasma membrane. The membrane breaks releasing the contents into the extracellular fluid.

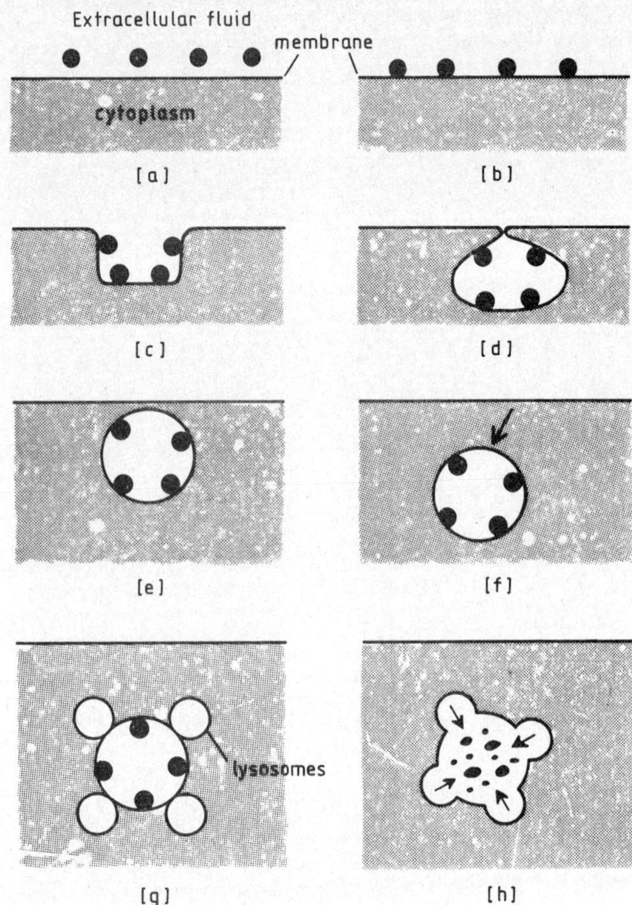

Figure 17.9 Mechanism of endocytosis:
(a) macromolecules such as proteins approach the plasma membrane,
(b) macromolecules are adsorbed onto the membrane
(c) an invagination of the plasma membrane results,
(d) the invagination forms an enclosed sac or vesicle,
(e) the sac separates from the plasma membrane,
(f) the sac moves towards the centre of the cell,
(g) lysosomes attach themselves to the sac,
(h) lysosomes release their contents into the sac causing a breakdown of the macromolecules.

PHAGOCYTOSIS

Phagocytosis is the third mechanism by which substances are moved across the membrane by active processes. It differs from pinocytosis in that only specialized cells can perform this function. Phagocytosis involves the ingestion of large particles such as bacteria, viruses, other cells and cell debris by white blood cells. The process is similar to pinocytosis in that the particle must be able to be adsorbed to the cell membrane. Once adsorbed, a large invatination forms around the particle which is then ingested (Figure 17.10). The ingested material then is digested by lysosomes.

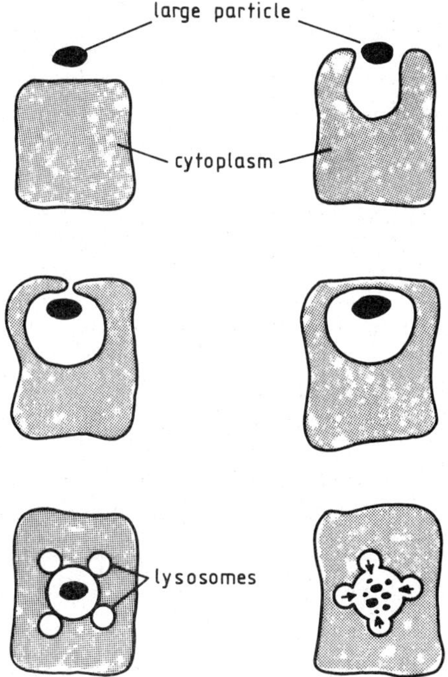

Figure 17.10 Mechanism of phagocytosis: (a) extracellular particle activates a specialized cell, (b) cell sends out arms to engulf the particle. (c) cell encloses the particle by forming a large vesicle or vacuole, (d) the vacuole moves into the cytoplasm, (e) lysosomes attach to the vacuole, (f) lysosomes digest the valuole's contents.

Summary

The plasma membrane protects a cell from unwanted substances and regulates the entry of wanted substances, that is, it can maintain cellular homeostasis.

The fluid-mosaic model proposes that a plasma membrane is composed of two layers of non-polar lipid which has mobile proteins interspersed throughout it.

Membrane pores probably act as channels for small substances such as electrolytes.

The selectively permeable nature of these membranes results from:

1. the non-polar lipid region only allowing non-substances to diffuse through it,
2. the specific nature of membrane carriers that transport polar substances across the membrane in the processes of facilitated diffusion and active transport,
3. the small diameter of pores,
4. the positive charges lining pores result in cations having a relatively lower permeability than anions,
5. the concentration of calcium lining the pores and thus conferring some of the positive charge to the lining,
6. the presence of substances that can alter membrane structure such as anti-diuretic hormone.

Net diffusion of non-electrolytes is influenced by a concentration gradient whereas electolytes are influenced by an electrochemical gradient. Facilitated diffusion differs from simple diffusion in that it is also dependent upon the relative amount of available carrier. When all the carrier is being utilized, diffusion of a substance has reached saturation point which is referred to as the transport maximum (T_m) for that substance. Active transport differs from facilitated diffusion by being independent of electrochemical gradients. Active transport also relies on the cell's energy for the movement of substances.

Both pinocytosis and phagocytosis differ from diffusion, facilitated diffusion and active transport. The latter are involved in the movement of substances through membranes whereas the former use energy to move substances via membrane-bound packages. On reaching their destination the vesicles are ruptured thereby releasing their contents across ruptured membranes. Pinocytosis occurs in most cells and involves the use of vesicles whereas phagocytosis occurs in specialized cells and involves the use of larger membrane-bound packages.

Chapter 18

Nucleus function

Objectives

At the completion of this chapter the student should be able to:
1. describe the functions of a nucleus,
2. describe the mechanism of mitosis and meiosis in terms of chromosomes,
3. relate chromosomal abnormalities to meiotic malfunctions,
4. define a gene,
5. relate genes to patterns of inheritance,
6. list the factors that can alter gene expression,
7. explain how DNA structure is related to gene function,
8. describe the genetic code and relate errors in the genetic code to molecular-based genetic disorders,
9. explain how molecular-based mitotic malfunctions can result in cancer and how cytotoxic drugs act in the treatment of cancer.

The nucleus performs a role analogous to the head office of an organization. It controls cell multiplication and the day to day functioning of a cell. The nucleus copies its programmed, coded messages during the formation of a new cell. This information ensures that the newly-formed cell will perform its correct functions in life. This hereditary information is packaged in structures known as chromosomes.

18.1 Microscopic structure of chromosomes

CHROMOSOME APPEARANCE

When using a microscope the nucleus is seen to contain thread-like structures which are called chromosomes. Chromosomes can only be observed with a microscope during periods of cell division. The chromosomes have been found to contain the hereditary information that is necessary for the growth and maintenance of a cell. In preparation for cell division this hereditary information is condensed from long, very thin molecules that are visible as a diffuse fine thread-like network, into the thick chromosomes. As these chromosomes are appearing hereditary information is being duplicated. The result of this duplication is the formation of chromosomes that are composed of two identical rods called Chromatids (Figure 18.1). The chromatids are joined by a centromere. When a chromosome consists of two chromatids it contains double the hereditary information that exists during periods when cell division is not occurring.

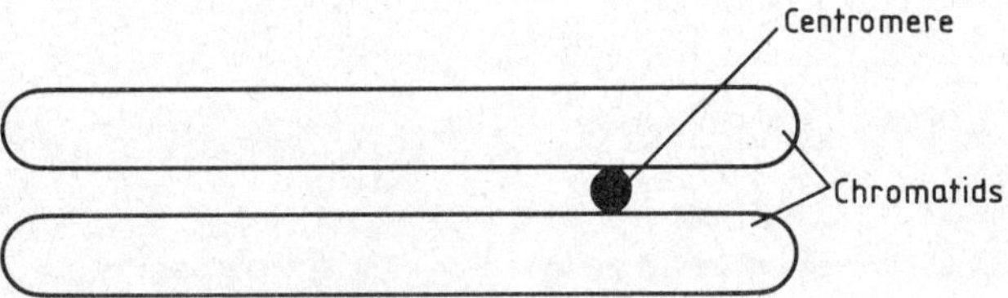

Figure 18.1 Representative diagram of a chromosome as observed with a microscope. The two identical chromatids are joined by a centromere.

CHROMOSOME PAIRS

All human cells contain twenty-three pairs of chromosomes, except mature sex cells which contain only twenty-three chromosomes. Chromosomes contain the hereditary information that is necessary for the growth and maintenance of a cell. Each pair consists of one maternal chromosome and one paternal chromosome, both of which contain similar but not identical hereditary information. The two chromosomes that belong to a pair are termed homologous chromosomes. Each pair of chromosomes has a different microscopic size and shape to the other pairs. In twenty-two pairs the homologous chromosomes have the same microscopic size and shape. These chromosomes are termed autosomes. The other pair can consist of two identical chromosomes or two different shaped structures. The twenty-third pair of chromosomes are called the sex chromosomes.

APPLICATION: KARYOTYPES

A karyotype is a formal arrangement of chromosomes used to identify chromosomal abnormalities and identify the sex of an individual. This arrangement designates a number to each homologous pair which is then used to identify any chromosomal pair. All human cells, apart from mature sex cells, contain the same karyotype. A description of chromosomal abnormalities that are detected by a karyotype is presented in 18.4.

SEX CHROMOSOMES

The sex chromosomes of a male contain a normal shaped chromosome in the shape of an X and a smaller chromosome with a shape similar to a Y. The Y chromosome contains hereditary information that codes for the formation of a testes. Normal males contain an X and a Y chromosome whereas normal females contain two X chromosomes.

APPLICATION: BARR BODIES

In instances where there is a need to determine the sex of an individual, cells can be taken from scraping the lining of the mouth and stained for the presence of Barr bodies. A Barr body is a small dark-staining area that occurs when more than one X chromosome is present in a cell, that is, in normal female cells.

18.2 Mitosis

Mitosis is the process of cell division where a new cell is formed that contains identical hereditary information to the parent cell. This form of cell division occurs in all body cells apart from mature sex cells. The function of mitosis is to duplicate cells for the growth and continued functioning of the body.

Mitosis consists of a series of continuous steps which are represented in Figure 18.2. The process of mitosis has been arbitarily divided into the following steps: prophase, metaphase, anaphase and telophase.

PROPHASE

Prophase is the earliest visible stage of mitosis. At the beginning of prophase, the chromosomes appear (Figure 18.2a) and by the end of this step the chromosomes are observed as two chromatids (Figure 18.2b). In this step, the chromosomes are firstly randomly distributed throughout the nucleus and then the cell as the nuclear membrane disappears. Clearly, a splitting of the cell at this stage would result in non-identical cells (Figure 18.2c). A need exists for the chromosomes to be oriented in a single line.

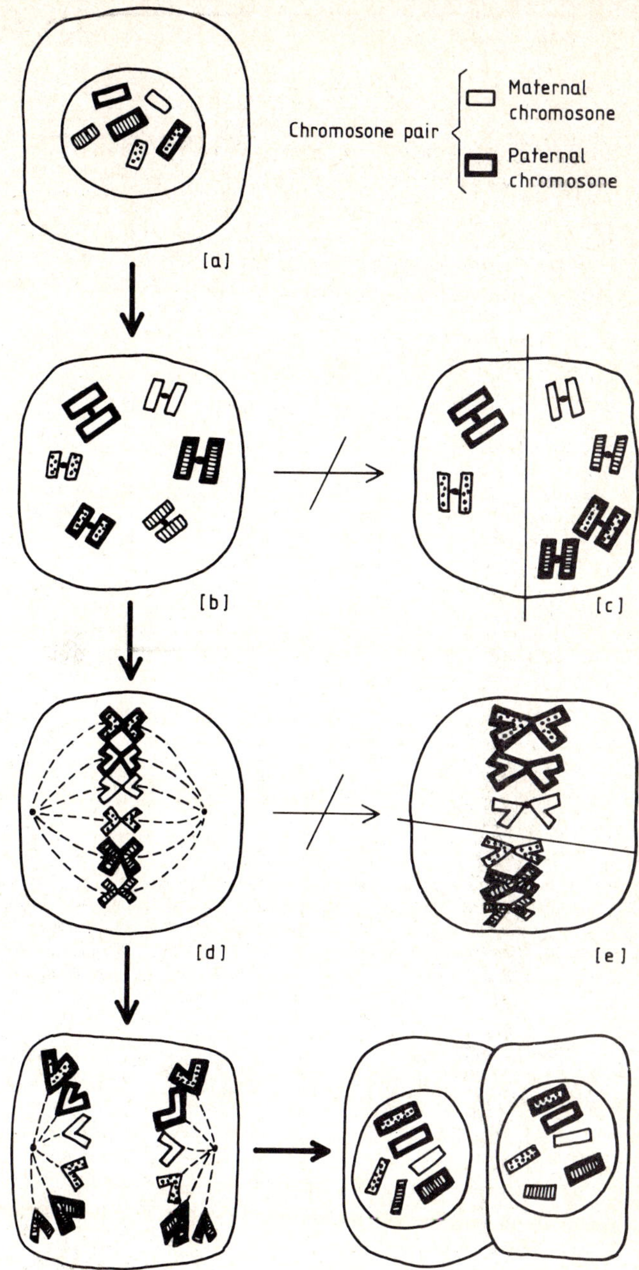

Figure 18.2 Mitosis: (a) prophase begins with the appearance of chromosomes, (b) prophase ends with the formation of two chromatids, (c) dividing the cell at this stage results in non-identical cells, (d) metaphase involves the orientation of chromosomes at the centre of the cell, (e) lack of a spindle results in non-identical cells being formed, (f) anaphase involves the spindle separating each pair of chromatids, (g) telephase and the formation of two identical cells.

METAPHASE

Metaphase involves the movement of chromosomes to the centre of a cell where they are oriented in a single line (Figure 18.2d). Each double stranded chromosome attaches by its centromere to a chain called a spindle fibril. The spindle fibril is connected at each end to a centriole. The centrioles pull the chromatids apart so that identical chromatids are pulled in opposite directions. The centriole–spindle system enables a cell to divide the duplicate hereditary information into two distinct regions. If this spindle system fails to function or is not present the chromosomes could be split to form two non-identical regions (Figure 18.2e).

ANAPHASE

The separated chromatids (now called chromosomes) move along the fibrils to the appropriate centriole (Figure 18.2g). The two complete sets of chromatids are now located at opposite sides of a cell.

TELOPHASE

This is the last step in mitosis. It involves the formation of nuclear membranes around each of the two sets of chromosomes. The chromosomes begin to become less dense and subsequently disappear from view. The cytoplasm divides, resulting in the formation of two identical cells (Figure 18.2g). These cells have the ability to undergo further mitotic divisions.

INTERPHASE

Interphase is the period between cell divisions. It is during this step that hereditary information is transcribed into instructions for cell function. The majority of cells spend most of their existence in interphase, that is, in a state where the normal physiological and chemical activities are present. Towards the end of interphase genetic information is duplicated in preparation for cell division.

18.3 Meiosis

In the last stages of sex-cell formation, it is necessary to halve the number of chromosomes in a cell. If the mature sex cells contain forty-six chromosomes, the combination of male and female sex cells (sperm and ovum) would result in ninety-two chromosomes.

Meiosis is the form of cell division that results in new cells with half the original number of chromosomes. In meiotic divisions, the new generation of cells contain one of each pair of chromosomes. This confers individual differences on offspring but maintains the essential characteristics of human-cell function.

Meiosis consists of two successive division sequences. The first division involves the organization of maternal and paternal chromosomal pairs along the centre of the cell (Figure 18.3a). These pairs are arranged in a random sequence. Each chromosome wraps around its partner and attaches to a spindle fibril. The centriole–spindle complex pulls the chromosome pairs apart to opposite sides of a cell. The cell is then divided into two separate cells with half the number of chromosomes (Figure 18.3b).

The second division sequence is identical to the steps shown for mitosis apart from the fact that only half the number of chromosomes are present. Briefly, the chromosomes are orientated in a line in the centre of a cell. The centromeres of each chromosome attach to a spindle, after which the chromatids are separated to opposite sides of the cell. Splitting each cell results in the formation of two cells. Four cells result from one parent cell instead of the two cells that occur in mitosis.

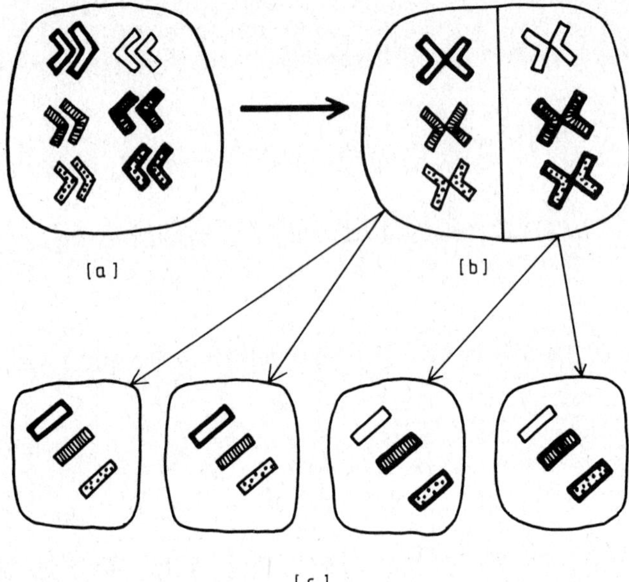

Figure 18.3 In meiosis the first division involves: (a) the alignment of chromosome pairs and (b) the subsequent separation of these pairs into two cells with half the number of chromosomes. (c) A further division occurs in which the chromatids of each chromosome separate resulting in four cells from the original single cell.

18.4 Chromosomal abnormalities

MEIOTIC MALFUNCTION

A malfunction in the meiotic process can lead to the formation of an abnormal cell which can be fertilized. Since an offspring originates from a fertilized cell, all the offspring's cells will be abnormal. The effect of a meiotic malfunction on the parent is minimal since meiosis only occurs in body cells that are involved in the formation of sex cells.

Chromosomal abnormalities can be classified according to whether a meiotic malfunction has resulted in the formation of cells with an abnormal number of chromosomes or chromosomes that have an altered structure.

Chromosomal Number

Normally chromatids separate or disjoin by the influence of the centriole-spindle complex. When a pair of chromatids fails to disjoin the event is referred to as a non-disjunction. EXAMPLE: The non-disjunction of two chromatids in meiosis results in the formation of cells that contain either twenty-two or twenty-four chromosomes. When the twenty-four chromosome sex cell is fertilized with a twenty-three chromosome cell, the resulting offspring cells will contain forty-seven chromosomes (Figure 18.4). These cells contain one chromosome pair that has been replaced by a triplet of chromosomes. When forty-seven chromosomes are present, the syndrome is referred to as a trisomy.

The incidence of non-disjunction increases with the age of the mother, especially after thirty-five years of age.

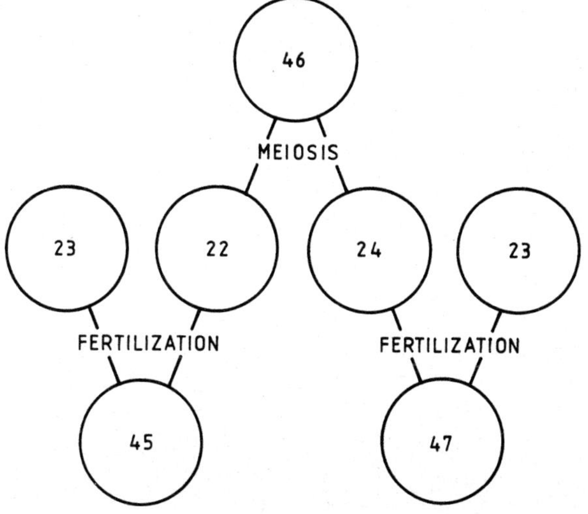

Figure 18.4 Relationship between meiotic non-disjunction and offspring chromosome number.

Different chromosome non-disjunctions are identified by using a karyotype. The karyotype number of chromosome non-disjunctions is used to distinguish different trisomies. Autosomal trisomies usually result in abortion of the foetus in the first trimester, although trisomies of the karyotype numbers eight, thirteen, eighteen, twenty-one and twenty-two are able to live. All of these autosomal trisomies result in offspring that have multiple defects with various degrees of mental retardation. The most common autosomal non-disjunction is trisomy twenty-one or Down's syndrome.

When the non-disjunction involves a sex chromosome, the result is an alteration to the sexual development of the child. With forty-seven chromosome trisomies, three syndromes can occur as a result of sex chromosome non-disjunction. A XXX female results from a non-disjunction of one of the X chromosomes, whereas the XXY (Klinefelter syndrome) results from a non-disjunction of the mother's sex chromosome. Klinefelter syndrome is one of the common forms of hypogonadism and infertility in males as it occurs approximately once in every four hundred live male births. Fathers that have a non-disjunction of their Y chromosome propagate offspring with a XYY syndrome.

Non-disjunction results in an equal number of sex cells losing chromosomes as those gaining chromosomes (Figure 18.4). A loss of one of the autosomal chromosomes results in an embryo that is generally incompatible with life. An embyro that has one less sex chromosome can be compatible with life when one X chromosome exists. Foetuses with forty-five chromosomes including only one X chromosome are usually aborted. However some are born and this is known as Turner's syndrome. These patients exhibit a short stature and poorly developed female sex characteristics.

Chromosomal Structure

Even though the total number of chromosomes is normal, the structure of one or more chromosomes could be abnormal. Structural abnormalities are mainly the result of chromosomal breakage where one part of the chromosome is lost. The loss of a section of hereditary information can result in clinical features similar to trisomies.

The term deletion syndromes is used to describe the group of syndromes that result from loss of a segment of a chromosome. Deletion syndromes can occur with the following chromosome numbers: four, five, thirteen, eighteen, twenty-one and twenty-two. EXAMPLE: The loss of a part of the karyotype number five chromosome results in the abnormal development of the larynx. Infants afflicted with this cry like a cat and thus this syndrome is often referred to as the Cri-du-chat syndrome (cry of the cat syndrome). These children also suffer from mental retardation and fail to thrive.

Detection of Chromosome Abnormalities

Patients with a suspected chromosomal

abnormality may have some of their cells cultured and then karyotyped. In families where a history of chromosomal disorders is prevalent, a karyotype of a developing foetus can be performed with the view of terminating the pregnancy if necessary.

Foetal cells can be obtained from the amniotic fluid as early as the fifteenth week of pregnancy. The procedure of collecting amniotic fluid is known as amniocentesis. This procedure involves the insertion of a sterile needle through a locally-anaesthetized region of the mother's abdominal wall and the subsequent withdrawal of amniotic fluid. The cells are then cultured until an adequate number exist for karyotyping.

18.5 Genetics

GENES

Chromosomes contain units of hereditary information called genes. A gene is the smallest unit which represents a single physical hereditary characteristic. EXAMPLE: Eye colour. Genes are arranged in a specific sequence along a chromosome. An understanding of meiosis assists in determining the chances of a particular gene being passed to offspring. The study of genes and their relationship to hereditary characteristics is termed genetics.

ALLELES

When a specific gene exists in several forms the different forms are referred to as alleles of that gene. EXAMPLE: The absence or presence of pigment in the iris of eyes, that is, blue eyes or brown eyes. Usually only two alleles exist for a gene, with one of the alleles being dominant over the other. Each gene is represented in both chromosomes of a homologous pair. The two chromosomes can contain the same or different alleles. EXAMPLE: Two alleles for brown eyes could be present or one allele for brown eyes and one for blue eyes. When a pair of chromosomes contain the same allele a person is said to be homozygous for that gene. Where a gene is represented by two different alleles a person is said to be heterozygous.

Usually a heterozygous representation of a gene involves the occurrence of a dominant allele and a recessive allele. Whenever a dominant allele is present, it overrules the recessive allele. EXAMPLE: A person who is heterozygous for eye colour, contains one brown allele and one blue allele. The brown allele is dominant and thus suppresses the blue allele which results in brown eyes.

A heterozygous distribution of a gene results in the recessive allele being masked by the dominant allele, that is, a person is a carrier of the recessive allele. This recessive allele can be passed onto the next generation where, if the fertilized egg contains two recessive alleles, the recessive characteristic will be expressed. Most genetic abnormalities result from a homozygous recessive allele combination. EXAMPLES: Colour blindness, haemophilia, muscular dystrophy and phenylketonuria. The dominant allele in these examples is normal function. Some genetic abnormalities result from the presence of a dominant allele. These abnormalities exist in people who contain either a homozygous allele combination or a heterozygous combination. EXAMPLES: The brain disorder Huntington's chorea, hypertension and diabetes insipidus.

Two terms are used to distinguish between the genetic makeup of an individual and their physical characteristics. Genotype describes the genetic composition of an individual. For any gene, a pair of chromosomes can contain the following combinations, that is, genotype:

dominant-dominant }
recessive-recessive } homozygous

dominant-recessive }
recessive-dominant } heterozygous

The expression of the genotype into a physical characteristics is termed the phenotype. The following different genotypes have the same phenotype:

dominant–dominant
dominant–recessive
recessive–dominant

In these three combinations the dominant gene expresses its physical characteristics. EXAMPLE: A brown-eyed person can contain any of the following allele combinations:

brown–brown
brown–*blue*
blue–brown

The genotype is used to predict the various phenotypic possibilities of offspring.

HEREDITY AND GENES

Meiosis results in the formation of mature sex cells that contain either allele but not both. EXAMPLE: Brown or blue. Each parent produces two different sex cells that can be used in fertilization. When predicting the possible inheritance of a genetic characteristic, each possible combination of both sets of sex cells must be taken into account. The relationship between hereditary and genes is illustrated by eye colour. What will the eye colour of offspring be when one parent has brown eyes while the other parent has blue eyes? The genotype of the brown-eyed parent can be either two brown alleles or one brown allele and one blue allele. Note that a general convention applies when abbreviating alleles. A dominant allele is denoted by a capital letter whereas a recessive allele is expressed by a lower

ALLELES BROWN B	ALLELES BROWN B	ALLELES BLUE b	ALLELES BLUE b	OFFSPRING GENOTYPE	OFFSPRING PHENOTYPE
B		b		Bb	BROWN
B			b	Bb	BROWN
	B	b		Bb	BROWN
	B		b	Bb	BROWN

OR

BROWN B	BLUE b	BLUE b	BLUE b		
B		b		Bb	BROWN
B			b	Bb	BROWN
	b	b		bb	BLUE
	b		b	bb	BLUE

Figure 18.5 Relationship between eye-colour alleles and inheritance.

case letter (brown-*B*, blue-*b*). The chances of offspring having brown or blue eyes is illustrated in Figure 18.5. If the brown-eyed person is homozygous, all offspring will have brown eyes, whereas a heterozygous brown-eyed parent would have an equal chance of producing a brown or blue-eyed offspring.

A simpler method of determining the possible allele combinations in an offspring involves the use of special charts called punnett squares (Figure 18.6). These charts are frequently used to determine the probabilities of offspring characteristics. A chart is completed by filling in the squares which immediately indicates the different genotypes that can be formed.

A knowledge of whether the genotype of the brown-eyed parent is *BB* or *Bb* would enable a more precise prediction of the eye colour of the offspring.

PATTERNS OF INHERITANCE

Determining the phenotypes of members of a family leads to the formation of a genetic family tree. This family tree can determine the genotype of most of its members. This enables a more precise prediction of possible child phenotypes. EXAMPLE: A brown-eyed parent would be heterozygous if one of their parents was blue-eyed.

Many genetic disorders follow the simple dominant–recessive predictions, whereas some disorders such as diabetes mellitus and high blood pressure have no simple pattern of inheritance. A history of a genetic disorder in a family requires the

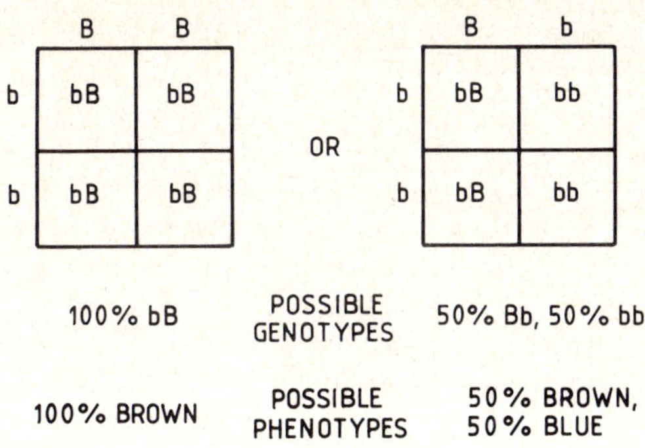

Figure 18.6 Inheritance of eye colour using punnett squares.

assistance of a genetic counsellor who will inform the parents of the risks and consequences of having children. The more complex patterns of genetic inheritance probably involve genetic predispositions, that is, an individual has the genetic potential for a specific disease but this potential is only realized when certain environmental conditions exist. EXAMPLE: Obesity may act as a stress on individuals genetically predisposed to diabetes mellitus which initiates a clinical expression of the diabetes.

SEX-LINKED CHARACTERISTICS

When an allele is located on a sex chromosome, the allele will manifest itself with respect to the sex that the chromosome represents. The X chromosomes contain many genes whereas the Y chromosome has lost most of the genes. This feature of sex chromosome genes produces a pattern of inheritance that is different to that described for autosomal chromosomes. An X-linked recessive disorder will only be expressed when the normal dominant gene is absent. A female is therefore only affected if she is homozygous for that gene, since a heterozygous female contains one X chromosome with a dominant normal allele. In males, whenever a recessive allele is present on the X chromosome, the abnormal characteristic is expressed since the Y chromosome lacks the gene. In effect, females act as carriers of sex-linked recessive characteristics.

APPLICATION: HAEMOPHILIA

Haemophilia is a sex-linked recessive disorder that results in either the failure of the blood to clot or the very slow clotting of blood following bleeding. The recessive allele is located only on an X chromosome which can be designated as X^h. The genotype for haemophilia in males is X^hY and for females X^hX^h. A female haemophiliac can only be produced from a haemophiliac male as a normal male confers a normal X^h allele to any female offspring (Figure 18.7).

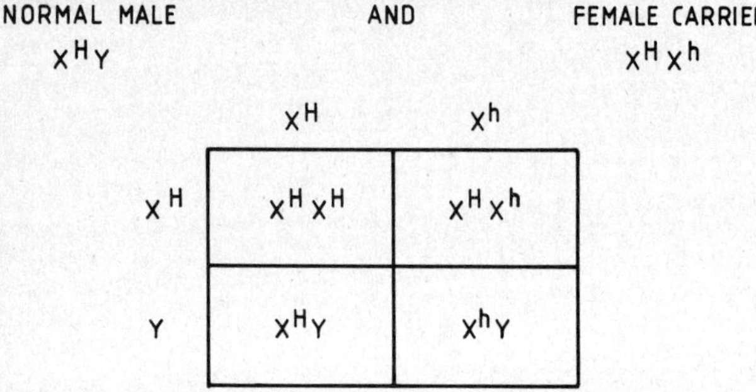

POSSIBLE GENOTYPES

25% $X^H X^H$, 25% $X^H X^h$, 25% $X^H Y$, 25% $X^h Y$

POSSIBLE PHENOTYPES

100% FEMALES NORMAL, 50% MALES NORMAL, 50% HAEMOPHILIACS

Figure 18.7 Inheritance of the sex-linked characteristic haemophilia. Note that a normal male is incapable of producing haemophiliac female offspring.

MULTIPLE ALLELES

Some genetic characteristics contain more than two alleles. In blood groups, *A* and *B* are equally dominant whereas *O* is the common recessive allele. The combination of *A* and *B* results in a third equally dominant allele called *AB*. These capital letters refer to substances on red blood cells called antigens. An antigen is a substance that stimulates the production of specific immune substances called antibodies. A combination of an antigen and its specific antibody results in agglutination.

The blood of individuals is distinguished according to the type of antigen present (Table 18.1). The plasma of each blood group contains antibodies that are different to the antigen, since complementary antibodies would cause agglutination of the blood. These antibodies are represented by lower case letters.

The common recessive allele for the equally dominant *A, B* and *AB* is identified as *i*. The homozygous recessive blood group *ii* is called blood group *O*.

Blood transfusions require the use of blood that is compatible with the recipient's blood. If the blood is not compatable, the antigens of the donor's red blood cells will agglutinate with the antibodies of the recipient's plasma. The agglutination of blood may block blood vessels and could prove fatal if the blocked vessels supply vital areas. Whenever possible, the donor blood is the same type and has similar properties as the recipient. This compatibility is determined by cross-matching.

In emergency situations, blood group O can be used since it lacks the red blood cell antigens that combine with a recipient's plasma. Blood group O is thus referred to as the universal donor. Conversely, group AB lacks anti-A and anti-B antibodies in the plasma and thus this blood group can receive any type of red blood cells, that is, AB is a universal recipient.

Table 18.1 ABO blood group

Genotype antigen alleles	Phenotype blood group	Plasma antibodies
AA	A	bb
Ai	A	bb
BB	B	aa
Bi	B	aa
AB	AB	–
ii	O	ab

FACTORS AFFECTING GENE EXPRESSION

Effect of Environment

The genotype determines the potential for growth and development of an individual. This potential is realized only when a suitable environment is present. If a person is exposed to a harmful environment the phenotype can differ from the genotype. The genotype for tallness and a large physique can be over-ridden by the effects of malnutrition.

APPLICATION: ENVIRONMENTAL DEAFNESS

If a person has a gene for normal hearing, this may not be expressed if factors such as infections, exposure to excessive noise or perforation of the ear drum occur. The most common cause of deafness in new born babies is infection of the mother with Rubella (German measles) during pregnancy. In this case a person has the genotype for normal hearing but has the phenotype of the recessive allele, deafness.

Multiple Genes

Multiple genes or polygenes are genes that combine to produce a specific phenotypic trait. These genes determine traits that are of a quantitative nature, such as height, weight and degree of pigmentation. Many common genetic disorders such as diabetes mellitus and hypertension probably also depend on several genes. Hereditary traits that are determined by polygenes differ from those where one gene determines one characteristic, since in the former a person may have a proportion of a trait expressed, whereas in the latter an individual either has it or they do not. The distribution of a polygenic characteristic in a population is usually exhibited as a curve similar to Figure 18.8.

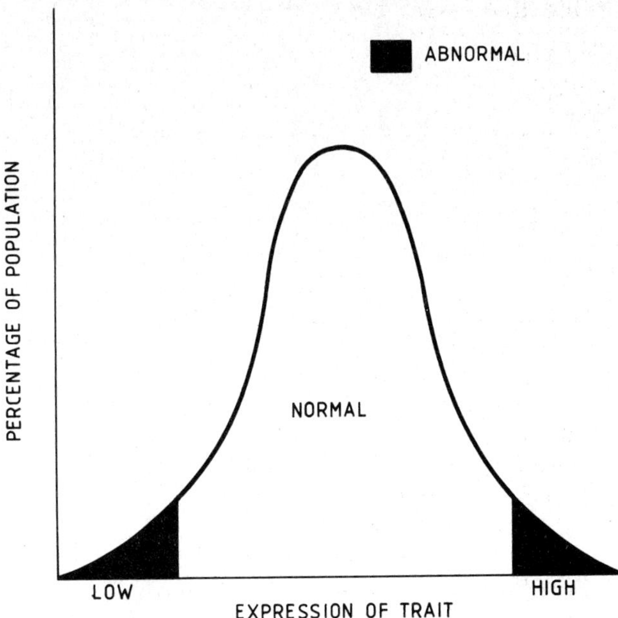

Figure 18.8 The distribution of a polygene characteristic in a population. The normal expression of the gene is usually considered to be the range of the trait that is expressed by ninety-five per cent of the population. A person who is in the remaining five per cent of the population is considered to be abnormal for that particular trait.

Modifier Genes

Most genes in the body are probably influenced to some degree by the presence of other genes. When one principal gene is involved in determining an inherited characteristic, its expression may be influenced by other genes with effects that can be so minor that they are difficult to detect. Modifier genes differ from multiple genes in that the former modify to a varying extent a single gene's expression, whereas the latter involve several genes that act in a cumulative manner to express a hereditary characteristic.

APPLICATION: HUMAN EYE COLOUR

Eye colour results from the presence (brown-eyed) or absence (blue-eyed) of the melanin pigment in the iris. Brown-eyed people contain the single dominant gene for the presence of melanin in the iris. The eye colour of humans can consist of various shades of blue and brown, that is, green, hazel, grey, light brown, dark brown and black. These variations in blue and brown eye colour result from the presence of modifier genes. These genes influence eye colour in the following ways:

1. amount of pigment in the iris,
2. affect the tone of the pigment,
3. distribution of the pigment throughout the iris.

It is still possible to use a punnett square when referring to a hereditary characteristic that is influenced by modifier genes. The punnet square can be used to predict the chance of brown eyes or blue eyes occurring in the offspring, but not the degree of shading of blue or brown eyes. The influence of modifier genes only alters the degree of expression of a genetic trait, they do not determine whether a genetic trait will be present.

In summary, the expressions of a gene depends on the other genes present (genetic environment) and on the physical environment.

18.6 Molecular basis of heredity

A chromosome is composed of the highly coiled molecule deoxyribonucleic acid, which is commonly referred to as DNA. Since a chromosome contains the units of hereditary information called genes, a relationship must exist between DNA structure and gene function.

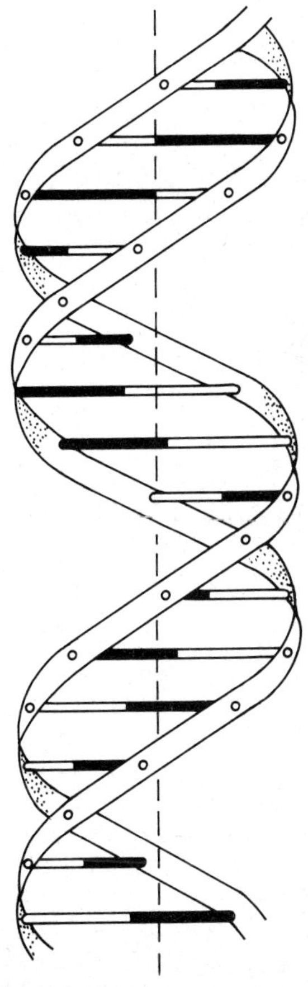

Figure 18.9 DNA molecule showing the double helix or spiral staircase shape of the two nucleotide strands.

DNA STRUCTURE

DNA is composed of sequences of chemical structures called nucleotides. Each nucleotide contains the sugar molecule deoxyribose, a phosphate group and one of a group of nitrogen related structures known as nitrogen bases. The four types of nitrogen base are adenine, thymine, cytosine and guanine. The arrangement of these nitrogen bases are the key to hereditary information. In a DNA molecule, the nitrogen bases form the "steps" of a spiral-shaped structure known as a double helix (Figure 18.9). The sugar molecules link with the phosphate groups to form a long DNA molecule. The double helix is a precisely-shaped molecule which requires a specific-sized nitrogen base to form the "steps" of the helix. The only nitrogen bases that can be linked together are adenine with thymine and cytosine with guanine. Any other combinations render the double-stranded helix unstable.

The only variable in a DNA molecule is the sequence of nitrogen bases. The restrictions of size results in a DNA molecule consisting of only four possible combinations (Figure 18.10). The hereditary information is contained in the nucleotide sequences along DNA molecules.

A gene is a group of about 1000 pairs of nucleotides. Each gene consists of a specific sequence of nucleotides, and no two genes contain exactly the same sequence.

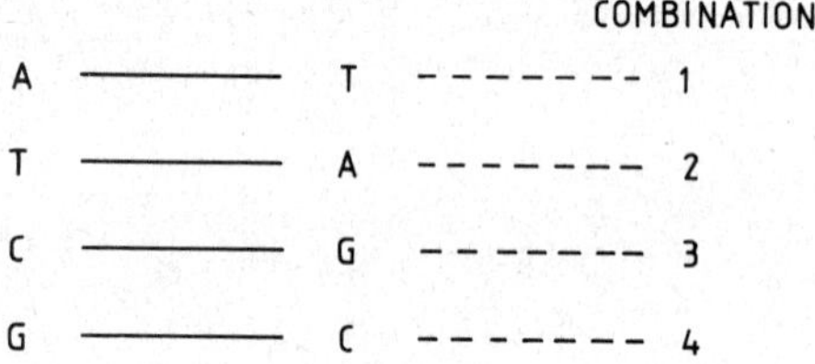

Figure 18.10 The four possible combinations of the nitrogen bases adenine (A), thymine (T), cytosine(C) and guanine (G).

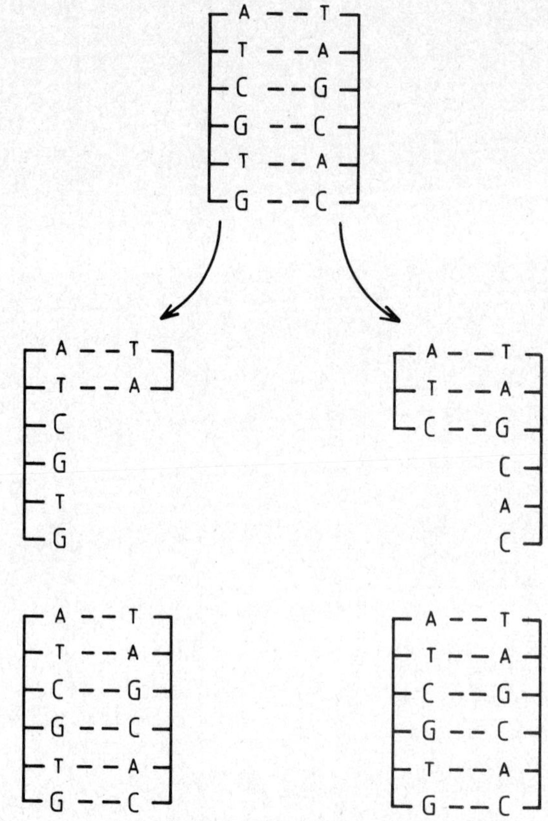

Figure 18.11 Replication of DNA. The two strands split exposing DNA nucleotides which act as a template for free nucleotides. As the free nucleotides attach, their close proximity results in these nucleotides being linked together and forming two identical DNA molecules.

REPLICATION OF DNA

The mechanism of DNA replication provides an understanding of the nature of life, since life depends upon the ability of chemicals to duplicate themselves.

Mechanism of Replication

DNA duplicates itself in the following manner:
1. hydrogen bonds that link complementary nucleotides break, resulting in the formation of two separate strands (Figure 18.11),
2. the exposed nucleotides in each strand act as a template for free nucleotides that are present in the nucleus,
3. complementary nucleotides form hydrogen bonds with exposed nitrogen bases in each single strand,
4. the physical proximity of the recently attached nucleotides results in these nucleotides linking together,
5. the two DNA molecules are passed onto the dividing cells resulting in each cell receiving identical genetic information.

Relationship of DNA Replication with Cell Division

The duplication of DNA occurs in the later stages of interphase. This duplication results in the formation of two identical chromatids. These very thin and elongated chromatids are unable to be observed with a microscope. As these chromatids condense in size, they appear as chromosomes in the stage described as prophase. Towards the end of prophase, the DNA molecules have moved to the centre of the cell. During metaphase the chromatids are oriented at the centre of a cell, after which each pair is separated. The separated DNA molecules are pulled via the spindle apparatus to the centrioles located at opposite sides of the cell. The two identical regions of DNA are then separated by the formation of a plasma membrane.

RNA STRUCTURE

RNA or ribonucleic acid provides the means by which genetic information can be transposed into actions within the cell. Since the manufacture of chemicals and energy occurs in the cytoplasm, the RNA must be capable of transmitting the genetic information from the nucleus to the cell factories if genes are to control cell function.

The two principal forms of RNA are termed messenger RNA (mRNA) and transfer RNA (tRNA). Both mRNA and tRNA display the following features:

1. they are similar to DNA in that they are composed of a sequence of nucleotides, that is, they are nucleic acids,
2. they consist of one strand of nucleotides compared with the double strand of DNA,
3. all RNA contains uracil in place of thymine and is the complementary nitrogen base for adenine,
4. RNA contains the sugar ribose in place of deoxyribose,
5. RNA is smaller than DNA which enables it to move from the nucleus to the cytoplasm.

mRNA differs from tRNA in terms of size and also function. mRNA is much larger than tRNA, and acts as a courier for genetic information from the nucleus to the cytoplasm, whereas tRNA carries the raw materials in the cytoplasmic matrix to the factories where these materials are assembled according to the specific genetic instructions.

RNA SYNTHESIS

For genetic information to be carried from the nucleus by RNA, a mechanism must exist where DNA is involved in the formation of RNA. DNA acts as a pattern by which RNA molecules are assembled. The formation of RNA from DNA is termed transcription.

In transcription, the DNA double helix splits, exposing the nucleotides. Complementary RNA nucleotides loosely attach via hydrogen bonds to their respective DNA nucleotides, that is, adenine RNA combines with thymine DNA, uracil–adenine guanine-cytosine, and cytosine-guanine. Since a specific RNA nucleotide can only attach to its complementary DNA molecule, a mirror image of the DNA nucleotide sequence results (Figure 18.12). The fixed position of assembling nucleotides results in adjacent nucleotides being linked by phosphate groups. As the phosphate groups form a nucleotide chain, the RNA-DNA linkages break. This break effectively separates the DNA from the RNA chain as the latter is being assembled.

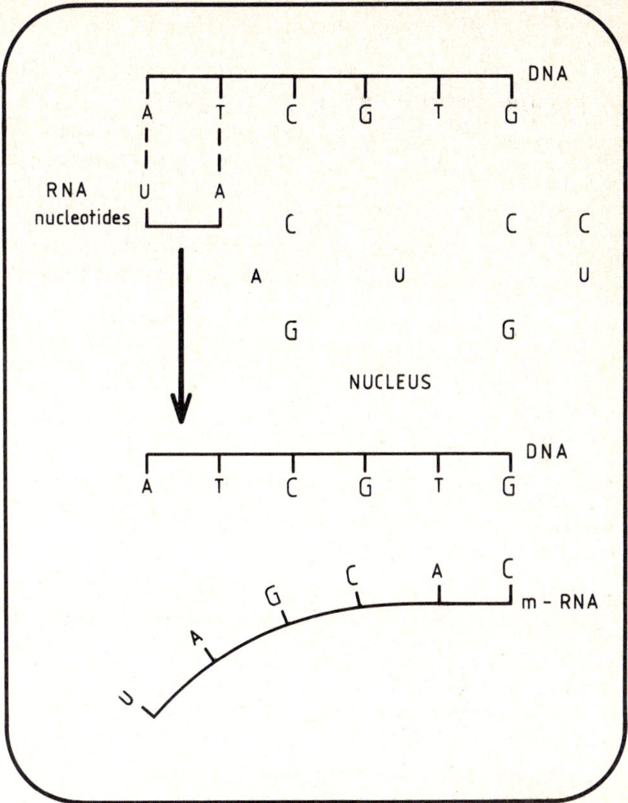

Figure 18.12 Transcription involves the linking of free RNA nucleotides to the exposed DNA nucleotides. The resulting RNA is a mirror image of the DNA nucleotide sequence.

Transcription effectively produces RNA molecules that are mirror images of the gene's nucleotide sequence. This process of transcription is necessary for the maintenance and growth of a cell. Since DNA is involved in cell division, the only stage that can be used for transcription of DNA is interphase. Most cells thus exist for the majority of their lives in interphase so that the genetic information on DNA can be used for the maintenance and growth of the cell.

GENETIC CODE

The transfer of hereditary information from one cell to another is accomplished through a series of specific nucleotide sequences called genes. These genes determine the properties of individual cells by controlling cell function, that is, the manufacture and breakdown of chemicals within cells. A knowledge of how genes control cell function is necessary for understanding such areas of cell biology as genetic disorders, virus function, and genetic engineering.

Relationship of Genes and Enzymes

The genetic code is the language that is used by a cell to transfer genetic information into chemical manufacture. Since genetic information is transcribed into RNA molecules, a form of communication is necessary between the RNA and the cytoplasmic substances that regulate chemical reactions. These substances are called enzymes.

Enzymes are a form of protein and, therefore, are composed of a sequence of amino acids. A specific enzyme consists of a unique sequence of amino acids which results in that enzyme only being able to regulate a specific chemical reaction (see 24.2).

For most cell chemical reactions to occur, a specific enzyme is required to regulate a specific reaction. Also, hereditary information from the nucleus is required to regulate specific cell chemical reactions. A mechanism must exist whereby hereditary information in the form of a gene controls the manufacture of the cytoplasmic chemical reaction regulators, that is, enzymes. By controlling the formation of enzymes, the gene is controlling the chemical reactions in the cytoplasm.

The function of a gene and an enzyme are determined by a sequence. In the case of the gene, a specific sequence of nucleotides determines its function, whereas with enzymes it is a specific sequence of amino acids. The mechanism of transferring genetic information to enzyme manufacture thus involves a relationship between nucleotide sequence and amino acid sequence. Parts of a nucleotide sequence must therefore correspond to individual amino acids if any correlation is to exist between the two sequences. The question then arises of how many nucleotides represent a particular amino acid.

Genetic Alphabet

It is often convenient to represent each nucleotide by a capital letter, that is, A—adenine, T—thymine, C—cytosine and G—guanine. The genetic code consists of three-letter words, that is, combinations of three nucleotides. EXAMPLES: TAC, CGA and ACC. ANALOGY: The English language is based on twenty-six letters which can be combined to form words of varying lengths. This variation in the number of letters that constitute a word and the relatively large number of letters results in virtually a limitless number of different words. In the genetic language, words are composed of only three letters. The small genetic alphabet of three letters restricts the number of genetic words to a maximum of sixty-four.

With any process where materials are to be assembled in a particular sequence, the following instructions are required:

1. words that signal the start of manufacture,
2. words that indicate the sequence in which the materials are to be assembled,
3. words that signal the completion of manufacture.

Transcription

Most of the sixty-four different nucleotide combinations represent the twenty forms of amino acid that are used as raw materials in protein manufacture. Some of the nucleotide triplets signal the start of protein manufacture whereas others stop the assembly of proteins. The genetic instructions are transcribed into complementary mRNA sequences of nucleotides. EXAMPLE: The DNA sequence TAC CGA ACC acts as a template to form the mRNA sequence AUG GCU UGG. The DNA instructions have been retained in the formation of a mirror image set of mRNA nucleotides. This process of transcription is necessary since the large DNA molecules are unable to move through the nuclear pores to the sites of protein manufacture. The mRNA acts as a courier of genetic information from the genetic master plan to the manufacturing plants. The nucleotide triplets in mRNA are referred to as codons. EXAMPLE: The mRNA sequence AUG GCU UGG consists of the three codons or words AUG, GCU and UGG.

A mRNA carries codons to initiate the formation of proteins, assemble amino acids in their correct order and stop at the end of the sequence when the protein is formed. Many of the meanings of the sixty-four codons have been resolved. EXAMPLES: The codons for the following amino acids: alanine—GCU, phenylalanine—UUU, tryptophan—UGG and tyrosine—UAU. An initiating condon is AUG whereas UAA stops protein manufacture.

A set of instructions carried by a mRNA to the protein factories could begin with the following nucleotide sequence: AUG GCU UGG UAU UAU UUU GCU . . . This gentic code sentence would be interpreted at the assembly site as follows: start assembling/alanine/tryptophan/tyrosine/tyrosine/phenylalanine/alanine/ . . .

Translation

The mechanism of converting genetic information into the formation of proteins is referred to as translation. The process of translation and thus protein manufacture occurs at the ribosomes. Ribosomes translate a sequence of codons into a sequence of amino acids with the aid of tRNA.

tRNA molecules are smaller than mRNA. They exist in approximately sixty different forms. The role of each form is that of transporting specific raw materials from the cytoplasmic matrix to the ribosomal factories. Each type of tRNA has a unique shape that is capable of carrying a specific shaped amino acid. In addition, each tRNA contains a combination of three nucleotides that are unique for that particular tRNA. These three nucleotides are termed the anticodon since they are able to form loose linkages with complementary mRNA triplets. EXAMPLE: A tRNA that contains the anticodon CGA can form a loose linkage with the mRNA codon GCU.

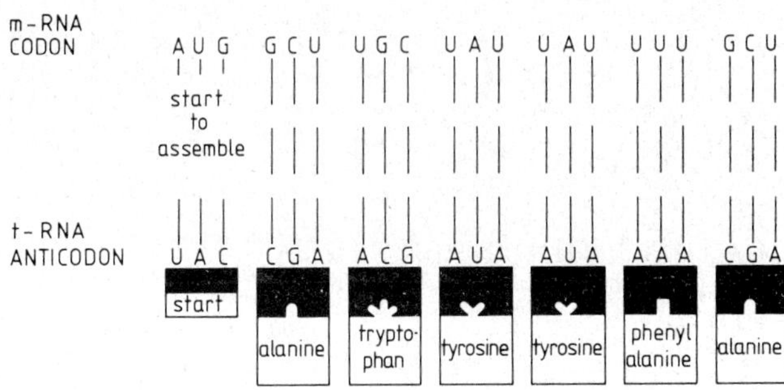

Figure 18.13 Translation of mRNA codons into an amino acid sequence.

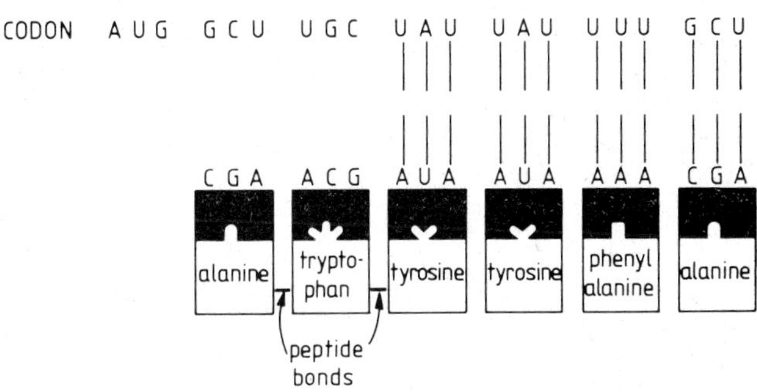

Figure 18.14 Assembling of an amino acid by linking of adjacent amino acids by peptide bonds.

A relationship exists between a codon and the amino acid carried by the complementary anticodon. EXAMPLE: The codon for the amino tyrosine is UAU. The tRNA that has the specific ability to carry tyrosine contains the anticodon AUA.

The mechanism by which the previously described mRNA nucleotide sequence AUG GCU UGG UAU UAU UUU GCU is translated into the amino acid sequence alanine/tryptophan/tyrosine/tyrosine/phenylalanine/alanine/ . . . can now be shown (Figure 18.13). As a sequence of tRNA forms loose linkages (hydrogen bonds) with their complementary mRNA nucleotides, a chain of amino acids is formed.

The close proximity of oppositely charged regions of adjacent amino acids results in the formation of bonds which are known as peptide bonds. A peptide bond links the amine group (NH_2) of one amino acid with the carboxyl group (COOH) of an adjacent amino acid. As peptide bonds link adjacent amino acids, an amino acid chain is formed. Simultaneously, the linked amino acids separate from the tRNA carriers (Figure 18.14).

When a chain is composed of many amino acids and thus peptide bonds, the molecule is referred to as a polypeptide. Proteins are an example of polypeptides. The manufacture of proteins in a cell is schematically summarized in Figure 18.15.

18.7 Molecular-based genetic disorders

THE CONCEPT OF ONE GENE–ONE PROTEIN

Errors in the molecular structure of a gene can result in a genetic disorder. The genetic disorder can result from an incorrect substitution of a nucleotide in the DNA molecule so that an incorrect amino acid is located within a protein molecule (Figure 18.16). Alternatively, a loss or addition of nucleotides in a gene will result in an altered instruction for the manufacture of the gene's protein. If the molecular error occurs at a functional part of the protein, a malfunction of the protein will occur. This malfunction is manifested as a genetic disorder. EXAMPLE: Sickle cell anaemia results from an alteration of only one amino acid in haemoglobin molecules.

Molecular-based genetic disorders require biochemical analysis for detection, since the chromosomes appear normal when karyotyped. This restriction in determining a molecular defect in a chromosome resulted in a delay of half a century before a linkage could be confirmed between hereditary diseases and single proteins.

The concept that certain diseases of lifelong duration arise because an enzyme governing a single biochemical step is reduced in activity or missing altogether was first proposed in 1909 by Sir Archibald Garrod. A half a century later Garrod's hypothesis that a block in a biochemical pathway can be inherited as a single recessive trait was proven. One of the first inborn errors of metabolism to be recognized was phenylketonuria.

APPLICATION: PHENYLKETONURIA

Normally the amino acid phenylalaline is converted into tyrosine which is then used for the manufacture of important body substances such as thyroid hormones and the neurotransmitters epinephrine (adrenaline), norepinephrine (noradrenaline) and dopamine. In patients suffering from phenylketonuria (PKU) a deficiency of the enzyme phenylalanine hydroxylase occurs. This enzyme is necessary for the conversion of phenylalanine into tyrosine, and thus a deficiency of this enzyme results in an accumulation of phenylalanine. The excessive accummulation of phenylalanine prevents the normal development of the brain. This abnormal development results in mental retardation of the affected infants.

The accummulation of phenylalanine can be tested in newborns by using a guthrie test. With an appropriate low phenylalanine diet during childhood, a PKU patient can be prevented from developing the serious effects that they would have normally suffered if they had not been treated.

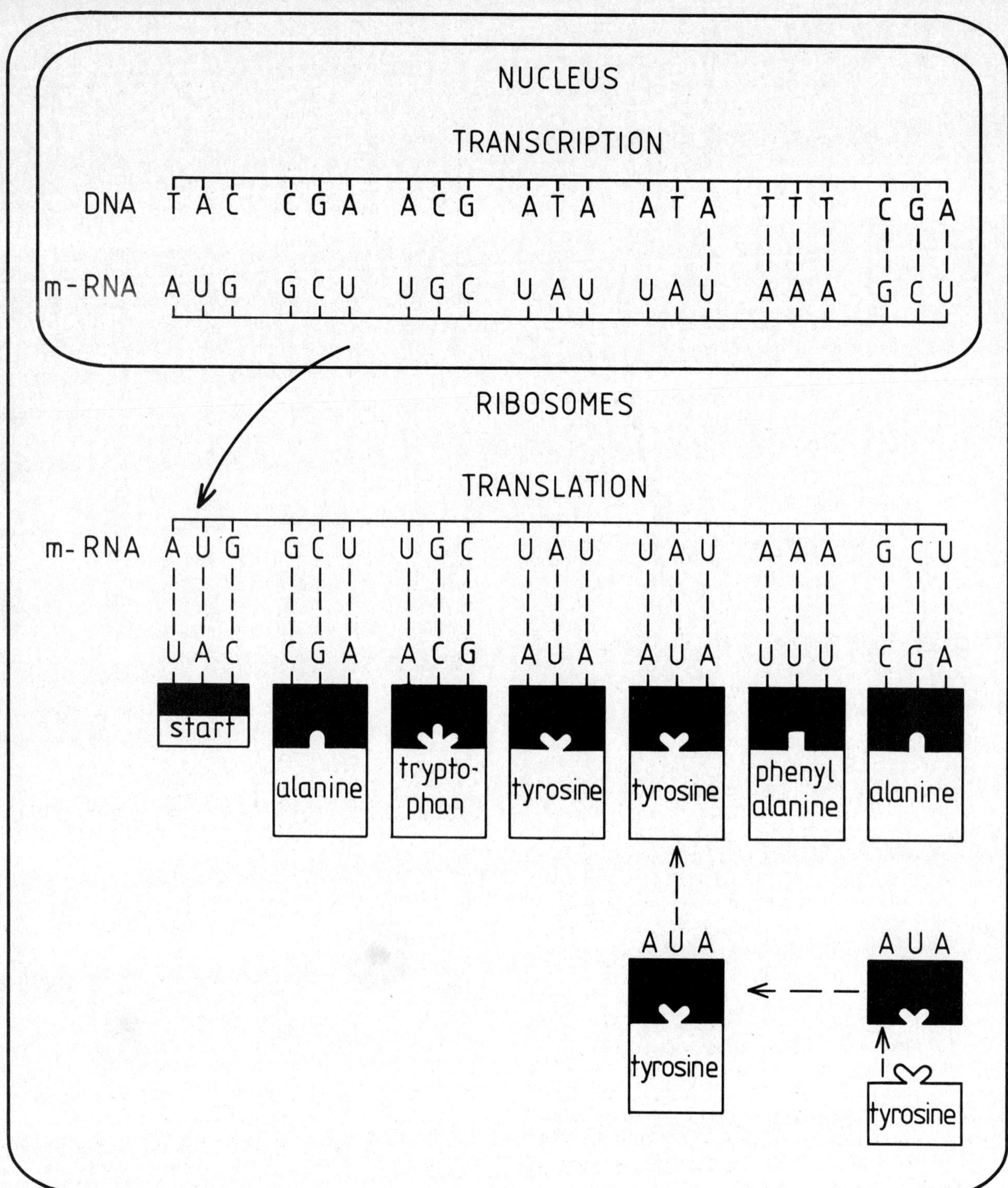

Figure 18.15 Summary of the mechanism of protein manufacture.

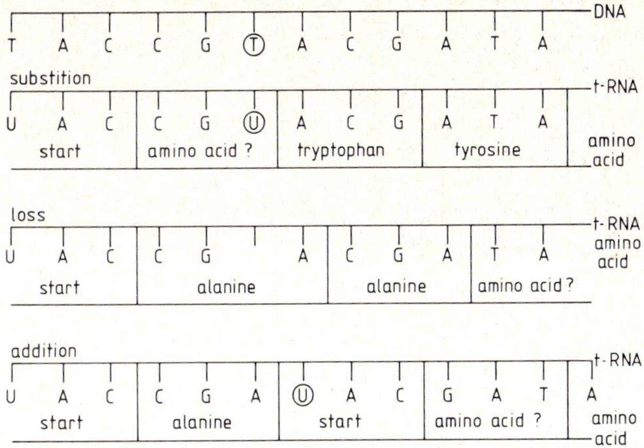

Figure 18.16 The substitution of an incorrect nucleotide into a DNA sequence can result in an incorrect amino acid being located in the amino acid sequence. A loss of addition of a nucleotide will result in many errors in the formation of a protein.

VARIED CONSEQUENCES OF GENE DEFECTS

At present, over 2500 genetic disorders have been attributed to an abnormality in the formation of proteins. The impact on the body of a particular genetic defect varies according to the functional role of the corresponding protein. The role of proteins in body function varies enormously. EXAMPLES: Membrane structure, body defence mechanisms, transporting oxygen, hormones and the control of biochemical pathways by enzymes. It is currently estimated that every individual is a carrier of between five and eight recessive deleterious genes that would have serious phenotypic effects if present in the homozygous state. As knowledge of disease mechanisms is enhanced, the number of genetic disorders present or carried by a person could be expected to be even higher than at present.

Many genetic defects may only be detected when a patient is exposed to an environmental agent that requires the normal functioning of that gene. In such cases the patient has a genetic predisposition. This predisposition may result in a patient having an abnormal response to a drug.

APPLICATION: PSEUDOCHOLINESTERASE DEFICIENCY

Patients with pseudocholinesterase or serum cholinesterase deficiency only manifest this deficiency in the presence of the muscle relaxant succinylcholine (scoline). In the presence of this drug a patient will remain in transient paralysis for several hours instead of minutes. This response is due to the deficiency of the enzyme that normally breaks down this drug.

MOLECULAR SITES OF GENETIC DISORDERS

The replication of DNA is the probable site of most genetic disorders. For instance, the pairing of two nucleotides that are not complementary results in the erroneous nucleotide directing the combination of nucleotides complementary to itself in subsequent generations (Figure 18.17). Another failure of the genetic mechanism is the incorporation of similar structured substances into a DNA strand. These substances can cause erroneous replication of a nucleotide sequence. EXAMPLE: A substance that is structurally similar to thymine but pairs with guanine rather than adenine. Subsequent generations would have cytosine incorporated in the DNA sequence instead of thymine.

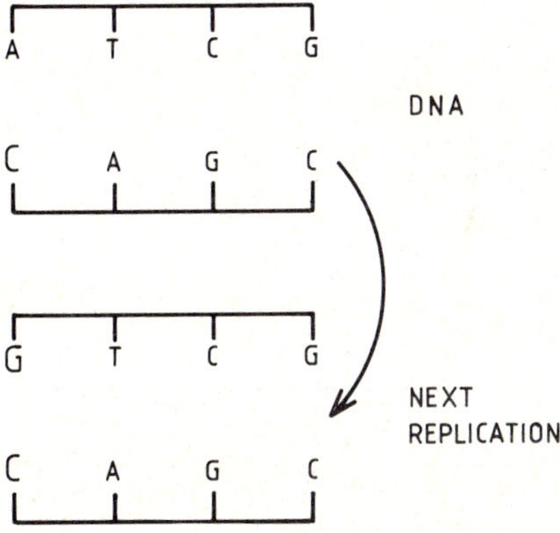

Figure 18.17 Genetic defect formed at one site in a DNA molecule where T has been replaced by a C.

18.8 Molecular-based mitotic malfunction

Difficulty arises in determining whether a malfunction in DNA replication in somatic cells is due to the direct effects of a harmful environmental agent or is a consequence of a genetic predisposition. Mitotic molecular malfunctions result in the formation of abnormal proteins in a certain type of body tissue. The genetic defects may result in the failure of the normal processes that control cell division. In such cases a cancer may occur. Various agents known as carcinogens can initiate this uncontrolled multiplication of cells. EXAMPLE: Pollution, ultraviolet radiation, cigarette tar and viruses. A current approach to the elimination of cancer is the use of drugs that hinder cell division, and thus the growth, of the rapidly dividing cancer cells.

APPLICATION: CYTOTOXIC DRUGS

An understanding of the mechanisms of cell division has enabled the development of a range of drugs that interferes with cell division mechanisms. Drugs that have a toxic action upon cells are known as cytotoxic drugs. Many of these drugs act by interfering with the mechanism of DNA replication. These drugs usually have a similar structure to one of the nucleotides. This similarity enables the drug to combine with a replicating DNA molecule and cause mutations that are incompatible with cell life. Since most cells divide at a relatively slow rate when compared with cancer cells, the cytotoxic drugs mainly influence the rapidly-dividing cancer cells. These drugs will also cause abnormalities in the frequently-dividing sex cells and foetal cells.

Summary

The nucleus controls cell multiplication and the day-to-day functioning of a cell. The hereditary material is packaged in chromosomes.

Human cells contain twenty-three pairs of chromosomes, apart from the mature sex cells which contain only twenty-three chromosomes. Each of the twenty-three homologous pairs of chromosomes contain one chromosome from the maternal parent and one from the paternal parent. One pair are involved in the expression of sexual characteristics and are thus called sex chromosomes whereas the other twenty-two pairs are termed autosomes.

Mitosis is the process whereby hereditary information (DNA) is doubled and separated to form two identical cells. The duplication of cells is necessary for the growth and continued functioning of the body. Meiosis is a form of cell division that results in new cells with half the original number of chromosomes. This process is used in the last stages of sex cell formation for the separation of the homologous pairs of chromosomes.

Malfunction of the meiotic apparatus results in cells being formed with either abnormal quantities of chromosomes or chromosomes that have an altered structure. These abnormalities can be detected in the foetus by the combined processes of amniocentesis, cell culture and karyotyping.

Individual units of hereditary information are represented as genes. Each gene represents a single physical hereditary characteristic which is usually expressed in two forms, although on some occasions more than two forms exist. These forms or alleles are usually either dominant or recessive. An allele of a particular gene is found in both chromosomes that constitute a homologous pair. When the two chromosomes contain an identical allele a person is said to be homozygous for that particular gene. When the two alleles are different the person is heterozygous and is often referred to as a carrier of the recessive allele.

The genotype refers to the genetic composition of an individual whereas the phenotype expresses the physical characteristics of that person.

Genetics is the study of genes and their relationship to hereditary characteristics. The patterns of inheritance for a particular gene can be determined by taking into account the possible combinations of the parents genotypes.

The potential genotypic combination's expression can be altered by the environment and the presence of other genes. The expression of a gene depends on the other genes present (genetic environment) and on the physical environment.

Each chromosome is composed of a long strand of DNA. The DNA is in the form of a double helix, with the two strands being linked by complementary nucleotides; adenine links with thymine and cytosine links with guanine. The sequence of these nucleotides determines the coding of hereditary information. A gene is represented by a specific sequence of appromixately 1000 pairs of nucleotides.

When the two strands of DNA are split, the exposed nucleotides link with complementary free DNA nucleotides which results in the formation of two identical strands of DNA. This doubling of genetic information is necessary for cell division.

During periods between cell division or interphase the DNA splits and the exposed nucleotides act as a template for the formation of mRNA in the process of transcription. This mRNA is single-stranded and contains uracil in place of thymine. The mRNA carries the genetic information in the form of nucleotide triplets or codons to the ribosome. These codons form loose linkages with complementary triplets or anticodons of tRNA. Each tRNA has a specific amino acid attached to it, and thus the aligning tRNA molecules line up their corresponding amino acids into a specific sequence. This process of translating mRNA genetic information into proteins is called translation.

Alterations to the nucleotide sequence result in malformation of proteins, such as in the genetic disorder phenylketonuria. Many of these molecular-based genetic defects can only be detected when a patient is exposed to an environmental agent that requires the normal functioning of the gene. Various agents can cause a malfunction in mitotic DNA duplication which can lead to cancer. Rapidly dividing cancer cells can be acted upon by cytotoxic drugs whose site of action is usually the interference of the molecular cell division mechanism.

Unit Eight
Extracellular structure and function

This unit relates the different specialized functions of cells to body homeostasis. The relationship between cellular organization and function is discussed, with particular reference to the impact of abnormal cell function on body homeostasis.

Chapter 19

Cellular organization

Objectives

At the completion of this chapter the student should be able to:
1. describe the similarities and differences between tissues, organs and systems,
2. list the types of tissue,
3. relate the structure of each tissue type to its function,
4. explain how the function of tissues maintains homeostasis,
5. compare the different functions of each type of tissue,
6. describe how cellular specialization and organization differ within each tissue type.

19.1 Cell forms

Human cells arise from a single cell. During the process of human development many cells are formed, and groups of these cells undergo cell differentiation. This process results in cells being altered into specialized types such as blood, nerve and muscle cells. These cells have each acquired specific functions from the same genetic makeup.

Following differentiation, specialized cells can be identified by their particular structure and function. The structure of different cells varies from the relatively simple red blood cell to the complex nerve cell (Figure 19.1). Some cells, such as the phagocytic white blood cells have a variable shape. This variable shape enables these cells to engulf nearby microbes and cell debris.

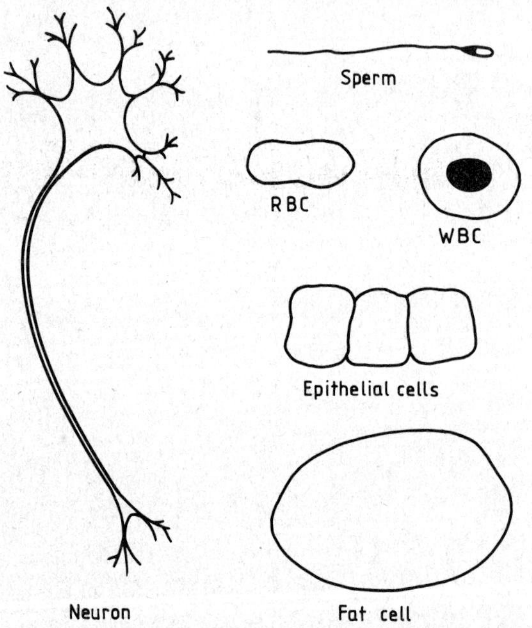

Figure 19.1 Diversity of size and shape of cells.

The size of cells varies from the relatively small red blood cell or erythrocyte with a diameter of 7.5 μm to the large fat cells with a diameter of 120 μm. A large variation of size can occur with the same cell type. EXAMPLE: The length of nerve cells varies from a few millimetres in the brain to over a metre in some spinal cord nerves.

Certain cells exhibit a specialized surface that consists of cytoplasmic extensions called microvilli (Figure 19.2). These extensions or folds increase the surface area of cells thus enabling a greater efficiency in the absorption of materials. EXAMPLE: Cells lining the surface of the small intestine.

The variety of cell types in the body indicates the high degree of specialization that exists in body structure and function (see Table 19.1).

19.2 Tissues

Cells are organized into tissues, organs and systems. Tissues contain cells with similar structure

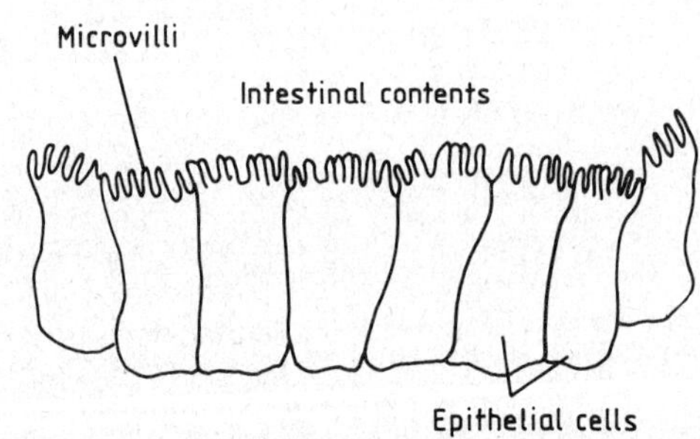

Figure 19.2 The surface of the small intestine showing microvilli.

Table 19.1 Examples of the specialized function of cells

Cell	Function
Epithelial	Absorption and secretion of chemicals; protective barrier to the external environment
Fat	Storage of energy
Goblet	Secretion of mucus
Muscle	Moving parts of the body
Nerve	Conduction of information
Red blood cell	Transporting oxygen and carbon dioxide
White blood cell	Helps protect the body against foreign material

and function whereas organs consist of several different tissues grouped together to form a structural and functional unit. EXAMPLES: Kidneys or liver. A system consists of a group of interacting organs that cooperate as a functional unit in the life of an organism. EXAMPLES: Endocrine and nervous systems.

A tissue is a combination of many cells that have a similar structure and function, that is, a tissue has a characteristic structure and function. EXAMPLE: Muscle tissue is composed of muscle cells that combine together to act as a structural and functional unit. Traditionally all animal tissues are divided into the four basic categories of epithelial tissue, muscular tissue, nervous tissue and connective tissue.

EPITHELIAL TISSUE

Epithelial tissue covers the surface of the body and organs. It also lines body cavities and organs and is found in secretory structures or glands.

The general functions of epithelial tissue are to act as a protective barrier for underlying tissue, secretion and absorption of substances. The function of this tissue is closely related to structure.

Epithelial cells provide the body with a protective barrier from the external environment. EXAMPLE: Surface of skin. Within the body these cells restrict the movement of foreign substances across the walls of the gastrointestinal tract, urinary tract, and respiratory tract. In order that they can function as a protective barrier, it is necessary for these cells to be packed so as to form an effective seal (Figure 19.3). The tight packing of cells results in only small amounts of intercellular substances being present. The exposure of these cells to wear, tear and injury necessitates frequent division and production of new cells.

Epithelial cells also perform the important functions of absorption and secretion, that is, the intake or removal of substances from the body. EXAMPLE: All digested nutrients are absorbed into the body across a layer of epithelial cells that line the gastrointestinal tract.

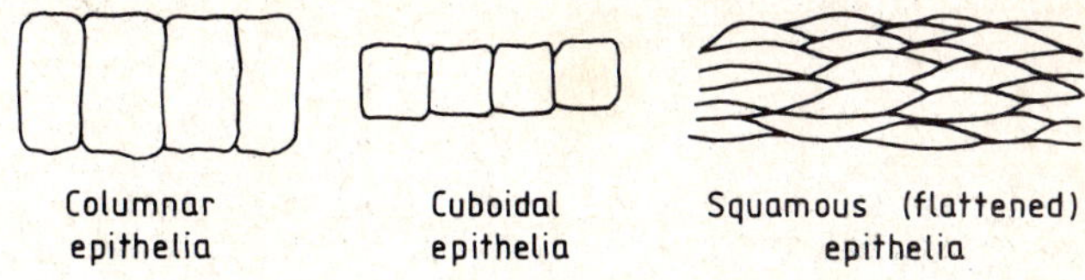

Figure 19.3 General structure of epithelial tissue.

The structure of epithelial tissue is arranged in such a way so as to maximize the movement of substances across it. This tissue is usually arranged in thin sheets of cells in structures that are involved in absorption or secretion. EXAMPLES: Cells lining the gastrointestinal tract, respiratory system, urinary and reproductive tracts.

The structure of epithelial cells varies from flat or squamous cells to cubes and column-shaped cells. The squamous cells are the most efficient cells for allowing substances to diffuse across them. Their thin shape offers minimal resistance to diffusing particles. These cells are therefore used to line capillaries where a minimum barrier to diffusion of oxygen and carbon dioxide is required.

All glands are composed of epithelial cells. A gland may consist of one cell or a group of highly-specialized epithelial cells that secrete substances into ducts or into the blood. All glands are classified as either exocrine which secrete their products into ducts or endocrine which secrete their products directly into the blood. Goblet cells (secrete mucus) and parietal cells (secrete hydrochloric acid) are examples of secretory cells that are often interspersed throughout epithelial layers. These single-cell exocrine glands enable epithelial layers to perform a variety of secretory roles.

Epithelial tissue has nerves penetrating it but not blood vessels. The lack of blood vessels requires epithelial cells to obtain their nutrients either from the region that they are lining or from blood capillaries in underlying connective tissue. Connective tissue also acts as a support for the epithelial tissue since the thin cell layers of epithelial tissue can be easily torn when not attached to a rigid structure.

MUSCULAR TISSUE

Muscular tissue has the ability to shorten or contract. It is responsible for locomotion and for movement of various parts of the body with respect to one another. Muscle cells are elongated, relatively large cells that can be up to 40 cm in length. The size and length of these cells has resulted in them being referred to as fibres. Muscle tissue is divided into three categories: smooth, skeletal and cardiac,

Smooth Muscle

Smooth muscle is the contractile tissue of all internal structures except for the heart. EXAMPLES: Walls of blood vessels, gastro-intestinal tract, respiratory tract and uterus.

Smooth muscle fibres may occur singly or in small groups. Each fibre is characterized by its long spindle shape (Figure 19.4). These fibres are able to contract spontaneously or they can be controlled by the internal nervous system which is known as the autonomic nervous system. This nervous system is an important component of the body's internal communication network as it links the brain to tissues (see Section 26.4). The contraction of smooth muscle in response to autonomic stimulation is usually not under voluntary control. For this reason, smooth muscle is often called involuntary muscle. The slow sequential contraction of smooth muscle along a tube such as the gastrointestinal tract results in the mixing and continual movement of materials.

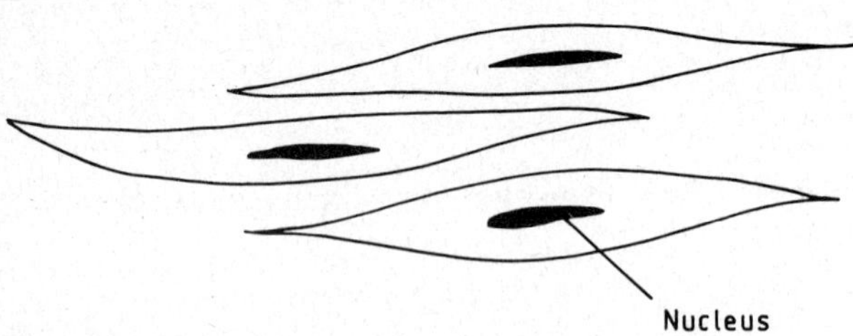

Figure 19.4 The structure of smooth muscle fibres.

APPLICATION: PERISTALSIS

Peristalsis is the rhythmic movement of smooth muscle contractions that mix and propel the contents of various tubular smooth muscle structures. EXAMPLES: Gastrointestinal tract, ureters and most glandular ducts. The usual stimulus for peristalsis is distention, that is the stretching of a tube wall due to the accumulation of fluid.

Various factors such as emotional stress can cause an elevated peristaltic rate in the small intestine. This increased frequency of waves propels the intestinal contents to the bowel and results in diarrhoea.

Cardiac Muscle

Cardiac muscle is the tissue that gives the heart its electrical and contractile properties. It is similar to smooth muscle in that it is automatic and can also be controlled by the autonomic nervous system.

Cardiac muscle is similar to skeletal muscle in that both exhibit cross bands or transverse striations along the length of their fibres (Figure 19.5). Cardiac muscle has a unique thickened region that connects the ends of adjacent muscle fibres. These thickened regions are called intercalated discs. They are responsible for the conduction of impulses to adjoining muscle fibres. The individual fibres contract in response to the spread of an electrical impulse along the surface of muscle cells. These impulses are similar to nerve impulses.

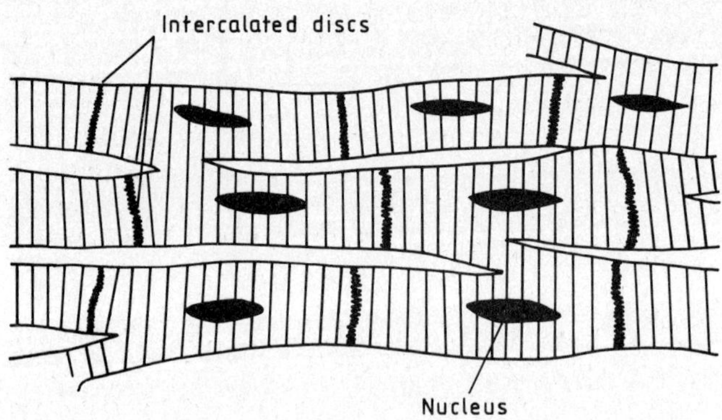

Figure 19.5 Cardiac muscle fibres showing the intercalated discs and the repetitive striations.

Under normal conditions the heart contracts in a rhythmic movement of about seventy-two times a minute. This rate can be altered by the autonomic nervous system in order to maintain homeostasis. The rhythmic contraction of muscle throughout the heart results in blood being forced from the heart and into the large arteries that emanate from it.

Cardiac muscle can exhibit three different physiological properties:

1. The automatic property of the heart spontaneously contracting at regular intervals is often referred to as autorhythmia. Cells that have highly developed autorhythmic properties are located at the pacemaker, the region that determines the rate of muscle contraction. Pacemaker muscle cells have little contractile ability, their main function being to initiate impulses that are conducted to the rest of the heart.
2. Coordinated contraction of the heart requires the rapid conduction of impulses throughout the heart. Since ordinary cardiac muscle cells can only conduct impulses at a slow rate, a network of specialized rapidly conducting cardiac cells exists. This is called the Purkinje network.
3. The vast majority of cardiac muscle cells exhibit the ability to contract. Both the pacemaker and Purkinje fibres have little contractile ability even though they are cardiac muscle fibres.

The efficiency of the heart to pump blood depends on a coordinated contraction that is initiated at the pacemaker at the top of the heart in a region called the sino-atrial (SA) node. The SA node sets the basic pace for the heart rate. Following the initiation of an electrical impulse, it spreads over the filling chambers of the heart or atria from top to bottom leading to a contraction of the atrium. This action forces blood into the two large chambers which are known as the ventricles. The spread of the impulse by the Purkinje fibres enables the ventricles to contract in a coordinated manner, thus forcing blood into the vessels that emanate from the ventricles.

APPLICATION: CARDIAC ARRHYTHMIAS

Efficient heart function relies on the development of electrical impulses and their subsequent conduction through the heart. Disturbances of impulse formation, conduction or a combination of both can result in a lack of rhythm in the hearts action. Abnormal rhythmic contractions are termed cardiac arrhythmias.

Asynchronous contraction and quivering of the atria or ventricle are referred to as atrial fibrillation and ventricular fibrillation respectively. Fibrillation results in regions of muscle contracting at different rates forming a quivering effect or rippling of the surface. This uncoordinated contraction reduces the effectiveness of the heart. In ventricular fibrillation failure to reverse the arrhythmia quickly will result in circulatory failure and death.

The process of defibrillation is achieved by applying strong electric shocks to the heart by defibrillators. If the normal pacemaker is still functioning and viable, it may take over the rhythm, and then normal coordinated cardiac contractions resume.

Skeletal Muscle

The muscles of the body that are attached to the skeleton are composed of skeletal muscle. It is similar to cardiac muscle in that it contains transverse striations. Large numbers of these long parallel fibres group together to form bundles (Figure 19.6). Many of these bundles are bound together by connective tissue to form large and small muscles.

Skeletal muscle differs from cardiac and smooth muscle by its inability to spontaneously contract, that is, it only contracts when it is stimulated by the nervous system. Skeletal muscle is controlled by the somatic nervous system. This nervous system involves the nerves that connect the spinal cord to the external regions of the body (see 26.4). This nervous system is under the influence of conscious control and thus skeletal muscle can be contracted at will.

Skeletal muscle fibres in a muscle will contract as one unit. They act in coordination with bones (levers) to move parts of the body. The rate of contraction can be rapid in the case of muscles used in locomotion, or slow in the case of muscles that are used for the maintenance of posture in which case they maintain long periods of sustained contraction.

Figure 19.6 A bundle of skeletal muscle fibres showing repetitive striations.

APPLICATION: DISORDERS OF SKELETAL MUSCLE

Disorders of skeletal muscle are known as myopathies. The dependence of skeletal muscle on nerve stimulation is shown when this stimulation is lost. The resulting disuse of the muscle results in some degree of muscle wasting which is referred to as muscular atrophy. EXAMPLE: Long-term immobilization of a limb will result in some muscle atrophy.

The most advanced cases of muscular atrophy are characterized by a substantial reduction in affected muscle size, along with a loss of tone of the muscle. This severe form of muscle atrophy occurs when the nerve supply to the muscle is disconnected. EXAMPLES: Poliomyelitis, peripheral neuritis and peripheral nerve injury cause the removal of nerve connections to regions of skeletal muscle.

Various forms of myopathies have a genetic origin and usually result in muscle weakness. EXAMPLE: The muscular dystrophies are a group of genetic abnormalities characterized by a progressive weakness and atrophy of muscles.

NERVOUS TISSUE

Nervous tissue is composed of cells capable of rapidly conveying information to different regions of the body. These nerve cells or neurons are an essential component of body homeostasis. The neuroglia (glial cells) comprise another group of cells which support, protect and assist the function of neurons.

Neuroglia

The neuroglia can be separated into three functional cell types. The first group form part of the blood–brain barrier, that is, the barrier which restricts the movement of certain substances into the brain. Another group help hold the neurons together. This group contains cells that produce a substance called myelin which forms a sheath around most neurons. Myelin serves the important function of increasing the velocity of nerve impulses. A third group of neuroglial cells are involved in the removal of microbes and cell debris from nervous tissue by the process of phagocytosis.

General Structure of Neurons

Neurons are cells that exhibit:

1. irritability—ability to respond to alterations in the environment,
2. conduction—ability to conduct impulses along their length,
3. neurotransmission—ability to transmit impulses to adjacent nerve, muscle or endocrine gland cells.

The structure of neurons is markedly different from all other cell types. A neuron consists of a cell body with projections extending from it. These projections are either numerous and relatively short or singular and up to one metre in length (Figure

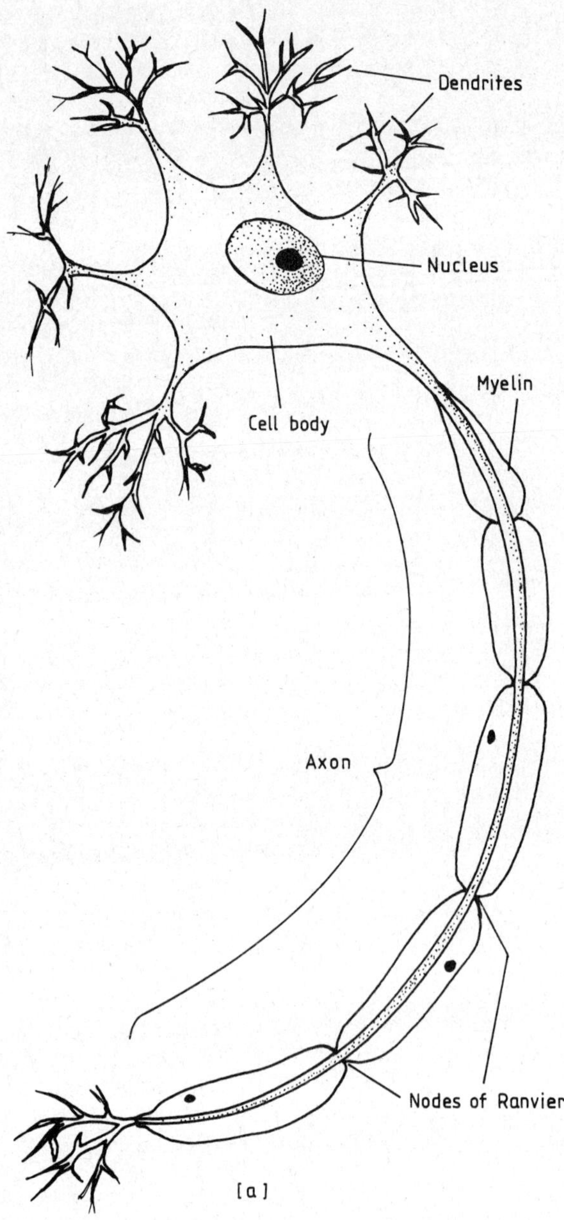

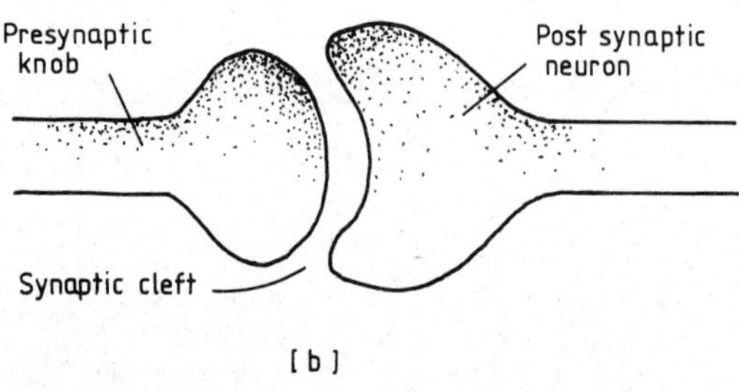

Figure 19.7 (a) Structure of a neuron, (b) a synapse.

19.7a). The former are referred to as dendrites while the latter is known as an axon.

The Cell Body

The cell body contains a nucleus and cytoplasmic organelles. It is the site of most of the cell's chemical reactions and thus damage to it results in the death of the neuron. The cell body is located in either the grey matter of the spinal cord and brain or in the ganglia of neurons associated with the autonomic and somatic nervous systems. The dendrites receive signals and transmit them to the cell body.

The Axon

The axon carries impulses away from the cell body and to other cells. These long structures are able to conduct nerve impulses to relatively distant cells. The conduction of these impulses can be assisted by covering the axon with a myelin sheath. The term nerve fibre is used to describe an axon and its sheaths. Axons containing these sheaths are myelinated whereas those lacking a covering are unmyelinated. Myelinated axons have an appearance similar to a string of sausages (Figure 19.7a). The myelin prevents sodium ions and potassium ions from entering and leaving the axon. This restriction results in impulses only being formed at points where the myelin sheath is absent, that is, at the nodes of Ranvier. A nerve impulse is thus required to jump from one node to the next node in a process called saltatory conduction. The importance of myelin to nerve function is highlighted in multiple sclerosis. This debilitating disease is characterized by a breakdown of the myelin sheath.

The axon terminates as a branched region with individual branches containing knobs at their ends (Figure 19.7b). These knobs form close associations with adjacent nerve, muscle or endocrine cells.

The Synapse

A synapse is the region where two neurons make close contact with each other. The axon located before the synapse forms the presynaptic region with the neuron being called a presynaptic neuron. The dendrites of another nerve form the postsynaptic region and the neuron is thus referred to as the postsynaptic neuron. The presynaptic knobs are separated from the postsynaptic dendrites by a minute gap of about 20 nm. This gap is called the synaptic cleft.

The region where a nerve and muscle make functional contact is known as a neuromuscular junction. The distance separating the two cells is also approximately 20 nm.

Neurotransmission

The electrical nerve impulse is unable to directly cross the 20 nm gap of synapses and neuromuscular junctions. A mechanism known as neurotransmission is used in both situations to indirectly transfer the nerve impulse across the cleft. This mechanism is initiated by the release of chemicals called neurotransmitters in response to a nerve impulse arriving at the presynaptic terminals. The released neurotransmitter diffuses from the presynaptic knob across the cleft to the adjacent cells membrane. The neurotransmitter alters the nature of this membrane so that a nerve impulse is either generated or is prevented from being formed. The effect of a neutrotransmitter is either the stimulation or inhibition of the postsynaptic membrane or muscle terminal. The main neurotransmitters are acetylcholine, norepinephrine, dopamine and serotonin. Each of these neurotransmitters can only exert their influence on specific membranes. Following their brief effect on the postsynaptic membrane, the neurotransmitters are removed or broken down by enzymes. Failure to remove or break down a neurotransmitter will result in it exerting a continued effect on the postsynaptic neuron. Any alteration of the neurotransmission mechanism by drugs can either enhance or inhibit the actions of nerves.

Forms of Neurons

Neurons vary in size and shape according to their function. One of the main functions of nerves is their ability to form nerve impulses from alterations in their extracellular environment. Some neurons contain highly specialized dendrites that are receptive to specific changes in the extracellular environment. These neurons are referred to as sensory neurons and the specialized dendrite region is called a sense organ or receptor. Except for pain receptors, each receptor responds to a specific stimulus, such as light. Receptors play a vital role in body homeostasis as they detect variations in such things as chemical concentration, pressure changes and tissue stretching.

Neurons that are required to conduct impulses over relatively large distances contain axons of up to one metre in length. Some neurons contain very short axons. These neurons usually link nerves in the grey matter of the spinal cord and brain. Neurons that contain specialized axon terminals for functional connection with muscles or glands are referred to as motor neurons.

White and Grey Matter

A nerve is a combination of many bundles of nerve fibres. The axons of a myelinated nerve appear white, and hence the term white matter is given to the regions of the brain and spinal cord that contain a predominance of axons. A cluster of cell bodies results in a slightly grey colour. The large regions of the brain and spinal cord that consist mainly of cell bodies are known as grey matter.

CONNECTIVE TISSUE

This is the most abundant tissue in the body. It binds, supports and protects other body tissue. The common features of connective tissue are the rich blood supply and the extensive intercellular material between the widely scattered cells. The intercellular substances form a matrix that may be a liquid, solid, or semi-rigid gel. The composition of the matrix determines the characteristics of that particular connective tissue. EXAMPLE: Blood is a connective tissue composed of a liquid matrix, whereas bone contains a solid matrix. The wide diversity in structure and function of connective tissue has resulted in the need to classify connective tissue into four categories which are known as connective tissue proper, cartilage, bone and vascular.

Connective Tissue Proper

Connective tissue proper consists of cells and extracellular fibres embedded in a semi-fluid matrix called tissue fluid.

Several types of cells are found in this tissue. Some remain in a relatively fixed position. EXAMPLE: Fibroblasts which are responsible for production of collagen and the maintenance of extracellular components. Other cells have the ability to move. EXAMPLES: Plasma cells for antibody production, mast cells for heparin and histamine production, and macrophages for engulfing cell debris and bacteria.

The three forms of fibres embedded in the matrix are:

1. Collagenous fibres—composed of the protein collagen. These fibres are flexible but offer great resistance to a pulling force, that is, they are difficult to stretch. They have a wavy appearance and are composed of intertwined units of collagen. When collagen is broken down by boiling, gelatin results.
2. Elastic fibres—composed of the protein elastin. These fibres are easily stretched and they will return to their original shape when the force is removed. They are thus similar to a rubber band.
3. Reticular fibres—composed of collagen. These fibres are very thin and form the framework of many soft organs such as the spleen and pancreas.

The relative quantity of various kinds of cells, fibres and matrix is found to differ greatly from one body region to another. The classification of connective tissue proper is thus difficult and inexact. The connective tissue proper is separated into two forms according to the relative density of fibres. These forms are termed loose connective tissue and dense connective tissue. In many regions of the body these two forms may overlap.

Loose connective tissue is continuous throughout the body and consists of a loose network of collagen and elastic fibres. It is located beneath the skin, around blood vessels, organs and nerves. The intercellular matrix is mainly composed of a gel, which forms a predominate proportion of the interstitial fluid. The loose spongy arrangement of collagen and elastic fibres is used to support body structures. This spongy nature enables this tissue to hold water. Most of the cells of loose connective tissue are involved in the repair and defence of body tissue. Examples of these cells are:

1. fibroblasts—formation of collagen following injury,
2. macrophages (histocytes)—phagocytic cells,
3. plasma cells—produce antibodies,
4. mast cells—produce histamine and heparin,
5. white blood cells—various functions (see 20.1),
6. adipose cells—fat storage.

Adipose tissue is a specialized form of loose connective tissue which is involved in the production and storage of fat and thus energy. As fat is a poor conductor of heat, the layers of fat under the skin prevent excessive heat loss. Adipose tissue also supports and physically protects various organs by being arranged as protective pads around these organs. About ten per cent of the body weight of an average person is fat which represents approximately forty days of reserve energy.

Dense connective tissue can be subdivided according to the arrangement of the fibres:

1. Dense irregular connective tissue is composed of the same cell types as loose connective tissue. However, these fibres are interwoven in an irregular dense framework between small amounts of intercellular matrix. This tissue forms the capsule that protects and supports organs. It is also located in the dermis and constitutes the sheaths of nerves and tendons.
2. Dense regular connective tissue consists of fibres arranged in a regular dense framework with little intercellular matrix. The orderly arrangement of the fibres is indicative of the mechanical requirements of the tissue. Several forms of arrangement and therefore function exist in the body. EXAMPLES: Ligaments—flexible, strong cords connecting bones at joints which consist of parallel bundles of elastic and collagen fibres. Tendons—strong cords connecting bones to muscles which consist of parallel bundles of collagen fibres.

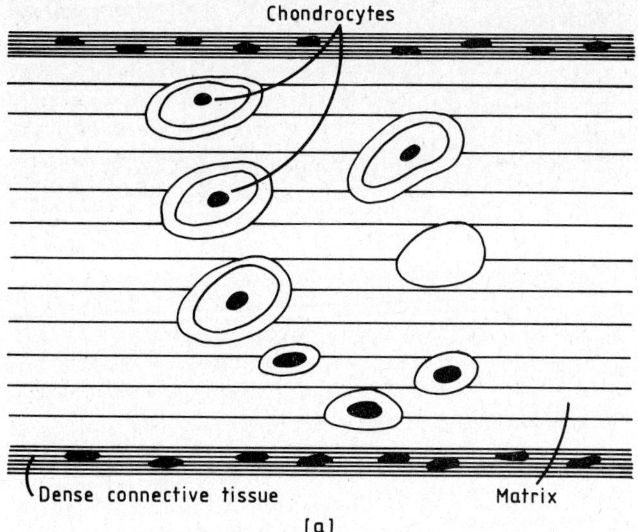

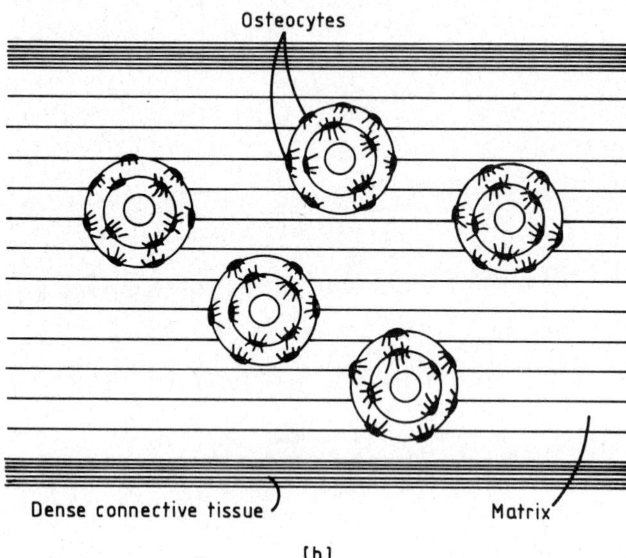

Figure 19.8 (a) Cartilage (hyaline) tissue
(b) bone tissue.

Cartilage

Cartilage consists of a gel-like matrix enclosed in a layer of dense connective tissue. The considerable degree of stiffness of this matrix enables cartilage to be used as a specialized supportive tissue. Embedded within the matrix are large quantities of fine collagen and elastic fibres and specialized cells called chondrocytes (Figure 19.8a). These cells are responsible for the production of collagen and the constituents of the intercellular matrix. This tissue lacks blood vessels, and thus the chondrocytes rely upon the efficient diffusion of nutrients and toxic waste products through the intercellular matrix to the blood vessels in the surrounding dense connective tissue.

The firmness and flexibility of cartilage varies according to the proportion of fibres embedded in it. These variations result in cartilage being located in a wide range of locations in the body. EXAMPLES: Large tubes of the respiratory tract, discs between vertebrae, menisci in the knee joint, articulating surfaces of bones, foetal skeleton.

Bone Tissue

Bone tissue differs from other forms of connective tissue since it contains calcified extracellular components. The calcified matrix confers the hard and unyielding nature to bone that makes it an ideal tissue to act as a skeleton. The rigid property of boney tissue enables bone to be used as a lever for mechanical movement.

Embedded within the calcified matrix are bone cells or osteocytes (Figure 19.8b). These cells perform the specialized function of producing the intercellular material. When diminished levels of calcium and magnesium occur in the blood, some of the matrix is broken down to release the required quantities of calcium and magnesium.

Bone is subdivided into one of two forms according to the nature of the matrix. Spongy bone contains relatively large spaces within the matrix. These spaces are occupied by bone marrow, the site of blood cell production (see 20.1). Compact bone appears as a solid mass with microscopic sized spaces. The proportion of spongy and compact boney tissue in a bone depends upon its function. A bone such as the femur requires great strength to support the body weight and is thus composed of a higher proportion of compact bone than the skull bones.

Bone has the ability to replace itself throughout the life of a person. This ability is shown in the healing of fractures.

APPLICATION: FRACTURES OF THE BONE

A fracture is a break in a bone. It results from a bone being unable to resist an excessive force being applied to it. It is important for the fragments of bone to be returned and then maintained in their pre-injury position. Failure to do this can lead to the healing of the fracture in an abnormal position and/or delayed healing. The process of fracture repair can take several months. Various immobilization methods such as traction and plaster are used to maintain the fragments in a relatively fixed position.

Fracture repair involves the following steps:
1. broken blood vessels at the fracture site form a clot,
2. dense connective tissue forms a collar around the gap between the bones,
3. the collar helps to stabilize the fragments as its surrounding boney regions are gradually replaced with new bone; at this stage the healing bone is not strong enough to bear weight,
4. the region inside the collar is joined and then the thickened collar region is remodelled; the complete healing process has resulted in the fractured bone being similar to its pre-fracture structure.

VASCULAR TISSUE

Vascular tissue or blood is composed of the liquid intercellular matrix known as plasma and is discussed in Chapter 20.

TISSUES AND HOMEOSTASIS

Tissues are combination of many cells with a similar structure and function. These similarities enable tissues to serve a variety of roles in maintaining body homeostasis such as:
1. Movement—muscles produce actions that assist the movement of substances throughout the body. EXAMPLE: Cardiac muscle.
2. Transportation—blood acts as a medium for the transportation of substances to different regions of the body.
3. Filtration—epithelial tissue acts as a filter for the movement of substances from the blood to cells.
4. Absorption—epithelial tissue acts as a selective barrier for the movement of substances into the body. EXAMPLE: Walls of the small intestine.
5. Secretion—glandular tissue secretes specific substances for the regulation of other cells. EXAMPLE: Endocrine glands secrete hormones.
6. Protection—loose connective tissue protects organs from both physical damage and infection.

19.3 Organs

An organ consists of several tissues that act in a unique, coordinated functional manner. EXAMPLE: The kidney consists of dense and loose connective tissue, epithelial tissue, vascular tissue, nervous tissue and muscle tissue. Each of these tissues individually contributes to the total function of the kidney. Muscle tissue in the arterioles regulates the arteriole diameter and thus flow of blood into and out of the kidney.

The unique specialized properties of different organs enables organs to play an important role in body homeostasis. EXAMPLE: The kidney has the ability to act as a filter for the body. It is thus able to regulate the removal and retention of important body substances such as water and electrolytes. It also acts as a site for the removal of toxic cellular waste materials.

19.4 Systems

The combined action of several organs is required for the effective function of many organs. EXAMPLE: The ability of the kidneys to function relies upon a mechanism for the physical excretion of kidney filtrate from the body. The kidney along with the ureters, urinary bladder and urethra form the urinary system. This system has the ability to remove water, electrolytes, waste products and other substances from the body.

A system relies upon the efficient functioning of all its constituent structures. EXAMPLE: Any factor that blocks the urinary tract will result in a diminished effectiveness of the entire urinary system.

Summary

The body consists of three levels of cellular organization:
1. Tissues are combinations of many cells that have a similar structure and function. The four main

tissue forms are epithelial tissue, muscular tissue, nervous tissue and connective tissue.

2. Organs consist of several tissues that act in a unique coordinated functional manner. EXAMPLES: Kidney, heart, liver.

3. Systems are composed of a group of interacting organs that operate as a functional unit. EXAMPLES: Endocrine and nervous system.

Cells are categorized into a particular type of tissue according to their structure and function. The four types of tissue exhibit specific characteristics.

Epithelial tissue consists of cells that form protective layers. Being packed closely together, these cells act as a barrier between the body and the external environment. These cells also line vessels, tubes and ducts within the body in order to prevent leakage. EXAMPLES: Cells lining the respiratory tract, blood vessels. Epithelial cells that can secrete substances are known as gland cells and exist as single secretory structures such as mucous membrane cells. Gland cells can also exist as multicellular structures that either secrete substances into ducts (exocrine glands) or blood vessels (endocrine glands).

Muscular tissue consists of cells that can contract. These cells are responsible for locomotion and for the movements of various body structures. This tissue can be divided into smooth muscle, skeletal muscle and cardiac muscle. Smooth muscle is the contractile tissue of all internal structures, apart from the heart which consists of cardiac muscle. Both smooth and cardiac muscle can contract automatically or be controlled by the autonomic nervous system. Skeletal muscle is controlled by the somatic nervous system. Skeletal muscle is attached to the skeleton and is used in locomotion, posture, facial expression and any other similar coordinated action.

Nervous tissue is composed of neurons and neuroglia. The neurons rapidly conduct information in the form of nerve impulses whereas the neuroglia support, protect and assist the function of neurons. A neuron consists of a cell body and projections from the cell body. The short projections receive information and are called dendrites, whereas a long projection or axon carries information away from the cell body. Axons form synapses with other nerves and neuromuscular junctions with muscles. These functional connections conduct information via a mechanism referred to as neurotransmission.

Connective tissue consists of cells, fibres and extensive amounts of intercellular material or matrix. This tissue binds, supports and protects other body tissue. The four main forms of connective tissue are connective tissue proper, cartilage, bone and vascular tissue. Each of these forms differs according to the intercellular matrix.

Chapter 20

Blood

Objectives

At the completion of this chapter the student should be able to:
1. list the constituents of blood,
2. describe the functions of the various blood constituents,
3. explain the effects of abnormal concentrations of these constituents,
4. explain the mechanisms of oxygen and carbon dioxide uptake, transport and release,
5. relate blood gas measurements to abnormal body function.

Blood is a body fluid that comprises about eight per cent of the total body weight. Blood is composed of a solution and a colloid of various substances which is called plasma, and a suspension of cells, the most predominate of which are red blood cells (erythrocytes).

Blood plays a vital role in the function of complex multicellular organisms such as humans. It is the medium of transportation of nutrients and chemical regulators to the cell, and for removal of cellular excretions. Blood also plays an important role in the defence of the body against foreign microorganisms (see 21.2). (The normal values of blood constituents are listed in the Appendix.)

20.1 Constituents of blood

PLASMA

Plasma is the fluid component of blood. It comprises about 55 per cent of total blood volume. Plasma consists of approximately 91.5 per cent water, 7 per cent protein and 1.5 per cent a mixture of a variety of substances such as electrolytes, lipids, hormones, carbohydrates and vitamins.

Plasma Proteins

The three major groups of protein present in plasma are albumin, globulin and fibrinogen. Their combined concentration is normally in the S.I. unit range of 66 to 82 g/L (6.6 to 8.2 g per cent). All plasma proteins perform the following functions in blood:
1. Exert osmotic pressure and so play an important role in the prevention of fluid loss from the capillaries (see 15.4).
2. Act as a mobile storage medium for amino acids. Whenever a particular tissue is depleted of protein the required amino acids can be readily supplied by the rapid breakdown of plasma protein.
3. Buffer increased concentrations of acid and base. Plasma proteins constitute the greatest quantity of plasma buffers and thus play an important role in the homeostasis of blood acid-base balance (see 11.5).

Albumin is the most abundant plasma protein: with the normal range of plasma concentration being 42 to 58 g/L (4.2 to 5.8 g per cent). It accounts for over ninety per cent of the colloid osmotic pressure and thus plays the principle role in blood capillary fluid regulation.

Plasma globulin is a collective term for three types of proteins known as globular proteins. These globulins constitute about forty-four per cent of the total protein concentration of plasma. They exist in a concentration of approximately 25 g/L (2.5 g per cent). The three forms of globulin are alpha (α), beta (β) and gamma (γ). The γ globulins are produced by cells in the reticuloendothelial system and form the antibody component of the body defence system (see 21.2). The α and β globulins are produced by the liver and bind with various substances for the purpose of transportation in plasma. These substances include electrolytes, EXAMPLES: Iron and copper; hormones, EXAMPLES: thyroid hormone thyroxin; lipids and fat-soluble vitamins.

Fibrinogen exists in small quantities in the order of 3.0 g/L (0.3 g per cent). This protein is essential for the formation of clots (see 20.1).

Non-protein Nitrogens

Non-protein nitrogens (NPN) are substances that contain nitrogen but are not proteins. Some of the important NPNs are urea, creatinine and uric acid. All of these substances are waste products of nitrogen metabolism.

Urea is the main end product of protein breakdown and it occurs in normal concentrations of between 3.2 to 7.5 mmol/L (19 to 45 mg/100mL). Urea is normally removed from the body by the kidneys. An elevated plasma urea concentration can indicate a decreased efficiency in kidney function.

The urea concentration will also be raised as a result of any factor that elevates the rate of protein breakdown. EXAMPLES: Low protein intake, injury and infection.

Creatinine is a waste product formed as a result of muscle activity. Its normal concentration range is 0.06 to 0.12 mmol/L (0.68 to 1.40 mg/100mL) in males and 0.04 to 0.10 mmol/L (0.45 to 1.10 mg/100mL) in females. Creatinine concentration is increased by muscle disuse, trauma and steroid therapy. The plasma level of creatinine depends directly on the efficiency of the kidneys. As the ability of the kidneys to filter creatinine falls, due to impaired renal function, the plasma concentration of creatinine rises.

Uric acid is a nitrogen-containing compound that results from nucleic acid breakdown. The normal plasma uric acid range is 0.18 to 0.47 mmol/L (3 to 8 mg/100mL). Uric acid has a low solubility in water and thus any elevation in the concentration of uric acid can result in precipitation. The formation of crystals in the kidney can lead to renal failure. Crystal formation of uric acid in joints results in gout, in which case foods such as meat that are rich in nucleic acids should be avoided.

Electrolytes

Plasma contains a variety of electrolytes. The predominate cations are sodium, potassium, calcium and magnesium. The principle anions are chloride, bicarbonate and phosphate.

Sodium (Na^+) is the most abundant plasma electrolyte, with normal plasma concentration ranging from 137 to 149 mmol/L (137-149 meq/L). Sodium plays an important role in nerve and muscle function, the homeostasis of fluid and electrolyte balance. Abnormally low sodium concentrations are referred to as hyponatraemia and may be caused by burns, certain diuretics and excessive perspiration. The effect of hyponatraemia is muscular weakness, decreased blood pressure, tachycardia and circulatory shock. Hypernatraemia causes cellular dehydration which causes thirst. The dehydration of the brain cells results in mental confusion and coma if not treated. The net effect of hypernatraemia is water depletion, and thus reversal is achieved by giving water orally and placing the patient on a low sodium diet. Another method is the intravenous administration of five per cent glucose in water.

Potassium (K^+) occurs in concentrations of 3.8 to 5.0 mmol/L (3.8 to 5.0 meq/L) in the plasma. It is essential for the normal function of nerve and muscle, especially cardiac muscle. Potassium depletion or hypokalaemia results in arrhythmias which may progress to cardiac arrest. Care must be taken when alleviating hypokalaemia by injecting intravenous potassium since hyperkalaemia may occur. Hyperkalaemia causes disordered cardiac function that may progress to ventricular fibrillation and cardiac arrest.

Calcium (Ca^{2+}) concentrations in plasma range from 2.25 to 2.60 mmol/L (4.5 to 5.2 meq, 9.0 to 10.4 mg/100 mL). About half this calcium exists as free dissolved ions while the remainder is bound to substances such as plasma proteins. Calcium plays an important role in blood clotting. It also influences nerves and muscle cells to sodium (see 17.2). The result is an increased activity of nerves and muscles which is manifested as increased nervous activity and twitching. In severe deficits, spasms of skeletal muscles and convulsions can occur. Hypocalcaemia may also inhibit the clotting process thus protracting any bleeding. Hypercalcaemia can cause vomiting, nausea, loss of muscle tone and confusion, and if untreated it may lead to coma.

Magnesium (Mg^{2+}) exists in normal plasma concentrations of 0.80 to 1.05 mmol/L (1.6 to 2.1 meq/L, 19.5 to 25.5 mg/L). Magnesium is essential for the functioning of many enzymes, particular those relating to carbohydrate metabolism. Magnesium also plays an important role in neuromuscular function. Hypomagnesaemia or magnesium deficiency in the blood causes symptoms similar to those of hypocalcaemia, that is, twitching and tremors which may lead to convulsions. Cardiac arrhythmias may also develop. Hypermagnesaemia depresses nerve and skeletal muscle activity.

Chloride ions (Cl^-) are the most predominate anions found in plasma. Its relatively large plasma concentration of 95 to 105 mmol/L (95 to 105 meq/L) indicates that it plays an important role in the maintenance of extracellular osmotic pressure and body water homeostasis. Chloride ions are also used in acid–base regulation. A close association exists between chloride and sodium ions. This is shown by many alterations of chloride concentrations being closely linked to changes in sodium concentration. A loss of sodium ions is usually accompanied by chloride depletion. Disturbances of the gastrointestinal tract can result in hypochloraemia or a chloride deficit which is independent of sodium concentration. In such cases, an elevated bicarbonate ion concentration develops along with alkalosis.

Bicarbonate ions (HCO_3^-) occur in normal plasma concentrations of 21 to 32 mmol/L (21 to 32 meq/L). Its principal role in blood is that of acid–base regulation and the indirect transport of carbon dioxide. Its function in acid–base balance is discussed in 11.5. The effect of altered bicarbonate ion concentration is described in 11.6.

Phosphate ions (PO_4^{2-}) exist in normal plasma concentrations of 0.80 to 1.5 mmol/L (2.5 to 4.6 mg/100 mL). They play an important role in energy production, and thus depleted plasma phosphate concentrations or hypophosphataemia results in lethagy and muscle weakness. A close association exists between phosphate and calcium. This association is shown in cases of hyperphosphataemia where plasma calcium concentrations usually fall and the resulting hypocalcaemia manifestations occur. The role of phosphate as a buffer is described in 11.5.

Non-electrolytes

The main non-electrolyte present in plasma is glucose. Its normal plasma concentration following fasting is 5.0 to 6.6 mmol/L (80 to 120 mg/mL). Glucose plays a vital role in the supply of energy to cells (see 22.5). Hypoglycaemia or depleted glucose results in a diminished supply of energy to the brain. This is manifested as dizziness, faintness or lethagy which can rapidly lead to coma and if untreated, death. Hyperglycaemia or elevated blood glucose results in loss of appetite, nausea, vomiting, thirst and drowsiness which can lead to acidosis and coma if not treated. A more detailed discussion of blood glucose concentrations occurs in Chapter 22.

Lipids

Lipids are a group of organic compounds that have the general property of insolubility in water (see 23.1). The four main forms of lipid that are present in plasma are fatty acids, triglycerides (fats), phospholipids and cholesterol. Most lipid is transported in blood by combining with protein to form a water-soluble lipoprotein.

Fatty acids are straight-chained organic acids that are either saturated or unsaturated (carbons linked by double or triple bonds—see 9.2). Fatty acids are mainly transported in combination with albumin. They are the main building block of many lipids. Their normal fasting concentrations are very low, but an increased breakdown of fat will result in elevated plasma levels.

Triglycerides or fats consist of fatty acids and the alcohol glycerol. Fat that is absorbed across the small intestine is transported as an emulsion. The emulsion consists of droplets of fat surrounded by proteins. These droplets are known as chylomicrons (see 10.4). Triglyceride can also be formed in the liver. This triglyceride is transported by lipoproteins to the adipose tissue. The normal range of plasma triglyceride concentration is 0.05 to 2.30 mmol/L (4.4 to 203 mg/100 mL). Elevated triglycerides would be expected to occur following ingestion of a meal and in situations where excessive amounts of fats are being formed by the liver.

Phospholipids are lipids that contain phosphate. These are the main constituent of cell membranes. They are transported throughout the body by lipoproteins.

Cholesterol is used by the body to form steroids. It is transported mainly by lipoproteins. The accepted cholesterol concentration ranges from 2.6 to 6.5 mmol/L (100 to 250 mg/100 mL). A diet containing saturated fatty acids will elevate plasma cholesterol, whereas polyunsaturated diets tend to lower cholesterol levels.

Lipoproteins play an important role in the transport of lipids. They are mainly involved in the transport of lipids from the liver to other tissues. Various types of lipoproteins occur, and each type carries specific lipids. EXAMPLE: β lipoproteins (low density lipoprotein—LDL) carry cholesterol.

APPLICATION: ATHEROSCLEROSIS

Atherosclerosis is a disease of large and medium-sized arteries where a lipid-protein deposit decreases the diameter of the artery. The deposit is called a plaque or atheroma and is mainly composed of cholesterol combined with proteins. Raised plasma cholesterol levels and thus elevated low density lipoprotein levels have been strongly correlated with atherosclerosis.

FORMED ELEMENTS OF BLOOD

Formed elements constitute the other major constituent of blood. They consist of blood cells and cell-like bodies suspended in the plasma. The formed elements can be placed into the following categories: erythrocytes (red blood cells), leucocytes (white blood cells), platelets (thrombocytes).

The formed elements enable the blood to perform important homeostatic functions such as transporting oxygen, combating inflammation, halting haemorrhages and fighting infection.

Erythrocytes

Erythrocytes are commonly called red blood cells (RBCs). These cells lack a nucleus and have a biconcave shape (Figure 20.1). The average life span of a red blood cell is 120 days and the production of erythrocytes in the bone marrow is approximately equal to the rate of destruction. The normal concentration of erythrocytes differs in women and men, with the normal female range being 3.9 to 5.6 x 10^{12}/L (3.9 to 5.6 x 10^{6} μL/mm^{3}) and the male range being 4.5 to 6.5 x 10^{12}/L (4.5 to 6.5 x 10^{6} μl/mm^{3}). These cells represent the greatest concentration of any formed element in plasma. They influence the viscosity of blood and thus the relative ease with which blood flows through the vascular system.

Red blood cells contain large quantities of haemoglobin which carries most of the oxygen in the blood. Haemoglobin therefore plays a vital role in body homeostasis, since normal cell function relies upon oxygen. Haemoglobin is also involved in the transport of carbon dioxide away from the tissues.

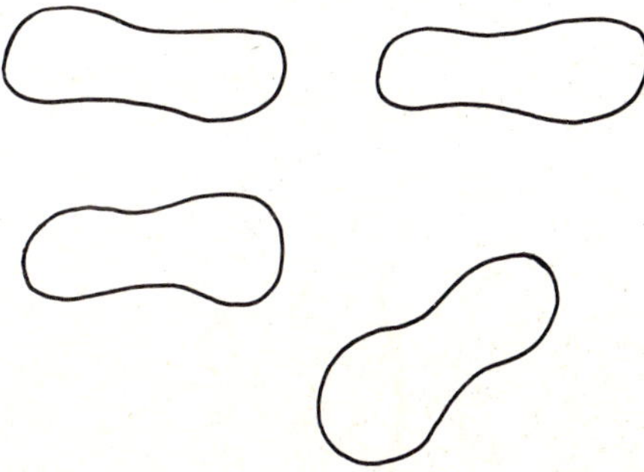

Figure 20.1 Erythrocytes showing their biconcave shape.

Related to this transport is the role of haemoglobin as a buffer in blood pH homeostasis (see 11.5). The relationship between haemoglobin and the transport of oxygen is discussed later in this chapter.

The production of erythrocytes is termed erythropoiesis. Red blood cell production depends upon the supply of raw materials such as iron, sufficient manufacturing sites (bone marrow) and correct production and release of red blood cells. Alterations to any of these factors can result in a relative deficiency of haemoglobin in the blood. Such a deficiency is called anaemia.

APPLICATION: ANAEMIA

Anaemia is a relative deficiency of haemoglobin that may result from either alterations in the production of red blood cells, excessive destruction in the blood or losses as a result of haemorrhage. The net effect of any of these alterations is a imbalance between the rate of formation and destruction of red blood cells. Anaemia results in a lack of oxygen being supplied to the tissues. This leads to fatigue, weakness and malaise and pallor. Anaemia can be classified according to the location of the abnormality (Table 20.1).

Leucocytes

Leucocytes or white blood cells (WBCs) occur in a variety of forms and perform functions associated with the body defence system. The total leucocyte concentration is less than that of the red blood cells, with the normal range being 4.0 to 11.0 x 10^9/L (4000 to 11,000/mm^3). Leucocytes originate from either lymphatic tissue or red bone marrow. The lymphatic tissue produces white blood cells that contain a non-granular cytoplasm and a regular-shaped nucleus. Red bone marrow produces some of these cells, but is mainly involved in the production of cells that have a granular cytoplasm and irregular-shaped nucleus. The latter cells exist in three forms: neutrophils (polymorphs), eosinophils and basophils. The non-granular white blood cells are classified as either lymphocytes or monocytes (Figure 20.2).

Neutrophils are phagocytic cells that respond to inflammation and tissue destruction. They account for sixty to seventy per cent of the total white blood cell count. They increase in number in response to infection and inflammation.

Eosinophils account for about two to four per cent of the leucocyte count. They appear to be involved in combating substances responsible for causing allergies. A high eosinophil count usually indicates an allergic condition, although some parasites such as hookworms can cause an elevated count.

Basophils account for only half to one per cent of the leucocyte count. Little is understood about the function of basophils although it is believed that they are involved in allergic reactions.

Lymphocytes comprise a group of different cell types, all of which appear similar when observed under a microscope. Most lymphocytes appear to play specialized roles in the immunity of the body to foreign agents. Lymphocytes represent twenty to twenty five per cent of a normal total white cell count. A comprehensive discussion of their functions in immunity is given in Chapter 21.

Table 20.1 Classification of the main forms of anaemia

Form of anaemia	Mechanism affected	Cause
Aplastic	Sites of production	Destruction or inhibition of red bone marrow caused by toxins and medications
Blood loss (haemorrhagic)	Demand exceeds production capability	Acute or chronic haemorrhage
Folic acid deficiency	Maturation of red blood cell altered	Dietary insufficiency, malabsorption and increased requirements during pregnancy
Haemolytic	Faulty red blood cell production and excessive breakdown	Faulty red blood cell membranes or enzymes results in a marked decrease in red blood cell life, agents such as parasites and toxins
Iron deficiency	Relative deficiency of raw material	Inadequate diet, malabsorption, increased body utilization such as occurs with pregnancy
Pernicious (vitamin B_{12} deficiency)	Maturation of red blood cell altered	Lack of gastric intrinsic factor (assists absorption), dietary insufficiency and disease in the intestine
Sickle cell	Faulty haemoglobin production leads to short life span, demand exceeds supply	Hereditary
Thalassaemia	Faulty haemoglobin production	Hereditary

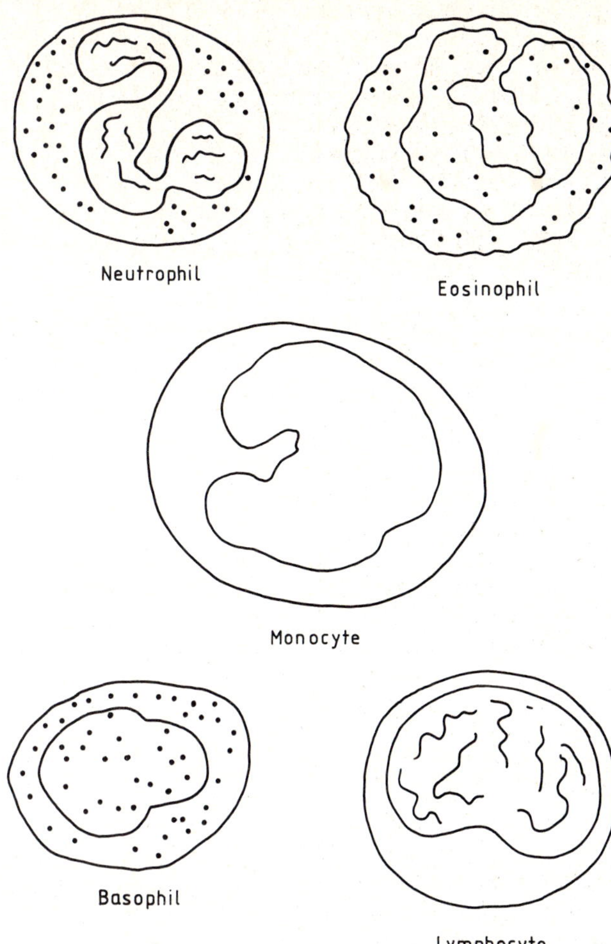

Figure 20.2 Different types of leucocytes.

Monocytes represent three to eight per cent of the total white cell count. Monocytes are able to move into tissues and phagocytose foreign material and are thus referred to as the scavenger cells of the body. They appear to clean up the remains left by the phagocytic neutrophils.

Generally, white blood cells remove debris and foreign material from the extracellular fluid. It would be reasonable to think that the more leucocytes present, the greater is our body's ability to fight infection and inflammation. However, in cases where an uncontrolled production of white blood cells occurs, the result is often infection!

APPLICATION: LEUKAEMIA

Leukaemia is a blood disease in which an uncontrolled production of white blood cells occurs. A marked increase of a specific white cell type occurs in the blood. EXAMPLE: Lymphocytes. Many of these white blood cells are abnormal, immature and lack the ability to perform their normal functions. Even though the number of white blood cells is elevated, the presence of large quantities of abnormal and immature cells restricts the body defence mechanisms, thus resulting in an increased susceptibility to infection. The relative proportion of other white blood cell forms decreases due to the cancerous spread of the abnormal leucocyte into their regions of formation.

The leukaemia tissues form new cells at a rate that places a drain on the nutrients circulating in the body. In particular, amino acids and vitamins are depleted, thus leading to a rapid deterioration of body proteins. This eventually results in death of the patient.

Platelets

Platelets (thrombocytes) are tiny disc-shaped cells that lack a nucleus. Platelets have a life span of about one week and thus a continual turnover of platelets is necessary for the maintenance of normal concentrations of 150 to 400 x 10^9/L (150,000 to 400,000/mm^3).

Platelets initiate the clotting mechanism by their ability to:

1. release vasoconstrictor substances,
2. provide a temporary plug capable of controlling flow in small vessels,
3. contribute to the coagulation mechanism which eventuates in a blood clot.

APPLICATION: HAEMOSTASIS

Haemostasis refers to the arrest of bleeding. Platelets are the key factor in the haemostatic mechanism. Following vascular injury, platelets near the damaged region adhere to the exposed collagen fibres of the vascular well. The platelets immediately release substances that cause the aggregation of new platelets and constriction of the vessel. The constriction performs the function of limiting the flow of blood through the damaged region. The aggregation of more platelets at the traumatized site results in the formation of a platelet plug. This platelet plug covers the vascular break. These platelets assist the formation of the enzyme thrombin from its inactive form prothrombin. Thrombin formation also requires the presence of various plasma coagulation factors and calcium.

Thrombin converts the plasma protein fibrinogen into long fibrous threads called fibrin. This is achieved by the process of coagulation, that is, the formation of suspended particles by combining colloid particles (see 10.4). These suspended particles coat the plug, giving it strength to resist being dislodged by the moving blood. The platelet plug-fibrin complex is referred to as a blood clot.

20.2 Function of haemoglobin

A knowledge of how oxygen enters the blood, is transported through the vessels and then released in the region of tissue, is important in gaining an understanding of the homeostasis of tissue oxygen supply. The homeostasis of oxygen is an essential component of body homeostasis since nearly all of the energy produced in the body requires the presence of oxygen.

MECHANISM OF OXYGEN TRANSPORT

The non-polar nature of oxygen results in oxygen having a low solubility in water and thus plasma. The amount of dissolved oxygen normally transported in blood is about 0.5 mL O_2/100mL blood or 15 mL/min. The body tissues usually require 20 mL O_2/100mL. Clearly an alternative method of transport is required for oxygen to travel from the lungs to the tissues. Nearly ninety-seven per cent of oxygen transported in the blood is carried by the substance haemoglobin which is the main constituent of red blood cells.

Haemoglobin is a pigment that consists of the protein globin wrapped around four haeme groups. A haeme group is a chemical structure that has iron located in its centre. An oxygen molecule is carried in the crevice between each haeme group and the globin molecule. Since four haeme groups exist, one haemoglobin molecule can therefore carry four oxygen molecules.

An important feature of haemoglobin is its ability to combine with oxygen in a loose, weak association. This enables haemoglobin to readily pick up oxygen and easily release oxygen without requiring the use of large amounts of energy. This ability is essential if a carrier is to pick up oxygen from the lungs and then readily release it in the tissue capillaries.

Each haeme group does not become oxygenated simultaneously. The binding of an oxygen molecule to one haeme group enhances further binding of oxygen with the remaining haeme groups. The effect of this binding mechanism is the sequential uptake and release of oxygen. This sequential binding plays an important role in the mechanism of oxygen release at the tissue capillaries.

The ability of oxygen to be sequentially combined with haemoglobin and sequentially released is determined by the difference in the partial pressure of oxygen (PO_2) in the blood. The high PO_2 in the pulmonary capillaries results in large quantities of oxygen binding with haemoglobin, whereas the low PO_2 in the tissue capillaries causes oxygen to be released from the haemoglobin.

UPTAKE OF OXYGEN

Oxygen uptake is dependant upon a high PO_2 in the blood. The site of oxygen intake is the lungs, therefore a high PO_2 is required in the pulmonary capillaries if oxygen is to be transported to the tissues. Diffusion of the oxygen from the lungs to the

Figure 20.3 The oxygen-haemoglobin dissociation curve.

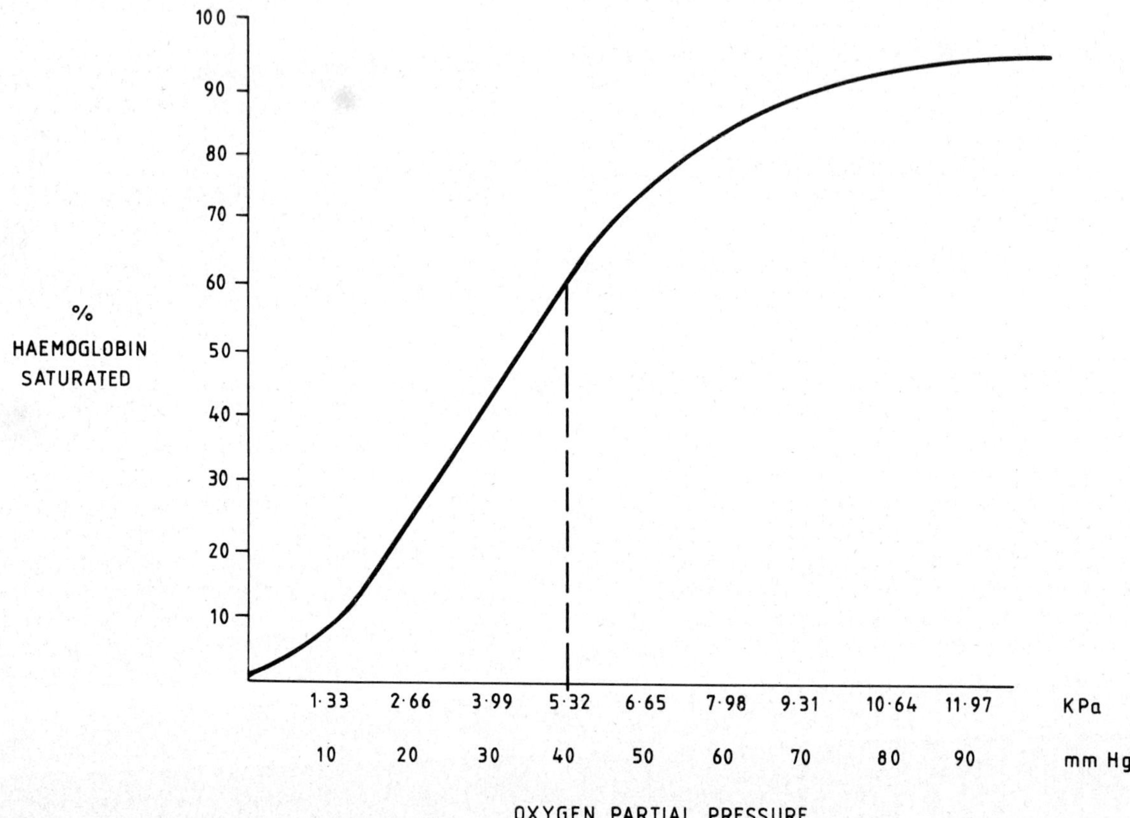

blood is dependant upon the relative difference in the concentration of oxygen between the lungs and the blood.

In 13.2 a relationship was described between PO_2 in the alveoli of the lungs and the pulmonary capillaries. The normal diffusion of oxygen from an alveolar PO_2 of 13.3 kPa (100 mm Hg) to the blood results in blood leaving the lungs with a PO_2 of 12.6 kPa (95 mm Hg). At this blood PO_2 all the haemoglobin is carrying its maximum amount of oxygen. The haemoglobin is said to be saturated.

Any condition which causes less oxygen to enter the pulmonary capillaries, thus a lower PO_2, can result in haemoglobin not being saturated. The relative uptake of oxygen is reflected in the degree of saturation of arterial blood, that is its PO_2. The measurement of blood arterial PO_2 is thus an important diagnostic aid in detecting conditions that result in the low uptake of oxygen in the blood.

RELEASE OF OXYGEN

Haemoglobin releases oxygen in accordance with tissue needs. Consumption of oxygen by tissue is reflected by a lower tissue capillary PO_2. A decrease in this capillary PO_2 results in the release of oxygen from the haemoglobin molecule. The normal relationship between oxygen release and capillary PO_2 is shown in Figure 20.3. The response of haemoglobin to PO_2 is often referred to as the oxygen–haemoglobin dissociation curve. The percentage of haemoglobin saturated is the proportion of oxygen bound to haemoglobin compared with the total amount of haemoglobin present. The PO_2 of tissue capillaries under normal conditions is 5.32 kPa (40 mm Hg). Under these conditions, twenty-five per cent of the total amount of haemoglobin-bound oxygen is released and seventy-five per cent is therefore retained. This retention acts as a readily accessible reserve of oxygen.

When oxygen utilization of tissues increases rapidly such as during strenuous exercise, the PO_2 in the interstitial fluid can decrease to levels of 2.0 kPa (15 mm Hg). These low PO_2 levels cause the release of most of the oxygen reserve. In the curve, a level of 2.0 kPa results in haemoglobin only being able to retain approximately twenty per cent oxygen saturation.

The result of normal levels of oxygen dissociation from haemoglobin is a venous PO_2 of 5.32 kPa (40 mm Hg). On arrival at the lungs, the difference in partial pressure results in a diffusion of oxygen from the alveoli to the blood and subsequently combination with haemoglobin.

20.3 Carbon dioxide transport

UPTAKE OF CARBON DIOXIDE

The PCO_2 of arterial blood is 5.3 kPa (40 mm Hg). The production of carbon dioxide by cells results in a tissue PCO_2 of 6.0 kPa (45 mm Hg). This relative difference in carbon dioxide concentration results in carbon dioxide diffusing from the tissues into the blood in the capillaries until the concentrations are equal. The venous blood thus has a PCO_2 of 6.0 kPa (45 mm Hg).

Carbon dioxide is carried in the blood by three mechanisms: dissolved in plasma (seven per cent), combined with haemoglobin (twenty-three per cent), converted into the bicarbonate ion (seventy per cent).

The non-polar nature of carbon dioxide results in a low solubility in plasma and therefore only about seven per cent of carbon dioxide is transported by this mechanism.

Approximately twenty-three per cent of carbon dioxide combines with haemoglobin. The weak bonding of carbon dioxide to haemoglobin is reversible, that is, the carbon dioxide can readily be released when a decrease in dissolved carbon dioxide concentration occurs.

CO_2 + haemoglobin $\rightleftarrows$ haemoglobinCO_2(Eqn 20.1)

Seventy per cent of carbon dioxide combines with water to form bicarbonate ions (Eqn 20.2).

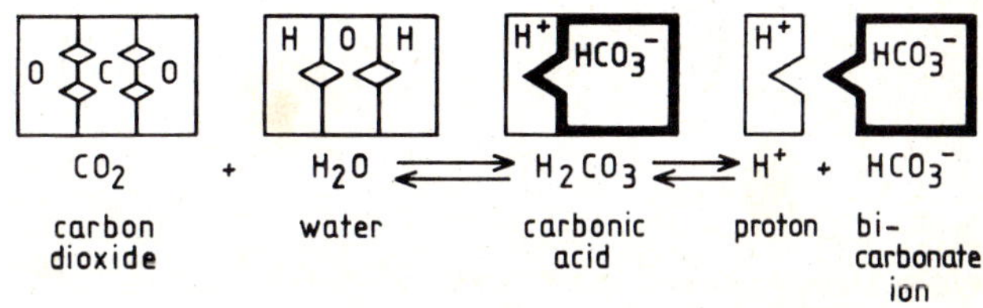

The reaction occurs in the plasma and in the cytoplasm of the red blood cells. Only about ten per cent of bicarbonate is formed in the plasma. In the red blood cells an enzyme called carbonic anhydrase effects the rapid formation of carbonic acid from carbon dioxide and water. About ninety-nine per cent of the carbonic acid dissociates into hydrogen and bicarbonate ions. The high concentration of bicarbonate ions in the cell results in large quantities diffusing from the cell into the plasma. This diffusion results in a loss of negative ions by the red blood cells. To maintain the negative charge, chloride ions move from the plasma into the cytoplasm in a process known as the chloride shift.

The hydrogen ions or acid formed as a result of carbonic acid dissociation are buffered by haemoglobin (see 11.5). Some hydrogen ions, however, remain in their free state, thus conferring greater acidity. For this reason the pH of venous blood (7.36) is lower than arterial blood pH (7.40).

RELEASE OF CARBON DIOXIDE

When blood arrives at the lungs, a difference in concentration exists between the venous carbon dioxide concentration and the alveolar carbon dioxide concentration. The alveoli have a PCO_2 of 5.3 kPa (40 mm Hg) compared with the returning blood carbon dioxide of 6.0 kPa (45 mm Hg). This difference results in carbon dioxide diffusing from the blood into the lungs. As carbon dioxide is removed from the plasma, an imbalance develops between the proportion of carbon dioxide, carbonic acid and bicarbonate ion. The lowered carbon dioxide levels results in a proportionate change of bicarbonate and hydrogen ions to carbonic acid resulting in decreased bicarbonate ions, and a proportionate conversion of carbonic acid to carbon dioxide and water.

The continued removal of carbon dioxide by diffusion thus results in a sustained conversion of bicarbonate ion to carbonic acid and thence carbon dioxide, that is, a reversal of Eqn 20.2.

The carbon dioxide combined with haemoglobin is removed as a result of the higher affinity of haemoglobin for oxygen and the relatively larger concentration of oxygen.

20.4 Use of blood gases

The determination of blood PO_2 and PCO_2 can be a useful aid in detecting abnormal respiratory function, blood function and metabolic function. Blood gas studies are also useful as an assessment of the therapy.

Blood gases are usually obtained from arterial blood. It is essential that the sample of blood is free of air bubbles since this results in false readings.

HYPOXIA

Hypoxia is a diminished amount of oxygen in the tissues. The general cause of this is hypoxaemia, that is, an insufficient concentration of oxygen in the blood. The effect of hypoxaemia ranges from decreased mental, emotional and muscular stability at 6.65 kPa (50 mm Hg) to unconsciousness at 4.3 kPa (32 mm Hg) and circulatory failure at 3.7 kPa (28 mm Hg). Hypoxaemia can occur with either a normal PCO_2 or an elevated PCO_2.

Some of the common causes of hypoxia are:

1. Decreased alveolar PO_2 due to either inhalation of low oxygen gases or inadequate respiratory function. EXAMPLES: Impaired lung movement and chronic obstructive airways disease (COAD).
2. Altered diffusion rates between the alveoli and pulmonary capillaries. EXAMPLES: Emphysema, pulmonary oedema, pneumonia and pulmonary fibrosis.
3. Decreased transport capacity of blood. EXAMPLES: Anaemia, inadequate circulation.
4. Abnormal tissue metabolism. EXAMPLES: Severe oedema, and toxicity due to substances such as cyanide.

ARTERIAL PCO_2 LEVELS

The normal arterial PCO_2 ranges from 4.8 to 5.8 kPa (36 to 44 mm Hg). Alterations in the PCO_2 can assist in the identification of the cause of hypoxia. An elevated PCO_2 suggests that either the lungs have difficulty in eliminating carbon dioxide or that abnormal metabolism results in the excessive production of carbon dioxide.

Factors that alter the ability of oxygen to diffuse from the alveoli to the pulmonary capillaries can have a normal PCO_2. This is due to carbon dioxide having a better ability to diffuse from the blood to the alveoli than oxygen diffusing from the alveoli to the pulmonary capillaries.

Measurements of arterial PCO_2 can also be used in conjunction with acid–base measurements to determine acid–base status. A description of the various forms of acid–base imbalance is given in 11.6.

Compensated metabolic acidosis has low PCO_2 levels, whereas uncompensated metabolic acidosis has normal PCO_2. In compensated and uncompensated respiratory acidosis an elevated PCO_2 occurs. Both uncompensated and compensated respiratory alkalosis cause a decreased PCO_2, whereas metabolic alkalosis usually has a normal PCO_2.

Summary

Blood consists of a fluid called plasma which contains a variety of suspended particles referred to as formed elements. Plasma consists of:

1. colloids in the form of proteins and lipoproteins,
2. aqueous solution of electrolytes and non-electrolytes.

The three major plasma protein groups are albumin, globulin and fibrinogen. They influence blood osmotic pressure and are important blood buffers. These proteins also act as a mobile storage medium for amino acids. Globulin proteins play important roles in body defence mechanisms (immunoglobulins) and in the transportation of substances in plasma. Fibrinogen is an essential component of the blood-clotting mechanism.

Non-protein nitrogens refer to substances in the plasma that contain nitrogen but are not proteins such as urea, creatinine and uric acid.

Electrolytes play important roles in body homeostasis. EXAMPLE: Sodium in fluid and electrolyte balance. Abnormal concentrations of electrolytes can greatly influence the health of a patient. EXAMPLE: Low levels of potassium causes cardiac arrhythmias which may progress to a cardiac arrest.

The main non-electrolyte present in blood is glucose which is vital for brain function.

Most lipids are transported in the blood as either lipid–protein complexes (lipoproteins) or as droplets known as chylomicrons.

Suspended particles in blood are referred to as formed elements. These formed elements are:

1. Erythrocytes or red blood cells are small haemoglobin-containing cells that constitute the greater proportion of formed elements. The haemoglobin carries ninety-seven per cent of oxygen transported in the blood, and thus a lack of haemoglobin or anaemia results in a lack of oxygen being supplied to tissues.
2. Leucocytes or white blood cells occur in a variety of forms and perform functions associated with the body defence system. Leucocytes that contain granules in their cytoplasm are called granular leucocytes. They exist as either neutrophils, eosinophils or basophils. Non-granular leucocytes exist as either lymphocytes or monocytes. Leucocytes either phagocytose unwanted material or they produce substances that are involved in body defence mechanisms. Leukaemia is the uncontrolled production of white blood cells.
3. Platelets or thrombocytes are tiny cells that lack a nucleus. They form an important part of the clotting mechanism.

The transport of oxygen in the blood is dependent upon the ability of haemoglobin to pick it up in the pulmonary capillaries and release it according to tissue demand in the tissue capillaries. Haemoglobin requires a lower PO_2 than the alveoli to pick-up oxygen and a greater PO_2 than the tissues to release oxygen. Any factor that alters the difference in PO_2 can alter the amount of oxygen transported in the blood.

Carbon dioxide is transported in the blood by being dissolved in plasma (seven per cent), combining with haemoglobin (twenty-three per cent), or converting into bicarbonate ion (seventy per cent).

The greater the concentration of carbon dioxide in the blood, the greater is the quantity of carbonic acid, bicarbonate ion and hydrogen ion. Release of carbon dioxide to a lower PCO_2 in the alveoli results in lowered carbonic acid, bicarbonate ion and hydrogen ion.

The determination of blood PO_2 and PCO_2 can indicate respiratory and cardiovascular function. Hypoxia or decreased amounts of oxygen in the tissues can result from inadequate respiratory function, restricted diffusion between alveoli and pulmonary capillaries, inadequate transport capacity in blood or abnormal tissue metabolism. Elevated PCO_2 suggests that lungs have difficulty in removing carbon dioxide or that abnormal metabolism results in excessive carbon dioxide production. Altered blood PCO_2 also influences the acid–base balance of blood.

Chapter 21

Cellular defence mechanisms

Objectives

At the completion of this chapter the student should be able to:
1. describe the different body defence mechanisms,
2. describe the stages of an immune response,
3. explain the differences between humoral and cell mediated immunity,
4. explain the mechanisms of actively acquired immunity,
5. explain how allergies, autoiummune diseases and tissue rejection relate to body defence mechanisms.

An important feature of body homeostasis is the need for homeostatic mechanisms to protect the internal environment from harmful foreign matter. Several homeostatic mechanisms known as immune systems serve the function of body defence, that is they attempt to give the body immunity against harmful foreign matter.

The harmful foreign material can be one of the following: chemical, EXAMPLE: Foreign proteins; bacteria, EXAMPLE: *Staphylococci*; viruses, EXAMPLE: *Herpes*; other microorganisms, EXAMPLE: *Rickettsia*.

There are several homeostatic mechanisms involved in the protection and removal of harmful foreign matter. These mechanisms involve the epithelial tissue, blood, reticuloendothelial system and the lymphatic system. The body's defence mechanisms can be separated into those restricting the entrance of foreign particles into the body, that is, external defence mechanisms, and mechanisms that neutralize and remove foreign particles that have entered the body, that is, internal defence mechanisms.

21.1 External defence mechanisms

The tissues and fluid of the body are surrounded by a wall of epithelial cells. These cells form the basis of the external defence mechanism.

EPITHELIAL TISSUE

The epithelial tissue lines the skin, respiratory tract, gastrointestinal tract and urogenital tract. This wall acts as the first line of defence. Breaking the skin surface often results in an invasion of microorganisms into the body. The epithelial cells form part of the body's non-specific defence mechanism which is unable to differentiate between different types of foreign matter.

There are several ways that epithelial tissue prevents foreign organisms from entering the extracellular regions of the body: arrangement of epithelial cells, secretions from epithelial cells, surface structures of epithelial cells.

Arrangement of Epithelial Tissue

Epithelial tissue consists of cells that are closely packed together. The tight packing results in little intercellular material between the cells (see 19.2). No intercellular material is present at the surface of these cells since the plasma membranes of adjacent cells are joined together. This joining of membranes effectively seals the surface of epithelial tissue. Patients suffering from burns are extremely susceptible to infection since the damaged epithelial barrier can no longer perform its protective function.

Secretions from Epithelial Cells

Some epithelial cells have the ability to secrete substances, that is, they act as glands. The type of secretion varies according to the location of the epithelial tissue.

The epithelial cells of the skin produce keratin which confers a tough and waterproof property to the surface of the skin. In addition, the skin contains sweat glands that secrete substances that are capable of killing bacteria. In the ear, some epithelial cells produce the protective substance, wax.

Some of the epithelial cells lining the respiratory tract, gastrointestinal tract, urogenital tract and conjuctivae secrete mucus. The mucus traps any foreign matter that has entered these regions of the body. The trapped material is retained by the mucus until the material is mechanically removed.

The enzyme lysozyme is secreted from epithelial cells that produce nasal mucus, tears, saliva and skin secretions. This enzyme is able to breakdown the cell walls of many bacteria.

A large proportion of foreign material that enters

the stomach is destroyed by the highly acidic nature of the stomach's hydrochloric acid. This acid is secreted by epithelial cells lining the stomach wall.

Surface Structures of Epithelial Cells

Nasal hairs prevent the entry of large particles into the respiratory tract. The smaller particles that pass through the hairs are entrapped by the mucus and moved towards the exterior by ciliated cells. The foreign material can be removed from the respiratory tract by the mechanical action of coughing and sneezing. Foreign material that enters the gastrointestinal tract is entrapped by mucus and moved to the highly acidic stomach by the action of epithelial cells containing cilia.

21.2 Internal defence mechanisms

The internal defence mechanism is brought into action when the first line of defence has been penetrated by foreign particles. Internal defence mechanisms can be divided into specific and non-specific.

NON-SPECIFIC DEFENCE MECHANISMS

Non-specific defence mechanisms can be thought of as the second line of defence. These defence mechanisms are unable to differentiate between different foreign objects.

Blood

Blood plays an important role in the defence of our body against foreign particles. The white blood cells, and in particular the neutrophils, have the ability to phagocytose foreign matter. The presence of foreign particles provides a stimulus which attracts white blood cells to the area of invasion. Phagocytosis of the foreign matter ensues, with the likely result being rapid removal and destruction of the foreign object.

The fibrous nature of fibrin enables it to assist white blood cells by entangling the foreign material for sufficient time to enable the phagocytes to act.

Reticuloendothelial System

The reticuloendothelial system is a functional system that consists of highly phagocytic cells broadly distributed throughout the body. These cells are located in such tissues as bone marrow, lymphoid tissue, loose connective tissue, spleen, lung and nervous tissue. The majority of these cells appear to originate from monocytes. The reticulo-endothelial cells have the ability to engulf and then digest foreign material. The ability to digest this material is due to the large array of powerful enzymes located in their lysosomes.

Lympatic System

The lymphatic system is composed of the spleen, tonsils and thymus, tissue called lymph nodes, and a network of lymphatic vessels. These vessels drain fluid from the interstitial regions and return it to the vascular system. This fluid is referred to as lymph. The lymph passes through lymph nodes on its journey from the interstitial regions to the vascular system.

Lymph nodes contain phagocytic cells of the reticuloendothelial system located in a mesh of fibrous tissue. The lymph nodes function as filters of unwanted particles. The fibrous mesh traps or slows down particles, after which the phagocytic cells engulf and break down the unwanted particle.

The nodes are strategically located so as to pick up microorganisms that get past the main entry barriers such as the respiratory tract.

Inflammatory Response

The inflammatory response to harmful foreign material involves the combined action of non-specific defence mechanisms and the repair of any damaged regions. Generally inflammation is the response of tissues to infection or injury. It involves the destruction and walling-off of the injurious agent along with any cells that the agent may have destroyed. Following neutralization of the agent, a complex series of events is initiated which results in the healing of any damaged tissue.

The symptoms of inflammation are heat, redness, swelling and pain. Loss of function may occur but it is dependent upon the magnitude and site of inflammation.

The process of inflammation can be summarized in stages:

1. Stage 1—the small blood vessels at the site of infection or injury dilate which results in an increased flow of blood carrying white blood cells to the area. The dilation is probably due to histamine. The increased blood flow is observed as a redness. Histamine also increases the permeability of blood vessels which leads to the movement of fluid and phagocytic white blood cells into the interstitial fluid. The increased flow of fluid into the interstitial region results in a localized oedema.
2. Stage 2—the accumulated white blood cells, principally neutrophils and monocytes, phagocytose the deleterious agent, resulting in the destruction or weakening of the foreign intruders. The end-product of this action is the formation of a fluid containing cell debris which is known as pus.
3. Stage 3—the increased permeability of small blood vessels results in the leakage of fibrinogen to the interstitial fluid. Conversion of the fibrinogen into fibrin results in the formation of a clot (see 20.1). The clot isolates the infected area thus helping prevent the spread of the infection.

In recent years, a group of substances known as prostaglandins have been found to be released in inflammation.

Prostaglandins are a group of lipid substances that are extremely potent physiological agents (see Section 23.2). It appears that prostaglandins contribute to the genesis of fever, pain, vasodilation and increased permeability of blood vessels. The production of many prostaglandins is inhibited by aspirin-like drugs. These drugs reduce the signs and symptoms of inflammation.

SPECIFIC DEFENCE MECHANISMS

The entry of a foreign particle into the circulation or tissues is usually successfully countered by the non-specific defence mechanism. A backup defence mechanism or third line of defence is sometimes necessary to overcome the foreign invasion. EXAMPLE: Some microorganisms have the capacity to multiply inside phagocytes and also destroy phagocytes. This third line of defence involves mechanisms that can recognize and overcome specific foreign particles. These mechanisms are thus referred to as specific defence mechanisms.

Foreign particles can be toxins, foreign proteins, microorganisms or tissue cells from another organism. All of these particles contain a region that will react with the specific defence mechanisms, that is, they contain a region which is known as an antigen. The antigen evokes an immune response which involves the production of special lymphocytes. Each different form of antigen has its own unique shape. This property is used by the body to identify and combat a particular type of antigen.

The basis of specific defence mechanisms is the ability of special cells to form moulds of the shape of the active part of the antigen. This is followed by the production of cells and gamma globulin proteins that contain the mould of the antigen's active region. These moulds can combine with the antigen, effectively deactivating the antigens effect on the body. The deactivated antigen is then destroyed by phagocytic activity.

The first step in specific defence mechanisms is the processing of an antigen by a macrophage. The macrophage phagocytoses the antigen and then the active part of the antigen is "processed" into a form that will be recognized by certain lymphocytes. The processed antigen is moved to the surface of the macrophage where contact with recognition lymphocytes occurs. These recognition lymphocytes respond to the processed antigen by initiating either of the following two mechanisms:

1. humoral immunity involving the production of antibodies,
2. cell mediated immunity involving the production of specialized lymphocytes that contain antigen receptors on their plasma membranes.

Humoral Immunity

In humoral immunity, the contact of a recognition lymphocyte with its specific antigen results in the following process.

The recognition lymphocyte rapidly divides resulting in the formation of large numbers of cells known as B-lymphocytes or B-cells. This process of B-cell proliferation usually occurs in the lymph nodes near the site of antigen entry. This rapid increase in cells results in the lymph node becoming grossly enlarged. The enlarged lymph node may become tender due to pressure and stretching on the surrounding tissue.

After five to seven days, some of the B-lymphocytes have been transformed into plasma cells. These latter cells increase in number and remain in the lymph nodes secreting gamma globulin proteins that have a complementary shape to the specific antigen. These proteins are called antibodies or immunoglobulins (Ig) and they are secreted into the lymph fluid. These immunoglobulins circulate throughout the body and combine with their complementary antigen thereby deactivating the antigen.

The different forms of immunoglobulin are identified by an alphabetic letter. Presently, there are five different forms or classes of immunoglobulins which are represented as IgA, IgD, IgE, IgG and IgM. All immunoglobulins of a particular class contain the same basic structure apart from the terminal region of the polypeptide. This variable terminal region is the site where an immunoglobulin binds with an antigen.

During the stage of transformation of B-lymphocytes into plasma cells some of the B-lymphocytes remain unchanged, that is, they remain as recognition lymphocytes for that particular antigen. These cells act as memory cells for any subsequent invasion of the antigen. These memory cells continually circulate throughout the lymphatic and cardiovascular systems in search of their specific antigen. They tend to accumulate in the spleen and peripheral lymph nodes.

Following the first encounter with an antigen, an individual is said to be sensitized to that antigen. The individual possesses a certain quantity of circulating immunoglobulin along with a reserve of memory cells. These cells are able to rapidly multiply and be converted into plasma cells in a short period of time. The individual is thus able to fight the antigen soon after it enters into the tissues and blood.

The immunity acquired as a result of exposure to an antigen can be acquired in several different ways:

1. When a person is exposed to the antigen through some natural process such as infection, the individual is said to have actively acquired immunity.
2. If a small sample of a antigen is introduced by artificial means into a person, the result is also active acquired immunity. This preparation of antigen is known as a vaccine. The result of vaccination is a small initial response to the antigen, and more importantly the production of antibodies and recognition lymphocytes which immunize the individual against any subsequent natural invasion of the antigen.

3. When an individual acquires antibodies and recognition lymphocytes without any exposure to the antigen the immunity is called passively acquired immunity. This form of immunity can also be acquired by either natural or artificial means. A foetus acquires immunity by obtaining the mothers antibodies through the placenta. These maternal antibodies provide a temporary immunity for the first few months of the life of the newborn infant. This natural passive immunity can be mimicked by injecting serum containing antibodies. These immunoglobulins offer a temporary immunity against the antigens of diseases such as hepatitis and measles.

Cell Mediated Immunity

Cell mediated immunity involves the production of specialized lymphocytes that mature in the thymus. These cells are called T-cells and they interact with antigens by means of specific structures on the surface of the cell referred to as receptors. Since T-cells combine directly with the antigen, the immune response is referred to as cell mediated immunity.

The T-cells perform their function at the lymph nodes, spleen, tonsils and blood where they constitute seventy to eighty per cent of all lymphocytes. The ability of T-lymphocytes to recirculate throughout the body enables these cells to recognize and bind with antigens shortly after infection.

When a specific shaped antigen encounters a recognition T-cell with a complementary shaped receptor, it binds to that T-cell. The T-cell is now activated or sensitized. Following sensitization, the T-cell rapidly divides to form enormous numbers of identical T-cells that circulate in the blood and lymph and lodge in the lymph nodes. The receptors on these cells combine with the complementary shaped antigen thus deactivating the antigen. The T-cell then secretes substances that digest the antigen.

Following either active or passive exposure to an antigen, the body acquires immunity by retaining an increased number of specific recognition and sensitized T-cells for very long periods of time. These cells are quick to respond to any further exposure to the antigen.

21.3 Disorders of the immune system

Disorders of the immune system vary from slight discomfort to life threatening situations.

ALLERGY

Allergy or hypersensitivity results from a person being oversensitive to a particular antigen which is referred to as the allergen. A wide range of allergens have been found to cause oversensitive responses in individuals. EXAMPLES: Penicillins, foods, dust, pollen and cosmetics.

The first exposure to the allergen results in the production of antibodies. Unlike a normal humoral response, in some individuals these antibodies attach to granular cells known as mast cells located in connective tissue.

The second exposure to the allergen results in the antibodies combining with the antigens. This reaction causes the breakdown of the granules in the mast cells which results in the release of large quantities of histamine.

The release of histamine results in inflammation, constriction of vessels with smooth muscle walls such as the respiratory and blood vessels, oedema due to histamine causing an increase in blood vessel permeability.

APPLICATION: ANAPHYLACTIC SHOCK

Anaphylactic shock is a rapidly developed allergic reaction which may result from either a systemic or local reaction. The excessive release of histamine can result in difficulty of breathing due to the constricted air passages. In addition, the increased blood vessel permeability results in widespread oedema which leads to a lowered blood volume. Failure to reverse the histamine effect by administering antihistamines or epinephrine (adrenaline) can result in death.

AUTOIMMUNITY

The immune system normally differentiates between foreign particles and its own particles. Autoimmunity is the formation of an antigen within the body and the subsequent formation and reaction of antibodies to that antigen. The antigen can be formed as a result of a non-antigenic foreign particle or altered cell function producing abnormal proteins and cells. A wide range of diseases have been attributed to autoimmunity. EXAMPLES: Addison's disease, rheumatoid arthritis, pernicious anaemia and myasthenia gravis.

21.4 Tissue rejection

Transplantation of an organ or tissue requires the suppression of the immune response.

Immunosuppressive drugs are required since the body recognizes the transplanted cells as foreign and will produce an immune response to attack the donor cells. This response is referred to as tissue rejection. The administration of immunosuppressive drugs results in the patient having an increased susceptibility to infection as the body's defence mechanisms are suppressed. The tissue rejection response and thus the dependence on immunosuppressive drugs can be minimized by using transplanted tissue that is similar to the recipient's tissue.

Summary

The defence of the body against harmful foreign matter involves the following mechanisms:

1. External defence mechanisms which rely on the arrangement, secretions or specialized surface structures of epithelial cells. The function of epithelial tissue in defence mechanisms is to restrict the entrance of harmful substances. It is regarded as the first line of defence and is non-specific, that is, it does not differentiate between different invaders.
2. Internal defence mechanisms exist in two forms, those that are non-specific and those that attach to a particular invader.

The internal non-specific defence mechanism involves white blood cells that phagocytose foreign matter. The series of steps in a non-specific response of tissues to an invasion is referred to as inflammation. An inflammatory response involves inflammation and the repair of any damaged region.

Specific defence mechanisms can be subdivided into humoral immunity involving the production of immunoglobulins (antibodies) and cell mediated immunity which involves the production of specialized lymphocytes that selectively attach different forms of foreign matter. Specific defence mechanisms rely on the recognition of a particular antigen (toxins, foreign proteins, microorganisms, and other organisms cells) and the subsequent immune response. The immune response involves the formation of large numbers of B-lymphocytes (B-cells) and T-lymphocytes (T-cells) from recognition lymphocytes.

The B-cells are transformed into plasma cells. The plasma cells produce large quantities of specific complementary immunoglobulin (antibody) which circulates throughout the body and neutralizes the antigen. The T-cells contain a specific receptor that combines with, and neutralizes, its complementary antigen.

Following the immune response the body contains recognition B and T-cells that will rapidly multiply after contact with their specific antigen. These sensitized cells enable the body to rapidly counter any further invasions of a particular antigen. The individual is said to have acquired an immunity to that antigen.

Actively acquired immunity can result from a natural process such as infection. It can also be artificially acquired by using a vaccine that initiates a small immune response to that particular antigen.

Passively acquired immunity involves the acquisition of immunoglobulins and recognition lymphocytes by natural or artificial means. EXAMPLE: The foetus naturally acquires immunity, whereas injecting serum containing immunoglobulins is an artificial acquisition.

Disorders of the immune system account for a wide range of health problems that vary from slight discomfort to life-threatening situations. An allergy results from a person being oversensitive to a particular antigen. The result of this hypersensitivity is the release of large quantities of histamine. In severe allergic reactions anaphylactic shock can occur. Another disorder is the misrecognition of some of the body's own cells as foreign. This results in autoimmune diseases. Similarly, the recognition of transplanted material as foreign, initiates an immune response referred to as tissue rejection.

Unit Nine
Digestion and metabolism of chemicals

In this unit, the digestion and metabolism of biochemicals is described and related to body function. The influence of vitamins and minerals in metabolism is also discussed.

Chapter 22

Carbohydrates

Objectives

At the completion of this chapter the student should be able to:
1. define the terms metabolism, anabolism and catabolism,
2. describe the general structure of carbohydrates,
3. .describe the digestion and absorption of carbohydrates,
4. list the main sites of glucose utilization,
5. describe the process of ATP production from glucose catabolism,
6. explain the difference between aerobic and anaerobic catabolism of glucose.

22.1 Metabolism — general concepts

Metabolism is the mechanism which produces homeostasis of body chemical processes. Normal body metabolism involves a balance between the manufacture or synthesis of chemicals and their breakdown. The synthesis of any biochemical is often referred to as anabolism whereas their breakdown into simpler structures is called catabolism. The balance between anabolism and catabolism varies according to the demands of individual tissues.

Body homeostasis is dependent upon metabolism since the rate of chemical synthesis and breakdown controls body function. The maintenance of normal metabolism is dependent on an adequate supply of nutrients and the biochemical regulators known as enzymes and hormones.

The principal metabolic organ is the liver. It metabolizes many of the body's biochemicals and influences the metabolism of other tissue by acting as a storage and distribution centre.

Most molecules in food are used by cells either to provide energy, maintain the body's structure and its growth, or maintain the body's homeostatic mechanisms. Three principal groups of molecules perform these functions. These groups are known as carbohydrates, lipids and proteins. Each group can be converted into the other groups.

22.2 Carbohydrate structure and function

STRUCTURE

Carbohydrates contain carbon, hydrogen and oxygen. The basic structural unit of a carbohydrate is the sugar molecule. Most sugar molecules important to humans contain six carbon atoms and are commonly referred to as hexoses. There are sixteen different hexose isomers with the molecular formula $C_6H_{12}O_6$ (see 8.5). Body function relies almost entirely on the glucose isomer of $C_6H_{12}O_6$ (Figure 22.1).

A large variety of carbohydrates occur in nature. A classification system has been devised to identify individual carbohydrates. Firstly, only carbohydrates contain the suffix -ose. The presence of -ose identifies the substance as a carbohydrate.

The following method is used to determine the quantity of sugar units in a particular carbohydrate: mono—one sugar unit, di—two sugar units, poly—many sugar units.

The suffix saccharide, which simply means sugar base, is added to the above prefixes:
monosaccharide, EXAMPLES: Glucose, galactose,
disaccharide, EXAMPLES: Maltose, lactose,
polysaccharide, EXAMPLES: Starch, glycogen.

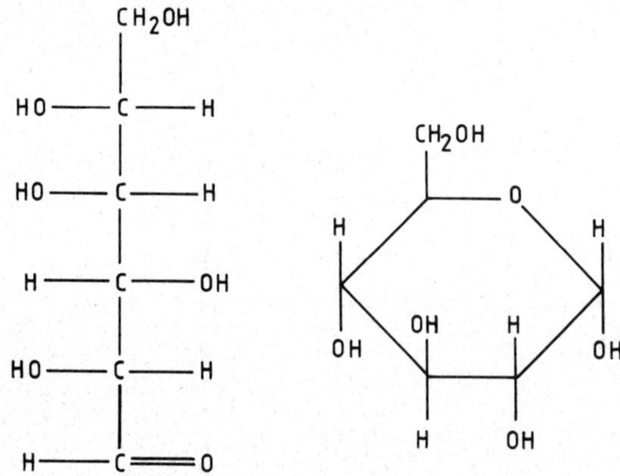

Figure 22.1 The structure of glucose can be represented as either a straight-chain molecule or as a cyclic molecule.

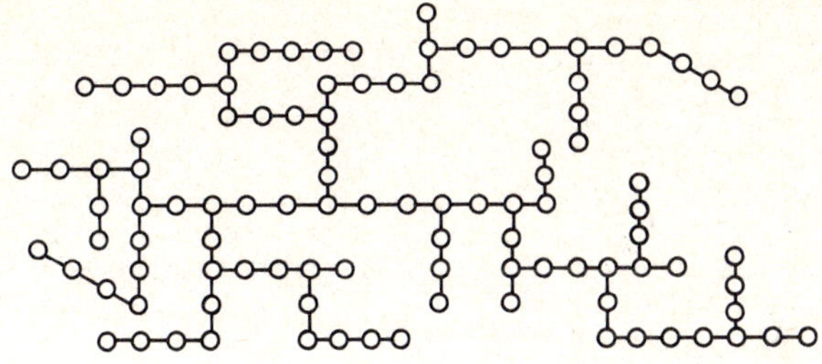

Figure 22.2 The structure of glycogen showing the chains and branches of glucose molecules.

FUNCTION

The principal function of carbohydrates is the supply of stored energy in the form of glucose to tissues such as nerve tissue. A temporary store of body carbohydrate exists in the form of the polysaccharide, glycogen (Figure 22.2). Glycogen is composed of large chains of glucose molecules. When blood glucose decreases, sufficient glucose is released from the glycogen to return the blood levels to normal.

22.3 Digestion and absorption of carbohydrate

Carbohydrate digestion and absorption involves the breakdown of polysaccharides to monosaccharides, since only monosaccharides are able to pass through the plasma membrane at the site of absorption.

DIGESTION

Mouth and Oesophagus
Saliva contains an enzyme (salivary amylase) that acts on polysaccharides to produce small quantities of the disaccharide maltose. The enzyme simply acts by slicing two glucose units at a time off the polysaccharide chain. The action of the enzyme continues in each bolus (ball) of food as it passes down the oesophagus.

Stomach
The highly acidic gastric contents (pH 1.2 to 3.0) inactivates the function of the enzyme. The stomach acts as a temporary reservoir for the carbohydrates and controls their rate of entry into the small intestine. This action of the stomach assists in maximum digestion of carbohydrate in the small intestine.

Small Intestine
The small intestine is the site of most carbohydrate digestion. The pancreas secretes an enzyme (pancreatic amylase) into the first part of the small intestine or duodenum via the pancreatic duct. This enzyme has a similar action to the salivary enzyme, that is, forms maltose from polysaccharides.

The small intestine is permeable to monosaccharides and impermeable to disaccharides. This membrane property thus necessitates the breakdown of maltose to monosaccharides for the absorption of carbohydrates into the body.

Two other disaccharides occur in the small intestine: sucrose which is common table sugar, and lactose which is derived from milk. Both these disaccharides are only digested in the small intestine.

The final step in carbohydrate digestion involves the breakdown of maltose, sucrose and lactose to their constituent monosaccharides. The specific enzymes involved in their digestion are released from the surface of the intestinal cells (maltase for maltose, sucrase for sucrose and lactase for lactose).

ABSORPTION

Absorption of monosaccharides across the small intestine is enhanced by its folded structure. The villi and microvilli markedly increase the surface area of available intestinal membrane (see 14.2). The enzymes responsible for breaking down disaccharides are located in the microvilli. The primary monosaccharides absorbed across the membrane are glucose, fructose and galactose. These are derived from maltose—two glucose molecules, sucrose—one glucose and one fructose, lactose—one glucose and one galactose.

There appear to be two processes by which monosaccharides are absorbed across the intestinal wall: facilitated diffusion which requires a concentration gradient, and active transport which is independent of concentration gradients but requires energy.

Both these transport mechanisms involve the use of carriers since monosaccharides are too large for diffusion through membrane pores and they are relatively insoluble in the lipid component of the plasma membrane (see 17.2 and 17.3). All carriers have a specific preference for a particular substance, with other similar structured substances being transported at slower rates. The carriers involved in the transport of monosaccharides appear to be glucose carriers since glucose is transported at a faster rate than galactose or fructose.

APPLICATION: LACTOSE INTOLERANCE

A deficiency in the enzyme lactase results in an inability to digest and absorb milk. The undigested lactose causes:

1. an osmotic imbalance between the intestinal

contents and the body which results in additional water entering the intestinal contents,
2. bacteria break down the lactose to lactic acid which causes an increase in intestinal mobility.

Both these factors cause pain and diarrhoea. Elimination of milk from the diet usually overcomes the pain and diarrhoea.

22.4 Sites of absorbed carbohydrate

Following ingestion of a carbohydrate, the glucose concentration in blood rises for about thirty minutes. After three hours the blood glucose concentration has usually returned to the original pre-meal level. Monosaccharides absorbed into the body are mainly used for energy supply. The intermittent dietary supply of sugar necessitates a reserve of body sugar. Without this reserve, an insufficient supply of glucose would occur between meals. The carbohydrate reservoir in humans is in the form of glycogen which is stored in the liver and skeletal muscle. Once this reservoir is full, any excess sugar absorbed into the body is converted into fat. The carbohydrate is thus indirectly stored in another form of energy.

When entering the body, monosaccharides are either utilized by tissue immediately or stored as glycogen or fat. The liver, brain, blood cells, skeletal muscle and adipose tissues are the most important utilizers of glucose. The uptake and use of glucose is controlled by hormones (see 28.2).

LIVER

When glucose, fructose and galactose enter the blood stream after absorption, they are transported via the portal vein to the liver. The liver is the major metabolic organ and its blood supply is closely linked to that of the small intestine.

The liver is a major site of monosaccharide uptake. Most of the monosaccharides that have been taken up by the liver are converted into glycogen. Since glycogen is only composed of chains of glucose molecules, fructose and galactose must be converted into glucose before being stored as glycogen.

BRAIN

The brain requires a continual supply of glucose. It uses about 120 grams of glucose per day which is approximately twenty-five per cent of the energy requirement of a normal resting adult. An interruption to the supply of glucose will result in brain damage. Following a meal, a proportion of the glucose will be immediately utilized by the brain. During periods between meals, the brain receives glucose by the breakdown of glycogen and the subsequent release of glucose into the blood. The brain relies upon liver glycogen since no glycogen is formed in the brain.

APPLICATION: GLYCOGEN STORAGE DISEASE

The reliance of the brain on liver glycogen is shown in glycogen storage disease. Children with glycogen storage disease are able to form glycogen but are unable to reconvert glycogen into glucose. Without constant feeding, brain damage will occur as a result of decreased blood glucose levels (hypoglycaemia) during intervals between meals.

BLOOD CELLS

Blood cells have a small but constant demand for glucose. The total demand on the body is approximately a quarter that of the brain.

SKELETAL MUSCLE

Skeletal muscle can either use glucose immediately for energy or it can store glucose as muscle glycogen. The amount of glucose used by skeletal muscle varies according to the work performed by the muscles. At rest or with mild exertion this muscle only utilizes about a quarter of the quantity used by the brain. During times of vigorous exercise the demand for glucose is elevated. The increased glucose is supplied by the increased blood supply to the muscle, along with the conversion of glycogen to glucose. Muscle glycogen represents the major reserve of body carbohydrate.

ADIPOSE TISSUE

Adipose tissue converts excess sugar into fat. Excess sugar will only exist when both the immediate tissue demands are satisfied and the glycogen reserves are replenished to their normal levels.

KIDNEYS

The kidneys actively reabsorb glucose from the kidney filtrate back into the blood. When the blood sugar level exceeds this capacity, glucose appears in the urine and its presence is referred to as glycosuria.

The digestion and absorption of large quantities of carbohydrate can result in blood sugar concentrations that exceed the ability of adipose tissue to convert glucose into fat. The inability of adipose tissue to utilize all the excess glucose results in greater concentrations of sugar being filtered by the kidneys. If the blood sugar level exceeds 10 mmol/L, the threshold or T_m is exceeded and glycosuria results (see 17.3). This form of glycosuria is referred to as temporary or alimentary glycosuria. It is not considered to be pathological in the short term.

In diabetes mellitus the ability of tissues to remove sugar from the blood is decreased. This can result in blood sugar levels far in excess of what would be expected for a particular carbohydrate intake. The elevated blood sugar level will often exceed the T_m following meals, and thus glycosuria occurs more frequently. The measurement of glycosuria and a knowledge of the carbohydrate intake can be used to diagnose diabetes mellitus.

APPLICATION: TESTING FOR GLYCOSURIA

Simple, quick methods of testing urine for sugar exist. Tablets, powder and strips of paper can all have a specific colour changing reagent impregnated into them. EXAMPLES: Clinitest tablets and clinistix paper strips. The presence of glucose causes the reagent to change colour. The intensity of colour change is indicative of the amount of glucose present. A semi-qualitative measurement can be performed by comparing the colour intensity with the colour comparison chart that accompanies the product. If glucose is found, a blood sugar test may be ordered to confirm the observation.

22.5 Production of energy

The production of energy in the body involves the removal of hydrogen atoms, mainly from carbohydrates and fats. The energy stores in the hydrogen bonds originates from the sun in the form of electromagnetic radiation (see 5.4).

The energy released as a result of removing hydrogen from a molecule is not in a form that can be immediately used by the cell. The released energy is mainly used to form the substance ATP (Adenosine TriPhosphate).

ATP

This triphosphate molecule contains large amounts of energy stored in the third phosphate bond. When ATP is broken down to the two phosphate molecule Adenosine DiPhosphate (ADP) and phosphate, large amounts of energy are released:

$$\text{ATP} \rightarrow \text{ADP} + \text{phosphate} + \text{energy}$$

This equation is reversible, that is, the formation of ATP requires ADP, phosphate and energy.

The function of ATP is to provide energy for membrane transport, the synthesis of chemical compounds, and mechanical work. Cells are unable to operate on glucose or fat without them being converted into ATP. ANALOGY: An electric stove relies on electricity for its supply of energy. The stove is unable to operate on coal, or water in a dam.

GLYCOLYSIS

The chemical pathway for glucose breakdown or catobolism terminates in carbon dioxide plus hydrogen ions. The hydrogen ions combine with oxygen to form the neutral substance water.

Glycolysis is the first step in glucose catabolism. It results in the formation of two three-carbon molecules called pyruvic acid from each glucose molecule. Glycolysis occurs in the cytoplasmic matrix of cells.

The series of reactions in glycolysis results in a net loss of four of the twelve glucose hydrogen atoms. The loss of these hydrogen atoms results in the release of energy that is used to combine ADP and phosphate to form the high energy molecule ATP.

When oxygen is present, a net production of six ATP molecules occurs from the breakdown of one glucose molecule. The absence of oxygen, that is, anaerobic conditions, results in the net production of only two ATP molecules. The formation of the pyruvic acid molecules is independent of oxygen, however the net production of energy in glycolysis varies according to the presence or absence of oxygen.

BREAKDOWN OF PYRUVIC ACID

Krebs Cycle

In the presence of oxygen, that is, aerobic conditions, pyruvic acid is converted to carbon dioxide and water. The initial step in this biochemical pathway is the formation of a substance called acetyl CoA. Pyruvic acid loses a carbon and forms a two carbon group (acetyl) that attaches to a substance referred to as coenzyme A. The coenzyme A or CoA acts as a carrier and assists enzyme function. It transports the acetyl groups to a location on the mitochondrial membrane where the acetyl group is broken down in the presence of oxygen. The catabolism of the acetyl group is referred to as either the Krebs cycle, citric acid cycle or the tricarboxylic cycle (Figure 22.3).

The acetyl group enters the cycle by combining with a four carbon molecule (oxaloacetic acid) on the inner mitochondrial membrane. A series of reactions then occurs which result in the removal of the two carbon atoms from the acetyl group as carbon dioxide and also their associated hydrogen atoms. When the carbon, hydrogen and oxygen are removed from the acetyl group, the original oxaloacetic acid molecule is left. This molecule is then free to combine with another acetyl group and thus oxaloacetic acid in effect acts as a carrier.

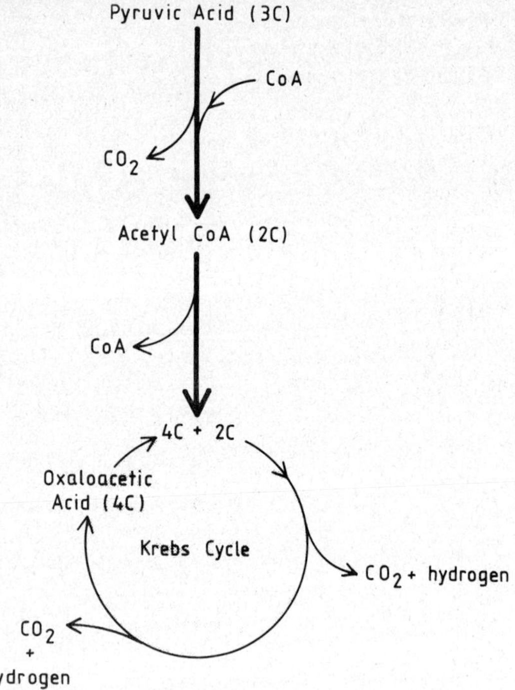

Figure 22.3 The catabolism of pyruvic acid and its relationship to the Krebs cycle.

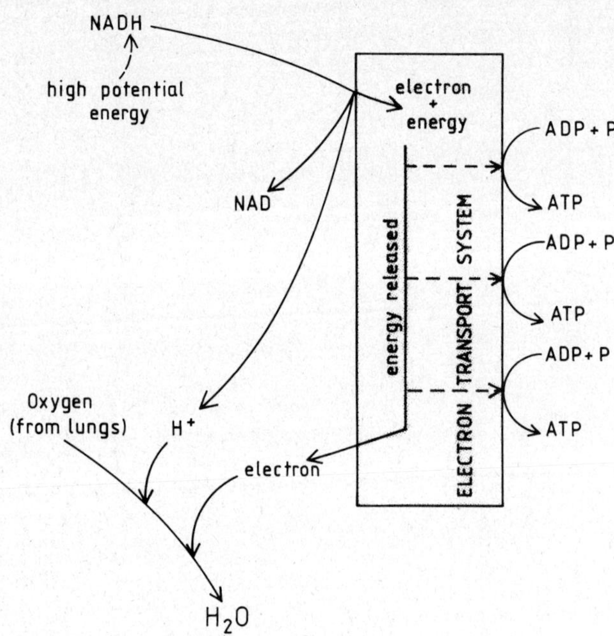

Figure 22.4 The relationship between hydrogen, the electron transport system and ATP formation.

Formation of ATP

The removal of hydrogen in the Krebs cycle releases large amounts of energy. The released energy is used to combine hydrogen with a coenzyme known as NAD. The NAD combines with two hydrogen atoms. This results in the formation of $NADH_2$:

$$NAD + H_2 \rightarrow NADH_2$$

The effect of the formation of $NADH_2$ is the transfer of the potential energy from a carbon–hydrogen–oxygen molecule to a $NADH_2$ molecule. At this stage the energy is unable to be used by the cell as cells need the energy to be supplied by ATP molecules.

The $NADH_2$ is broken down, resulting in the formation of NAD and hydrogen. Some of the released energy is used to form a molecule of ATP. A series of reactions then follows which results in the release of hydrogen ions and the transfer of their electrons and the energy along a chain of substances known as cytochromes. As the electrons are passed along the cytochrome chain or electron transport system the energy required to bind the electrons decreases. This results in a corresponding increase in the amount of energy available for ATP production (Figure 22.4).

The end products of the removal of hydrogen from the Krebs cycle are hydrogen ions (acid), electrons and energy. A substance is required to neutralize the hydrogen ions otherwise the body would be unable to cope with the excessive quantities of acid. These protons are combined with electrons and oxygen to produce the neutral substance water. *The principal reason for the body requiring oxygen is the neutralization of protons and the removal of electrons formed as a result of energy production.* The energy released during the electron transfer system is used to form ATP.

The transfer of the energy to form ATP is oxidative phosphorylation. This process requires the bonding of a phosphate to ADP. The energy stored in the third phosphate bond is able to be utilized by the cell. Approximately ninety per cent of all ATP is produced by this mechanism.

The formation of ATP is dependent upon the release of energy in the electron transport system. A breakdown in the electron transport system results in no ATP being formed by oxidative phosphorylation. EXAMPLE: Cyanide blocks ATP production by preventing the electrons being transferred to oxygen. This leads to a blockage of the electron transport system and thus oxidative phosphorylation.

The function of hydrogen and oxygen have been shown to be closely related. Without either, homeostasis would be impossible, and therefore death would occur. In the body, the formation of ATP from carbohydrate catabolism is an important mechanism for the functioning of organs such as the brain. The brain thus relies upon an adequate supply of glucose and oxygen. Without either, it will fail to function.

The overall mechanism of ATP production due to aerobic glucose catabolism is summarized in Figure 22.5. Each turn of the Krebs cycle results in the net production of fifteen ATP molecules. Since two acetyl groups are formed from glucose, a net

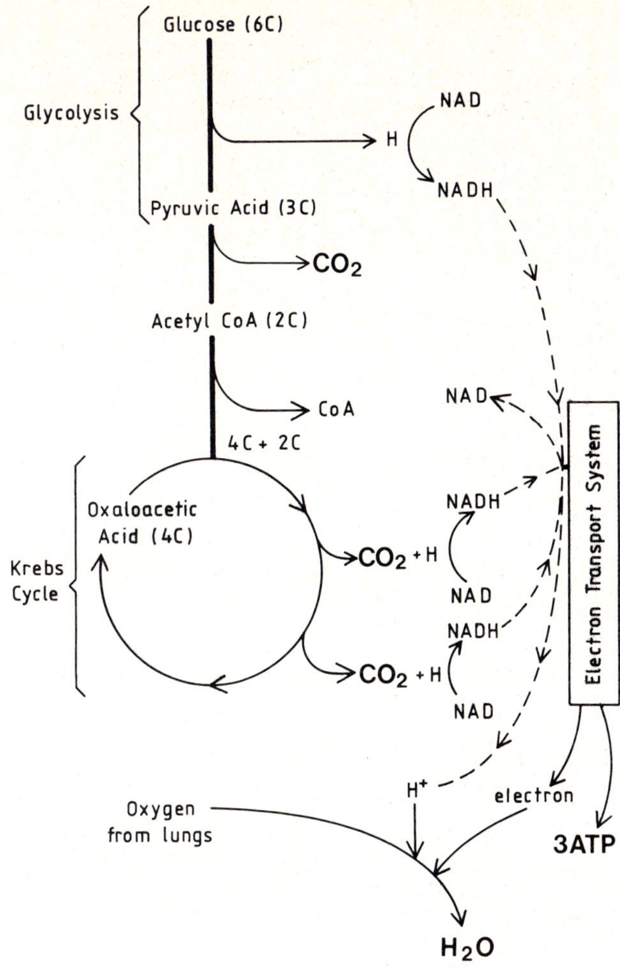

Figure 22.5 Aerobic catabolism of glucose to carbon dioxide, water and ATP.

production of thirty ATP occurs in the Krebs cycle from one glucose molecule. When combined with the six ATP produced during glycolysis a net quantity of thirty-six ATP molecules results from aerobic catabolism of glucose.

Lactic Acid Formation

In the absence of oxygen, that is, anaerobic conditions, pyruvic acid is converted into lactic acid without the formation of any additional ATP molecules. The net result of the catabolism of one glucose molecule is the formation of two molecules of the three carbon acid called lactic acid and two molecules of ATP.

Lactic acid formation can occur as a result of oxygen supply not meeting the tissue demands. EXAMPLE: The large increase in oxygen requirements in vigorous exercise can result in insufficient oxygen being supplied to the relevant muscles. Moderate amounts of lactic acid can be buffered by the body preventing acidosis. In the presence of large amounts of oxygen, lactic acid can be converted in the liver to glycogen which can be utilized at a later time by the body.

Comparison of Energy Produced During Aerobic and Anaerobic Glucose Catabolism

The net total quantity of usable energy produced in the presence of oxygen is eighteen times that produced by anaerobic glucose catabolism (Table 22.1). The supply of oxygen is essential for the efficient use of glucose. Organs such as the brain require large amounts of energy to function. Without oxygen, these organs are unable to produce sufficient energy anaerobically to meet their energy requirements.

Table 22.1 Comparison of the energy produced during aerobic and anaerobic glucose catabolism

	Aerobic (no. ATP molecules)	Anaerobic (no. ATP molecules)
Glycolysis	6	2
Lactic acid	—	—
Krebs cycle	30	—
Total	*36*	*2*

Summary

Metabolism is the mechanism which produces homeostasis of body chemical processes. Metabolism can be divided into anabolism (chemical synthesis) and catabolism (chemical breakdown).

Carbohydrates are compounds that contain carbon, hydrogen and oxygen arranged into structural units known as sugars.

The principal function of carbohydrates is the supply of stored energy in the form of glucose to cells. Glucose can be stored in the liver and muscles in the form of the polysaccharide glycogen.

Carbohydrates are partially digested in the mouth due to a salivary enzyme, and completely digested in the small intestine by the action of pancreatic and intestinal enzymes. The primary monosaccharides absorbed across the small intestine are glucose, fructose and galactose. They are absorbed by facilitated diffusion and active transport.

Following absorption, monosaccharides are transported to the liver where any depleted glycogen reserves are replaced. Since glycogen consists of only glucose molecules, fructose and galactose are converted into glucose. When the requirements of liver and other tissues are met, the remaining glucose is converted into fat and stored in adipose tissue.

Carbohydrates are an important source of energy for the body, and in particular the brain. The energy released from breaking hydrogen bonds is used in a series of reactions to form ATP from ADP and phosphate. Energy in the form of ATP is used by cells for membrane transport, synthesis of chemical compounds and mechanical work.

The release of hydrogen from carbon and oxygen and the production of ATP involves the following steps:
1. Glycolysis—the six carbon glucose is broken down to form two three-carbon pyruvic acid molecules. In the presence of oxygen (aerobic), six ATP molecules are produced from each glucose molecule. Anaerobic conditions result in only two ATP molecules per glucose molecule.
2. Formation of acetyl CoA—each pyruvic acid molecule is converted into a two carbon molecule known as acetyl CoA.
3. Krebs cycle—the two carbon acetyl group combines with a four carbon molecule, and after a series of reactions two carbon atoms are removed as carbon dioxide. The four carbon molecule is then free to combine with another two carbon acetyl group. Hydrogen is removed as a consequence of carbon dioxide removal. The removal of hydrogen releases large amounts of energy.
4. ATP production—following the release of hydrogen atoms and its associated energy, a series of reactions occurs which results in the release of hydrogen ions, transfer of electrons and energy along an electron transport system, and release of energy from the electron transport system which is used to form ATP.
5. Formation of water—oxygen combines with the hydrogen ions and electrons to form water.

In the absence of oxygen, pyruvic acid is converted into lactic acid. This is a relatively inefficient energy producing system since only two ATP molecules are produced per glucose molecule compared with thirty-six ATP produced when oxygen is present.

Chapter 23

Lipids

Objectives

At the completion of this chapter the student should be able to:
1. list and briefly describe the features of the different forms of lipid,
2. describe the process of fat digestion and absorption,
3. list the major sites of fat utilization,
4. explain how the production of energy from fat catabolism is linked to the pathways involved in glucose catabolism,
5. describe how fat can be changed into glucose,
6. relate ketosis to fat catabolism.

Lipids are a diverse group of organic substances which share the property of being relatively insoluble in water and readily soluble in organic solvents such as ethanol, ether and chloroform. The diverse nature of lipids in both structure and function has necessitated the classification of lipids into several groups. The important lipid groups found in the body are fats (triglycerides), phospholipids, prostaglandins, steroids and waxes. These lipid groups have a range of functions such as the storage of energy, membrane structure and chemical stimulants and inhibitors. Fats constitute the major form of dietary lipid and body lipid. This chapter will therefore place an emphasis on their metabolism.

23.1 Fats

STRUCTURE

Fats or triglycerides consist of three fatty acids bonded to a molecule of alcohol called glycerol (Figure 23.1). These substances vary in size according to the length of the hydrocarbon chains of the fatty acids. Animal fats contain saturated fatty acids whereas plant fats are polyunsaturated (see 9.2).

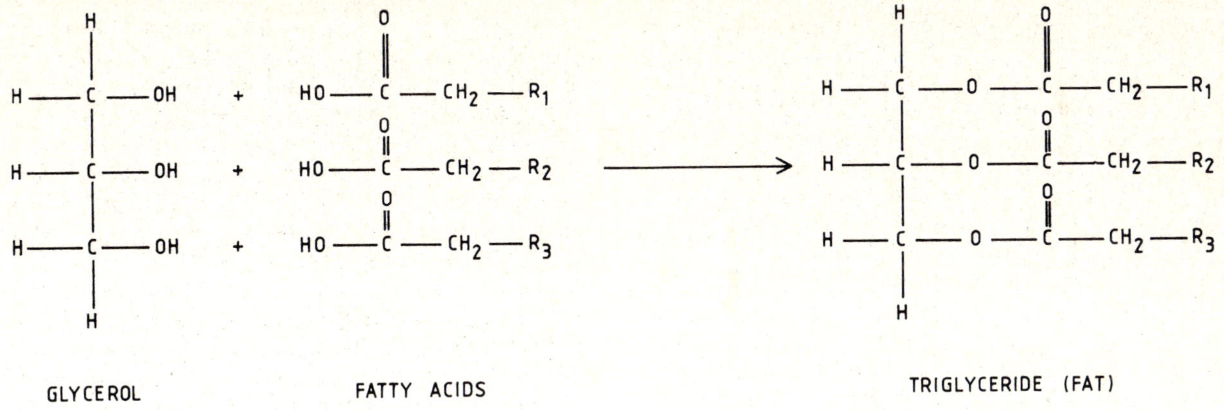

$$H-\underset{|}{\overset{H}{\overset{|}{C}}}-OH \quad + \quad HO-\overset{O}{\overset{\|}{C}}-CH_2-R_1 \qquad\qquad H-\overset{H}{\overset{|}{C}}-O-\overset{O}{\overset{\|}{C}}-CH_2-R_1$$

$$H-\underset{|}{C}-OH \quad + \quad HO-\overset{O}{\overset{\|}{C}}-CH_2-R_2 \quad \longrightarrow \quad H-\underset{|}{C}-O-\overset{O}{\overset{\|}{C}}-CH_2-R_2$$

$$H-\underset{H}{\underset{|}{C}}-OH \quad + \quad HO-\overset{O}{\overset{\|}{C}}-CH_2-R_3 \qquad\qquad H-\underset{H}{\underset{|}{C}}-O-\overset{O}{\overset{\|}{C}}-CH_2-R_3$$

GLYCEROL FATTY ACIDS TRIGLYCERIDE (FAT)

Figure 23.1 The structure of fats showing glycerol and three fatty acid molecules.

DIGESTION

Large quantities of triglyceride are unable to move across the wall of the gastrointestinal tract. It is therefore necessary for fat to be digested.

Digestion begins in the stomach where a small quantity of fatty acids are liberated from the glycerol. This breakdown of triglyceride is due to the action of an enzyme (gastric lipase). The greater the acidity of the gastric contents, the lower is the ability of this enzyme to release fatty acids.

The small intestine is the major site of lipid digestion. The undigested triglycerides are exposed to bile and pancreatic juice when they enter the first section of the small intestine or duodenum.

Bile is a secretion of the gall bladder which enters the duodenum via the bile duct. Bile emulsifies fat, that is, forms an emulsion by breaking down the ingested fat into droplets. This formation of the emulsion is assisted by the churning effect of peristalsis. Both obstruction of the bile duct and the removal of the gall bladder result in incomplete lipid digestion.

The formation of an emulsion of tiny fat droplets assists the action of a pancreatic enzyme (lipase) in breaking down triglyceride into fatty acids and glycerol. The enzyme acts on the surface of a fat droplet. The greater the number of small droplets, the greater is the surface area, and thus the available lipid, for enzymatic digestion.

ABSORPTION

The products of fat digestion, that is, glycerol and fatty acids are absorbed across the walls of the small intestine. Fatty acids appear to be absorbed by active transport. In addition some undigested droplets are absorbed as a temporary emulsion (see 10.4). These large molecules are absorbed across the membrane by pinocytosis (see 17.3). The rate of digestion and absorption of fat is slower than that of carbohydrates and proteins. However, most ingested fat is absorbed in normal circumstances. Only a small amount of undigested fat should be present in the faeces.

APPLICATION: STEATORRHOEA

The occurrence of large amounts of fat in the faeces is known as steatorrhoea. This can indicate maldigestion due to either deficient secretion of pancreatic lipase or bile salts. If digestion is normal, the steatorrhoea may be due to malabsorption of fat as a result of disease to the small intestine wall. EXAMPLE: Coeliac disease.

SITES OF ABSORBED FAT

Following absorption, most fat is transported in the plasma as droplets of fat known as chylomicrons or as fatty acids attached to albumin.

Muscles

Muscles usually use fat as their source of energy. During periods of vigorous activity they stop fat use and switch over to glucose utilization. Some of the absorbed fat is immediately utilized by muscle.

Liver and Body Tissue

The liver can immediately utilize fat to produce energy or form other substances according to the metabolic needs of the body.

Fat is found in most cells in the body. In adipose tissue the stored fat occupies most of the cell. In other cell types fat is normally dispersed throughout the cytoplasm in combination with phospholipid and proteins.

Adipose Tissue

Most absorbed fat is stored in adipose tissue in various regions of the body. Most fat is stored in fat depots in subcutaneous tissue, around the kidneys and in the abdomen. The fat depots are the major sites for the storage of energy, that is, they act as a reservoir of potential chemical energy. This depot fat does not remain stationary for long periods. It is continually being mobilized and shifted from one depot to another. During periods of either fasting or poor nutrition, sufficient quantities of mobilised fat are released and sent to the liver. In the liver the fat can be converted into other substances such as glucose. In addition, sufficient quantities of fat are sent to tissues such as muscle so as to maintain their production of energy. When excessive amounts of nutrients are taken into the body, the nutrients are stored as fat in adipose tissue.

APPLICATION: OBESITY

Obesity can be defined as a condition where more kilojoules are taken in than the body requires over a period of time. The excess energy is stored as fat. As the obese person more frequently develops metabolic disease than a person who is the same age and has a normal body weight, obesity is now considered to be a disease. This disease is likely to start in middle age. Some of the diseases that an obese person has a greater chance of developing are mature onset diabetes mellitus, atherosclerosis, hypertension, heart disease and phlebitis. The enlarged deposits of fat can also impair organ function by crowding the organ, restraining circulation or placing the organ in an abnormal position.

PRODUCTION OF ENERGY

The body has a preference for glucose as a source of potential energy. Fat which is more difficult to mobilize is the body's next preference. This order of preference can be altered by hormone action (see Section 28.2). Mobilized fat can produce more ATP than glucose since the former contains a greater number of hydrogen atoms.

Fat is the major storage depot of energy in the body. The hydrophobic nature of fat (water repelling) results in fat being able to be stored free of water. Glycogen is surrounded by large quantities of water and is therefore relatively heavy when compared with fat. Fat can form ATP because its breakdown products can directly enter the Krebs cycle. It can also produce ATP indirectly by being converted to glucose which is subsequently catabolised to produce ATP (Figure 23.2).

The breakdown of fat into its constituent molecules results in the formation of glycerol and three fatty acids. Both these substances consist of only carbon, hydrogen and oxygen atoms. The principal of fat energy production is simply the rearrangement of the quantities and proportions of carbon, hydrogen and oxygen into molecules that are found in the previously mentioned pathway used for glucose energy production (see 22.5).

The three-carbon-alcohol glycerol is converted in the liver into a three carbon compound found in the process of glycolysis. This latter substance can then be either converted to glucose by a reversal of glycolysis, and if necessary glycogen, or it can be catabolized to produce ATP via pyruvic acid formation, and if oxygen is present via the Krebs cycle (Figure 23.2).

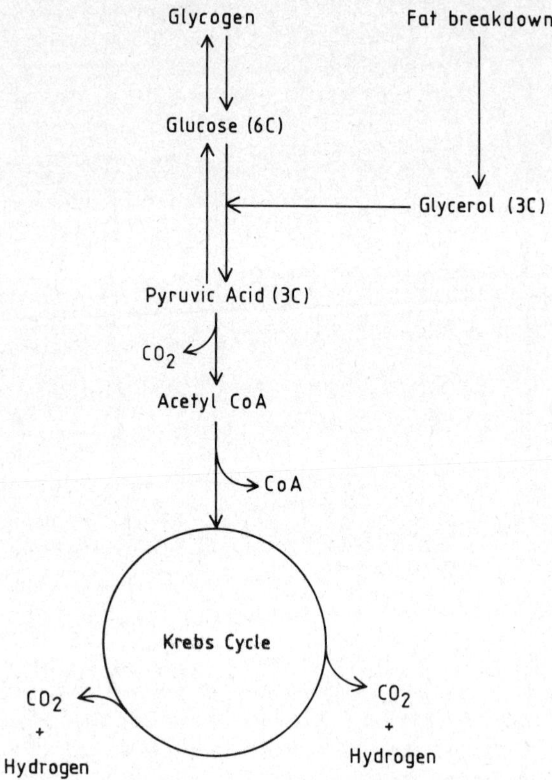

Figure 23.2 The catabolism of glycerol showing its entry into glucose metabolic pathways.

The catabolism of fatty acids readily takes place in tissues such as liver and muscle. An enzyme slices off two carbon units of the hydrocarbon fatty acid chain (Figure 23.3). Each two carbon unit or acetyl group is then combined with CoA to form acetyl CoA. The acetyl CoA is identical to that formed from glucose catabolism. The acetyl group is transported by the CoA to the site in the mitochondrial membrane where the Krebs cycle breaks down the two carbon unit to form ATP (see 22.5). Alternatively, when low glucose concentrations exist, glucose is able to be formed from one of the substances in the Krebs cycle which is known as oxaloacetic acid (Figure 23.3).

The action of slicing the fatty acid continues until the fatty acid is completely broken down. The process of forming acetyl CoA by removing acetyl groups from a fatty acid is referred to as β oxidation.

A problem exists with odd numbered hydrocarbon chains since the last carbon atoms in the chain have to form either a three or one carbon molecule. In fact, three carbon units are formed. These three carbon molecules are converted by a series of reactions, and are then able to either enter the Krebs cycle or yield glucose by a reversal of glycolysis.

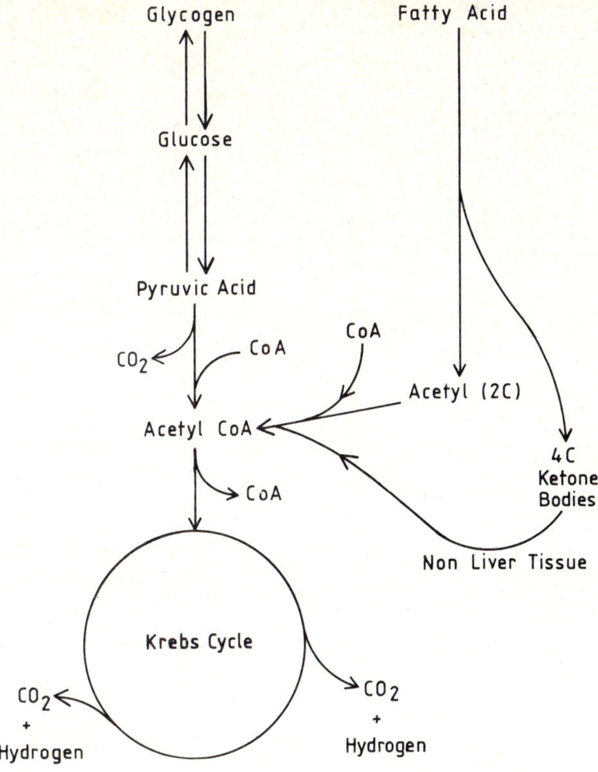

Figure 23.3 The catabolism of fatty acids showing the entry of acetyl CoA into the glucose metabolic pathways.

FORMATION OF KETONE BODIES

The breakdown of long chain fatty acids can result in the last four carbon atoms remaining together to form a four carbon acid. These acids are called ketone bodies. EXAMPLE: Acetoacetic acid.

In humans, the liver is the only organ that produces significant amounts of ketone bodies. However, the liver lacks the enzyme that converts ketone bodies into acetyl CoA. Ketone bodies leave the liver and are transported to most tissues where the ketone bodies are converted into acetyl CoA.

Ketone bodies rely on blood for transport to their breakdown sites, and the acidic nature of these substances can thus influence acid–base balance. During normal metabolism, the quantities of ketone bodies in blood are easily buffered by the various blood buffers. Factors that elevate fat catabolism will increase the concentration of ketones in the blood. Excessive fat catabolism leads to an acidosis of the blood which is referred to as ketoacidosis.

APPLICATION: KETOSIS

Ketosis is a form of metabolic acidosis (see 11.6). The most common causes of ketosis are starvation and diabetes mellitus. In both these conditions a lack of glucose in the body tissues results in large amounts of fat and protein being catabolized. The excessive breakdown of fat and protein results in large quantities of ketone bodies being released into the blood. The abnormally high concentrations of these acids are unable to be successfully buffered and thus acidosis occurs.

The onset of ketosis is associated with malaise, rapid breathing and thirst. If untreated, vomiting, dehydration and finally coma will occur. The large quantities of ketones in the blood also results in ketonuria, that is, the presence of ketone bodies in the urine.

OTHER FUNCTIONS OF FAT

Fat acts as an insulator and thus conserves the body's normal temperature. Adipose tissue also protects organs from mechanical injury and determines the figures of women and men. It also serves the function of an "overflow reservoir" since excess dietary intake of carbohydrates is converted to fat for storage.

A special form of adipose tissue is used to produce energy in the form of heat in new-born babies.

23.2 Other forms of lipid

PHOSPHOLIPIDS

Phospholipids consist of glycerol combined with two fatty acids and one phosphate group. All living cells contain a plasma membrane composed of phospholipid. The phosphate region is polar whereas the remainder of the molecule is non-polar. The non-polar nature of plasma membranes acts as a barrier for many ions and polar substances (see 17.2).

All tissues apart from blood and skin are capable of synthesizing their own phospholipids. The liver acts as a major source of phospholipid for the body. Unlike fats, phospholipids appear not to be involved in the storage and release of energy.

PROSTAGLANDINS

Prostaglandins are twenty carbon fatty acids containing a five carbon ring (Figure 23.4). They appear to be present in virtually all tissue and are released from membrane phospholipid. Their actions are diverse and only just beginning to be discovered. EXAMPLES: Induction of labour, decreasing gastric secretion of hydrochloric acid, altering blood vessel diameter. Prostaglandins are presently categorized into the following groups: PGA, PGB, PGC, PGD, PGE, PGF, PGG and PGH. In the future this list is likely to expand even further. Most of these groups appear to be derived from a substance known as arachidonic acid. The

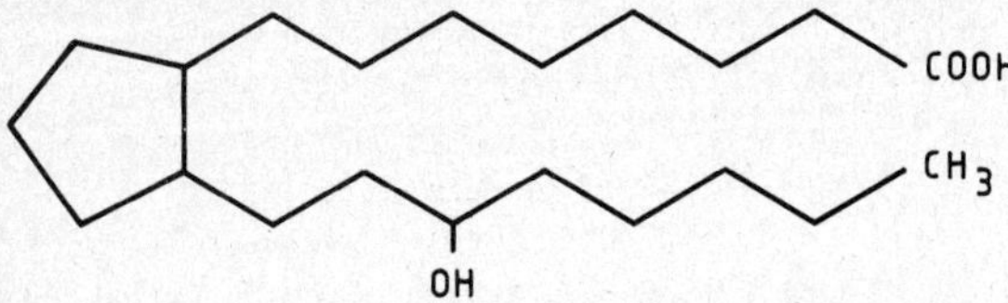

Figure 23.4 The general structure of prostaglandins.

effect of prostaglandins on the body is shown by their inhibition. The marvels of aspirin and indomethacin are attributed to their ability to block the formation of prostaglandins from arachidonic acid.

Prostaglandin groups often act in opposition to each other. EXAMPLES: PGF inhibits platelet aggregation whereas PGB accelerates aggregation, PGF causes vasoconstriction whereas PGE causes vasodilation. The role of prostaglandins in body function is only just beginning to be understood. In future years, an understanding of their complex variety of actions is likely to lead to these substances playing an important role in arresting disease.

STEROIDS

Steroids are a group of substances that include cholesterol, bile salts, sex hormones and hormones secreted from the adrenal cortex. They contain four carbon rings as illustrated in Figure 23.5.

Cholesterol

Cholesterol plays an important role in body metabolism. It is an essential component of membranes and is also used to form bile salts, steroid hormones and vitamin D. It is normally abundant in nervous tissue, the adrenal glands and the skin. About ninety per cent of cholesterol is made by the liver and regions of the intestine. The liver also serves as the chief agent for removing cholesterol by converting it into bile salts.

Bile Salts

Bile salts are formed from cholesterol in the liver and secreted via the gall bladder and bile ducts into the duodenum. Bile salts act as the emulsifying agent for facilitating lipid digestion.

Steroid Hormones

Steroid hormones have a variety of actions in the body. They are secreted from the adrenal cortex and gonads. EXAMPLES: Cortisol and aldosterone from the adrenal cortex and oestrogens and androgens from gonads. The actions of some steroid hormones is described in Chapter 28.

WAXES

Waxes are hydrophobic substances that are less greasy than fats. They are secreted by the skin to

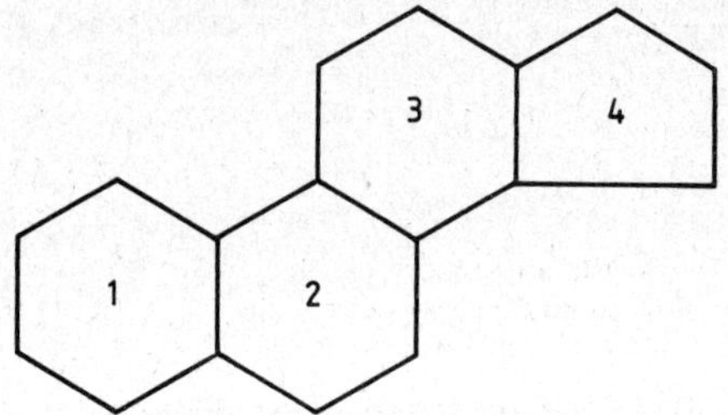

Figure 23.5 The general structure of steroids.

reduce water loss. They are used in ointments, cosmetics and a variety of pharmaceutical preparations.

Summary

Lipids are a diverse group of organic substances which share the property of being relatively insoluble in water. There are several categories of lipids:

1. Fat or triglycerides consist of three fatty acids and glycerol. They are the major reserves of energy and are stored in adipose tissue.
2. Phospholipids contain two fatty acids, a phosphate group and glycerol. They are the main constituent of plasma membranes.
3. Prostaglandins are twenty carbon fatty acids containing a five carbon ring. They appear to have a wide variety of actions on the body and are sometimes referred to as tissue hormones. Their actions on the body are blocked by aspirin.
4. Steroids contain four carbon rings. They are groups of substances that include cholesterol, bile salts and steroid hormones.
5. Waxes are hydrophobic substances secreted by the skin to reduce water loss.

Fat digestion begins in the stomach where a small quantity of fatty acids are released from glycerol. The major site of fat digestion occurs in the small intestine where the actions of bile and pancreatic juice complete the digestive process. The products of fat digestion, that is fatty acids and glycerol, are then absorbed across the wall of the intestine. Some fat is absorbed as a temporary emulsion by the process of pinocytosis.

Absorbed fat is transported as chylomicrons and fatty acids are transported by attaching to albumin. The sites of absorbed fat are muscles, liver, adipose tissue and most other cells. The principal site of stored fat is adipose tissue.

The catabolism of fat for the purpose of producing energy involves the following steps.

1. Fat is broken down into glycerol and fatty acids.
2. Glycerol is converted in the liver into a three carbon compound used in the process of glycolysis. This latter substance can be converted into either glucose, and if necessary glycogen, or it can be

broken down to pyruvic acid and subsequently enter the Krebs cycle.
3. Fatty acid catabolism readily takes place in the liver and muscle. An enzyme slices the hydrocarbon chain into two carbon acetyl molecules which combine with CoA to form acetyl CoA.
4. The acetyl group enters the Krebs cycle and produces ATP in the presence of oxygen. Alternatively, low glucose concentrations stimulate the formation of glucose from a four carbon Krebs cycle compound known as oxaloacetic acid.
5. Catabolism of long chain fatty acids can result in the formation of four carbon molecules known as ketone bodies. These substances are mainly formed in the liver but are only broken down in other tissues. The acidic nature of most ketone bodies and their dependence on the vascular system for their breakdown in other tissues, can lead to ketoacidosis in cases of excess ketone body formation.

Chapter 24

Proteins

Objectives

At the completion of this chapter the student should be able to:
1. describe the structure of amino acids,
2. relate the structure of proteins to their function,
3. list the groups of proteins based on function,
4. describe the process of protein digestion and absorption,
5. explain the difference between essential and non-essential amino acids,
6. describe transamination and deamination and the part that they play in protein metabolism,
7. explain how carbohydrates, lipids and proteins can be converted into each other.

Proteins are one of the most important groups of body chemicals. They have many functions within the human body, notably as enzymes responsible for regulating biochemical processes, as structural proteins (such as in cell membranes and collagen), as hormones (such as insulin) and as antibodies for body defence. Even though proteins perform a wide variety of body functions, they all are composed of the same fundamental chemical units which are known as amino acids.

H

R — C — COOH

NH_2

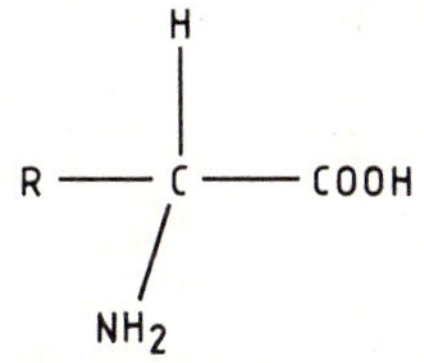

Figure 24.1 General formula of amino acids where NH_2 represents an amine group, COOH an acid group and R varies according to the amino acid.

24.1 Amino acids

AMINO ACID STRUCTURE

There are twenty different amino acids that commonly make up proteins. All of these amino acids contain an amine group ($-NH_2$) and an acid group (-COOH) (Figure 24.1). These amino acids differ only by the atoms that are present in the R group, that is, side chain. The names of these amino acids are listed in Table 24.1. Note that the amino acids have been grouped according to their acid/base properties and hydrophilic/hydrophobic properties. Remember that hydrophilia is water attracting whereas hydrophobia is water repelling.

PROPERTIES OF AMINO ACIDS

The properties of amino acids play an important role in determining the structure and function of a protein.

Ionization

Amino acids ionize in aqueous solutions by the weak acid (COOH) donating its proton (H^+) to the weak base, amine (NH_2). The result is the formation of an amino acid that contains a negatively charged region (COO^-) and a positively charged region (NH_3^+) (Figure 24.2). This negatively and positively charged compound is referred to as a

Table 24.1 The twenty common amino acids that are present in proteins

Group	Name
Basic	Lysine
	Argine
	Histidine
Acidic	Aspartic acid (aspartate)
	Glutamic acid (glutamate)
Hydrophilic	Glycine
	Serine
	Threonine
	Asparagine
	Glutamine
	Cysteine
	Methionine
Hydrophobic	Alanine
	Valine
	Leucine
	Isoleucine
	Proline
	Phenylalanine
	Tyrosine
	Tryptophan

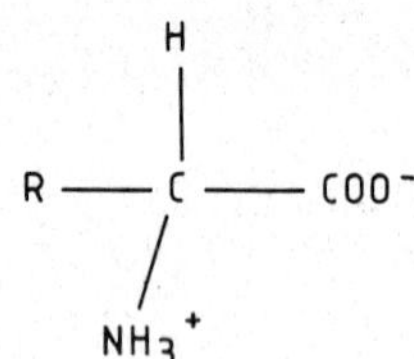

Figure 24.2 The ionized form of an amino acid.

zwitterion. When amino acids exist as zwitterions they can function as buffers (see 11.5).

Amino acids such as aspartic acid and glutamic acid contain as the R group an additional carboxyl (COOH) group. These acids can donate the additional proton into the solution, thus confering the title of acidic amino acids. Some amino acids contain an amine as the R group. EXAMPLES: Lysine and histidine. The ability of these amino acids to accept protons from the solution and combine them with the extra amine group renders these amino acids as bases. Since proteins are composed of amino acids, some of which are weak acids or bases, proteins are able to perform an important role as buffers.

Affinity for Water

The nature of the R group or side chain can influence the affinity of an amino acid to water. Side chains with a high affinity to water are polar and are described as hydrophilic. EXAMPLES: Glycine and cysteine. In contrast, hydrophobic side chains are non-polar and repel water. The relative affinity of these side chains to water plays an important role in determining the physical shape of proteins.

24.2 Protein structure and function

STRUCTURE

Proteins consist of large chains of amino acids that are joined together by bonds that are referred to as peptide bonds. Hence proteins are called polypeptides. The arrangement and number of amino acids determines a protein's structure. Proteins can be divided into the two groups fibrous proteins and globular proteins.

Fibrous Proteins

Fibrous proteins are composed of long helical chains of amino acids entwined with each other. ANALOGY: The twines in a rope wrap around each other in a similar manner to that of the polypeptide chains. In addition, the polypeptide chains are bound to each other by side chain bonds (Figure 24.3a).

These proteins have great strength, and as the name implies they form the fibres in connective tissue and muscle. EXAMPLES: Collagen and fibrinogen in connective tissue, and myosin in muscle.

[a]

[b]

Figure 24.3 The general shape of polypeptide chains in: (a) fibrous proteins, and (b) globular proteins.

Globular Proteins

Globular proteins consist of a folded polypeptide chain (Figure 24.3b). ANALOGY: A tangled piece of fishing line has a unique folded shape which usually appears as a mess and any effort to untangle the line will result in an alteration to the shape of the tangle. In the body, each form of globular protein has a unique shape which is maintained by the interaction of side chains between nearby parts of the polypeptide chain. The function of the protein is directly related to the shape of the tangle. Alterations to this shape can interfere with the proteins function.

There are an infinite number of possible tangled or globular shapes. A large range of different shaped globular proteins exist in the body. EXAMPLES: Enzymes, immunoglobulins and the globin part of haemoglobin. The unique shape of individual enzymes and immunoglobulins enables them to adhere with a complementary shaped structure, that is, a biochemical with enzymes and antigens with immunoglobulins.

FUNCTION

Proteins can be classified by their function into seven broad groups.

Nutrient Proteins

Nutrient proteins or dietary proteins are the bodys source of amino acids. Since some amino acids are unable to be formed in the body, these proteins are an essential raw material for the synthesis of other proteins.

Structural Proteins

Structural proteins are fibrous proteins found in connective tissue. EXAMPLE: Collagen. Structural proteins are used to support body structures forming a loose mesh in loose connective tissue and a dense framework in dense connective tissue.

Contractile Proteins

Contractile proteins are responsible for muscle contraction. The proteins actin (fibrous) and myosin (globular) effect a contraction by sliding over each other.

Blood Proteins

The four main types of blood proteins are albumin, globulin, fibrinogen and the globin part of haemoglobin. All of these proteins act as buffers. The specific roles that each of them play in blood is described in 20.1.

Immunoglobulins

Immunoglobulins or antibodies are globular proteins that are produced by plasma cells. Each specific immunoglobulin acts on a complementary shaped antigen. Antibodies are circulated throughout the cardiovascular and lymphatic systems (see 21.2).

Hormones

Hormones are regulators of cell function. They play a vital role in the coordination of body function. EXAMPLES: Insulin, antidiuretic hormone. Note that not all hormones are proteins. Some are steroids and others are small amino acid based molecules. A comprehensive description of their actions is described in Chapter 28.

Enzymes

Enzymes are catalysts of chemical reactions, that is, they alter the rate of the chemical reaction without being changed themselves. Without enzymes, most body chemical reactions would procede at a rate that is too slow to support life. Enzymes are able to exert their influence in minute concentrations.

The globular nature of enzymes enables each enzyme to act upon a specific chemical reaction. Any alteration to the shape of the enzyme will result in diminished enzyme function. Factors such as acidity, temperature and organic solvents can alter enzyme shape and thus hinder enzyme function. EXAMPLE: A specific enzyme is designed to work in a specific pH range. When exposed to an acidity or alkalinity outside that range the enzyme's action is diminished. In the gastrointestinal tract, salivary enzymes are unable to function in the stomach as a result of the change in acidity.

24.3 Digestion and absorption of proteins

The inability of most proteins to be absorbed across the intestinal membrane and into the body necessitates the breakdown of protein into a form that can be absorbed.

DIGESTION

Stomach

Protein digestion begins in the stomach. Under the action of the enzyme pepsin, large protein molecules are broken down to smaller more soluble molecules known as peptides. Pepsin operates most efficiently at a pH of between 1.5 and 2.0 and thus can operate effectively in the highly acidic gastric contents. If gastric hydrochloric acid production is unable to maintain a satisfactory pH, protein digestion in the stomach is minimal.

Small Intestine

Pancreatic enzymes split the peptides down to units that contain two or three amino acids. These small peptides are broken down by enzymes secreted by the epithelial cells lining the small intestine.

ABSORPTION

Active Transport

Following digestion of the protein into individual amino acids, the amino acids are actively transported across the small intestine wall. There are four different carriers present in the membrane for the active transport of amino acids. Each of these carriers transports a specific group of amino acids. EXAMPLE: One carrier is used for the transport of the acidic amino acids while another is used for the basic amino acids.

Most amino acids rely on active transport for their movement across membranes. Without this active transport, an amino acid imbalance will occur in the body.

APPLICATION: CYSTINURIA

Cystinúria is an inherited disorder of membrane transport. It is due to the genetic loss of the mechanism by which cysteine and similar structured amino acids are actively transported across membranes. It appears that the genetic error is a failure to produce the carrier associated with the cysteine group of amino acids in the membranes of the intestine and kidney. The disease is expressed clinically by the formation of calculi in the urinary tract. Calculi result from the elevated levels of urine amino acid.

Pinocytosis

Small amounts of undigested protein are absorbed by pinocytosis (see 17.3). The combination of active transport and pinocytosis results in about ninety per cent of ingested protein being absorbed.

24.4 Protein metabolism

AMINO ACID POOL

Following the absorption of amino acids across the small intestine wall, the amino acids enter a common amino acid pool located in the blood and tissues. This pool also contains amino acids that have resulted from the breakdown of tissue proteins during normal metabolism (Figure 24.4). Once the amino acids are in this common metabolic pool the body cannot distinguish from where a particular amino acid has been derived.

The amino acid pool serves as the distribution centre for amino acid based biochemicals such as hormones and nucleotides. The principal site of this amino acid metabolism is the liver. During starvation or when an inadequate intake of protein occurs, the liver loses protein to a greater extent than other tissues. The liver is more susceptible to injury by such agents as alcohol when a situation of depleted protein exists.

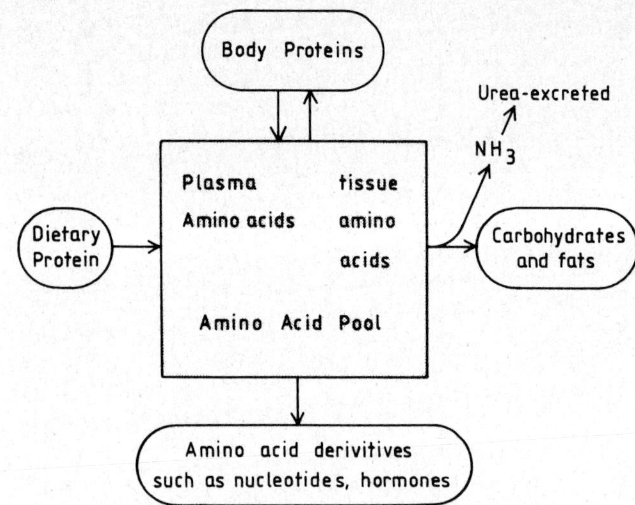

Figure 24.4 The amino acid pool and its relationship with biochemicals.

TRANSAMINATION

Within the amino acid pool, some amino acids can be converted into other amino acids. Transamination is the process where an amine group is transferred from one amino acid to form another amino acid. The chemical that combines with the amine group is usually a chemical formed in the breakdown pathway of carbohydrates and fats. Transamination assists the homeostasis of protein metabolism by being able to alter the concentration of particular amino acids according to the needs of the body.

Not all amino acids can be formed by transamination. Amino acids that are unable to be manufactured by the body have to be supplied by the diet.

ESSENTIAL AND NON-ESSENTIAL AMINO ACIDS

The nutritional value of any protein depends upon the composition and proportion of its constituent amino acids. Approximately half of the

Table 24.2 Essential and non-essential amino acids for man

Essential	Non-essential
Arginine	Alanine
Histidine	Asparagine
Isoleucine	Aspartic acid
Leucine	Cysteine
Lysine	Glycine
Methionine	Glutamic acid
Phenylalanine	Hydroxyproline
Threonine	Proline
Tryptophan	Serine
Valine	Tyrosine

amino acids are readily manufactured in the body. The presence of these amino acids in the diet is therefore not essential (Table 24.2). The remainder of the amino acids are not readily made in the body. These essential amino acids are either not synthesized at a rate that is required for normal metabolism or they are not manufactured at all.

Essential amino acids must be supplied in the diet and thus the dietary value of any protein depends upon its content of essential amino acids.

APPLICATION: THE BIOLOGICAL VALUE OF PROTEINS

The biological value of a particular protein is determined by its content of essential amino acids. Generally, animal proteins contain all of the essential amino acids and thus have a high biological value. Most plant proteins are of a low biological value because they are deficient in one or more of the essential amino acids.

DEAMINATION

The removal of amine groups from the amino acid pool is termed deamination. This process is the main pathway for the removal of nitrogen from the body. On leaving the amino acid pool, the amine forms ammonia. Most of this relatively toxic substance is combined with carbon dioxide to form non-toxic urea which is excreted in the urine. Urea is the main end product of protein catabolism and it accounts for eighty to ninety per cent of the nitrogen that is excreted in the urine. Some of the ammonia is used in the formation of the nucleotides, adenine and guanine. The breakdown of these substances yields the excretory product uric acid.

The fate of the amine free part of a deaminated amino acid depends on the type of amino acid that has been deaminated. The removal of amine groups from most amino acids results in the remaining part being composed of only carbon, hydrogen and oxygen. These amine free substances thus contain the essential ingredients for entrance into carbohydrate biochemical pathways.

FORMATION OF CARBOHYDRATES AND FATS FROM AMINO ACIDS

Proteins are the last major reserves of energy, since carbohydrates and fats are preferred as energy sources. Proteins can produce energy by the amine free parts of amino acids being incorporated into the Krebs cycle (Figure 24.5). Amino acids can be grouped according to the end-products formed as a result of the metabolism of their amine free part.

1. Glucogenic amino acids enter the Krebs cycle as oxaloacetic acid (one of the substances in the cycle). The oxaloacetic acid can then either be catabolized to form ATP, carbon dioxide and water or it is anabolized to form glucose, and if necessary glycogen or fat. EXAMPLE: Alanine.

2. Ketogenic amino acids form the ketone body acetoacetic acid. The ketone bodies leave the liver and are transported to cells where the acetoacetic acid is catabolized to form ATP, carbon dioxide and water. EXAMPLE: Leucine.

3. Deamination of the third group of amino acids results in the formation of glucose and ketone bodies. EXAMPLES: Tyrosine and phenylalanine.

Approximately sixty per cent of the body's protein can be converted in the liver to glucose. The breakdown of this protein is influenced by hormones such as cortisol (see Chapter 28).

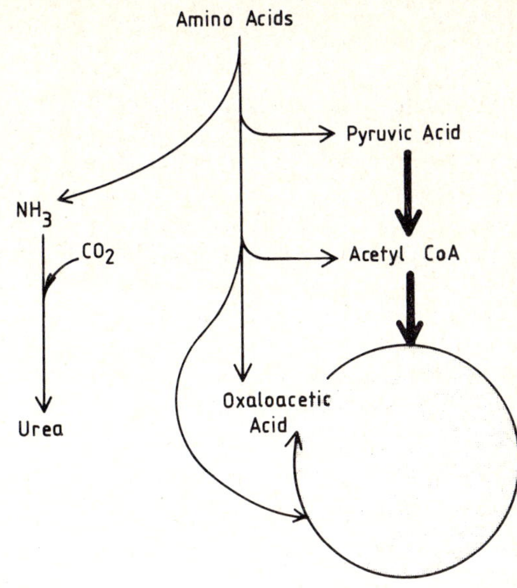

Figure 24.5 Incorporation of the amine free part of amino acids into the Krebs cycle.

NITROGEN BALANCE

Nitrogen balance refers to the amount of nitrogen in the urine compared to nitrogen intake over a period of time. When the intake of nitrogen exceeds its excretion, a person has a positive nitrogen balance. EXAMPLES: Children, pregnant women, and patients recovering from a prolonged illness. A negative nitrogen balance exists when more nitrogen is being excreted than is being taken into the body. This imbalance is due to some factor causing a relative excess of amino acid deamination. EXAMPLES: Starvation, wasting disease and diets lacking sufficient quantities of essential amino acids.

24.5 Interconversion of carbohydrates, lipids and proteins

Carbohydrates, lipids and proteins all have the potential to enter the Krebs cycle in the presence of oxygen, and produce energy in the form of ATP (Figure 24.6). These substances can also be converted into each other. The ability of these important chemicals to form ATP and be interconverted is an important mechanism in the maintenance of body and cell homeostasis. In cases of disease where an imbalance in body metabolism exists, interconversion of body chemicals is essential for a healthy life. EXAMPLE: People with diabetes mellitus convert fat and protein into glucose.

APPLICATION: LOW CARBOHYDRATE DIETS

A person on a low carbohydrate diet converts fat and to a limited extent protein into glucose and glycogen. Without this formation of glucose, body function would be severely impaired. The conversion of fat to glucose results in decreased levels of body fat and a decrease in body weight. The conversion of fat to glucose can be monitored by measuring the concentration of ketones in the urine.

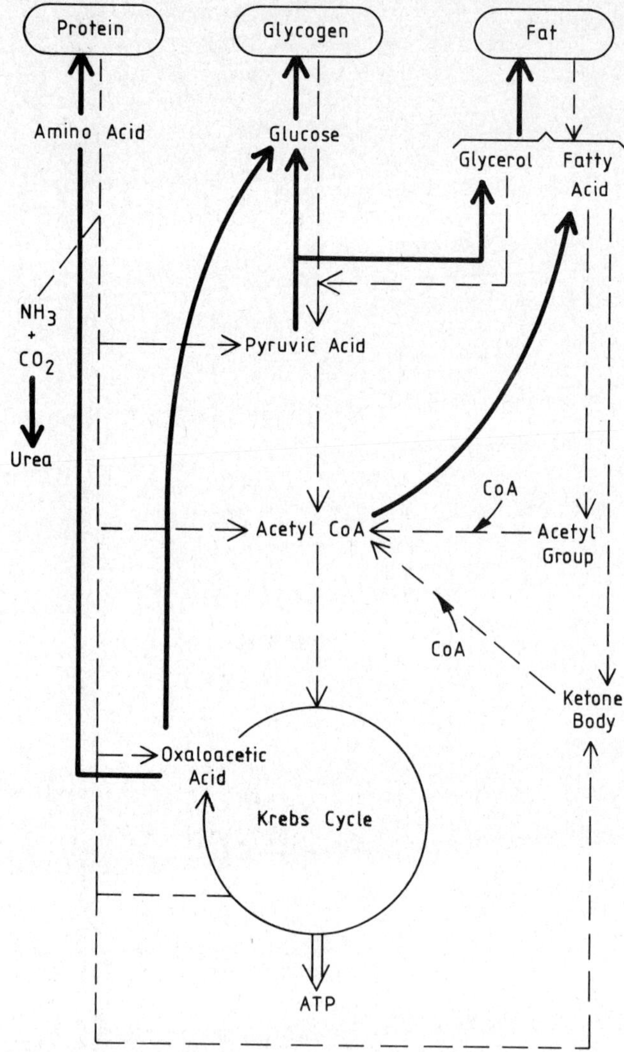

Figure 24.6 The relationship of the Krebs cycle to carbohydrate, lipid and protein catabolism. Note that these biochemicals can be converted into each other by the reversal of catabolic pathways (——live catabolic pathways,—— anabolic pathways).

Summary

Proteins are molecules consisting of large chains of amino acids. There are twenty common amino acids used to form proteins, of which ten are unable to be synthesized by the body. These latter amino acids are called essential amino acids.

The long chains of amino acids (polypeptides) can be arranged as either long helical chains (fibrous proteins) or they can be considerably folded (globular proteins).

Proteins can be classified according to their function:

1. nutrient proteins are the body's source of amino acids,
2. structural proteins are fibrous proteins used to support body structures,
3. contractile proteins are responsible for muscle contraction,
4. blood proteins refer to a group of proteins found in blood and the main types are albumin and globulins; fibrinogen and the globin part of haemoglobin,
5. immunoglobulins or antibodies are proteins capable of neutralizing specific antigens,
6. hormones are regulators of cell function; note that not all hormones are proteins,
7. enzymes are catalysts of chemical reactions.

Digestion of protein begins in the stomach where an enzyme breaks the polypeptides down into smaller molecules. In the small intestine, the pancreas and intestine secrete enzymes that complete the digestive process by forming amino acids. The amino acids are then actively transported across the wall of the small intestine. Small amounts of undigested protein are absorbed by pinocytosis.

Following absorption, the amino acids enter a common amino acid pool located in the blood and tissues. This pool serves as the distribution centre for all amino acid based biochemicals. Within the pool, a non-essential amino acid can be converted into another non-essential amino acid by a process known as transamination.

The process of removing an amine group from an amino acid is called deamination. The amine group is converted into the relatively non-toxic substance, urea. The remaining amine-free part enters the pathway involved in the production of ATP from glucose. These amine-free groups can either form ATP, carbon dioxide and water, or in the presence of low glucose concentrations form glucose and glycogen. The catabolism of some amino acids results in the formation of ketone bodies.

The relative utilization of nitrogen is referred to as nitrogen balance. A positive nitrogen balance exists when the intake of nitrogen exceeds its excretion in the urine. A negative nitrogen balance indicates a greater use of amino acids by the body than are being ingested.

Proteins, carbohydrates and fats can all enter the Krebs cycle and all have the potential to be converted into either of the other two biochemicals.

Chapter 25

Vitamins and minerals

Objectives

At the completion of this chapter the student should be able to:
1. list the important water soluble and fat soluble vitamins,
2. describe the deficiency effects of each vitamin,
3. list the important minerals and briefly describe their function.

25.1 Vitamins

Vitamins are organic substances that act in trace amounts to promote enzyme function. They play an important role in regulating biochemical and physiological processes.

Most vitamins cannot be synthesized by the body, they are mainly obtained from food. EXAMPLE: Vitamin C in citrus fruit. No single food can supply all the different forms of vitamins, and thus it is important to have a balanced diet. The daily requirement of each vitamin varies between individuals. Factors such as pregnancy can alter the vitamin requirement within an individual. Abnormally low and high levels of vitamin intake can lead to altered physiological and biochemical function.

Evidence for the actions of vitamins on the body is still conflicting in many cases. The effect of drugs on vitamin absorption and function is little understood at present. In the future, hopefully, a clear picture will emerge as to their precise role in the body and the impact that drugs have on vitamins fulfilling their role. This book will only attempt to introduce the topic of vitamins to students.

Vitamins are separated into two categories according to their solubility in water.

WATER SOLUBLE VITAMINS

These vitamins are dissolved in the gastrointestinal tract and body fluids. The body has difficulty in storing these vitamins and any excess is excreted in the urine. EXAMPLE: Vitamin C.

The water soluble vitamins consist of the B group of vitamins and vitamin C. The function of each of these vitamins and the effects of their deficiency is summarized in Table 25.1.

FAT SOLUBLE VITAMINS

This group of vitamins are insoluble in water but soluble in fat. Their ability to be stored in cells enables the body to keep an adequate reserve under normal conditions. They are found in most fatty foods. EXAMPLE: Vitamin A in cod liver oil.

The fat soluble vitamins are vitamins A, D, E and K. A summary of their functions and the effects of deficiencies on the body are listed in Table 25.2.

25.2 Minerals

Minerals are inorganic substances that function in the body in combination with other minerals or with

Table 25.1 A summary of water-soluble vitamins

Vitamin	Food examples	Function	Deficiency effects
B_1 (thiamine)	Most foods except white flour and sugar	Involved in catabolism of pyruvic acid to carbon dioxide and water	Elevated pyruvic acid and tactic acid levels, insufficient energy production, beriberi and polyneuritis
B_2 (riboflavin)	Dairy products, eggs, meat, whole-meal bread	Involved in carbohydrate and protein metabolism in tissues such as intestinal wall and skin	Digestive disturbances, dermatitis, eye soreness, cataracts
B_6 (pyridoxine)	Unpolished rice, whole-grain cereals, liver, yoghurt	Important in lipid and protein metabolism, maintenance of healthy skin	Dermatitis, nausea
B_{12} (cyanocob-alamin)	Not in vegetables, found in liver, milk, eggs, cheese, meat	Necessary for red blood cell formation, entrance of some amino acids into Krebs cycle	Pernicious anaemia, malfunction of nervous system
Nicotinamide (niacin)	Meat, whole-grain cereals, bread and yeast	Important in lipid catabolism for energy release, inhibits cholesterol production	Pellagra
Pantothenic acid	Green vegetables, liver, kidney	Essential in the conversion of lipids and amino acids to glucose, synthesis of cholesterol and steroid hormones	Fatigue, neuromuscular abnormalities, insufficient adrenal hormones
Folic acid	Green leafy vegetables	Important in the Krebs cycle	Fatigue, mental depression and nausea
Vitamin C (ascorbic acid)	Citrus fruits, green vegetables, tomatoes	Function is controversial but appears to promote many metabolic reactions, work with immunoglobulins, involved in steroid production and epithelial growth	Slow healing of wounds, anaemia, scurvy

organic substances. The complete function of minerals is still being determined, but the actions of some are already known to be essential for normal metabolism and the maintenance of homeostasis. EXAMPLES: Minerals such as calcium are essential for body structure whereas minerals such as sodium play important roles in fluid balance. The body appears to only use the ions of minerals, and thus minerals appear as electrolytes and salts in the body.

The required intake of each mineral varies. Some such as sodium and calcium are needed in relatively large amounts, whereas others such as copper and zinc perform their function from the intake of only trace quantities. A summary of the main minerals influencing the body are listed in Table 25.3. Additional information relating to electrolytes in blood occurs in 20.1.

Summary

Vitamins are organic substances that play an important role in regulating biochemical and physiological processes. Most vitamins cannot be synthesized in the body. To obtain the correct daily requirement of each vitamin it is important to have a balanced diet. Some vitamins are water soluble while others are fat soluble. The water soluble vitamins are the B complex vitamins and vitamin C. The body has difficulty in storing these vitamins, whereas the fat soluble vitamins are able to be stored in the body. The fat soluble vitamins are vitamin A, D, E and K.

Minerals are inorganic substances that function in the body in combination with other minerals or with organic substances. Some minerals such as sodium and calcium are needed in relatively large amounts, whereas others such as copper and zinc are only needed in trace amounts.

Table 25.2 A summary of fat-soluble vitamins

Vitamin	Food examples	Function	Deficiency effects
A	From carotene in yellow and green vegetables, also directly from fish, liver, oils, and dairy products	Maintains health and vigour of epithelial cells especially mucous membrane cells, important in vision	Night blindness, dry skin and hair, infections of the ear, sinus, respiratory, urinary and digestive tracts
D	Fish liver, oils, egg yolk, milk, produced in body by exposure to sunlight	Important in calcium and phosphorus absorption and metabolism	Poor muscle tone, retarded skeletal growth leads to rickets
E (tocopherols)	Fresh nuts, wheat germ, green leafy vegetables, seed oils	Suspected of being involved in the formation of DNA, RNA and red blood cells, inhibits membrane lipid catabolism, protects liver and lungs from toxic chemicals	Catabolism of lipids in membranes especially in red blood cells, which leads to haemolytic anaemia
K	Spinach, cauliflower, cabbage, liver and produced in large quantities by intestinal bacteria	Involved in clot formation and probably in the electron transport system	Prolonged bleeding and clotting times

Table 25.3 Summary of the main minerals that influence body function

Mineral	Food examples	Function
Calcium	Dairy products, egg yolk, green leafy vegetables	Necessary constituents of bone and teeth; essential for hormone manufacture, blood clotting, nerve and muscle activity
Chlorine	Salt in food processing and preparation	Important in fluid balance and hydrochloric acid formation in the stomach
Cobalt	Trace quantities required, sufficient in well-balanced diet	Essential part of vitamin B_{12} and thus plays an important role in red blood cell maturation
Copper	Trace quantities required, sufficient in well-balanced diet	Speeds up haemoglobin formation and is important in the function of some enzymes
Iodine	Trace quantities required, iodized salt, seafoods, vegetables grown in iodine-rich soils	Required for production of thyroid hormones
Iron	Meat, fish, vegetables, fruit, cereals, liver	Essential component of haemoglobin, important in production of ATP
Magnesium	Dairy products, flour, cereals, green leafy vegetables, nuts, soybeans	Required for normal nerve and muscle function
Manganese	Trace quantities required, sufficient in well-balanced diet	Important in the function of some enzymes, haemoglobin synthesis, required for growth
Phosphorus	Meat, fish, dairy products, nuts, maize	Important in acid-base balance, bone and teeth formation, nerve and muscle activity
Potassium	Present in many substances in a normal diet	Important in nerve and muscle function, essential constituent of intracellular fluid
Sodium	Found in all animal foods as salt, present in processed and prepared foods	Essential in blood volume control along with nerve and muscle function
Zinc	Trace quantities required, present in many foods	Constituent of many enzymes

Unit Ten
Control and regulation of body function: Neuroendocrine systems

This unit introduces the organization and function of the nervous and endocrine systems. The interrelationship between these systems is related to homeostatic mechanisms.

Chapter 26

General organization and function of the nervous system

Objectives

At the completion of this chapter the student should be able to:
1. describe the general organization and function of the nervous system,
2. list the principal components of the central nervous system,
3. explain the difference between white and grey matter,
4. describe where spinal nerves and sympathetic nerves connect with the spinal cord,
5. compare the organization and actions of sympathetic nerves with parasympathetic nerves,
6. describe the process of biofeedback.

26.1 Body communication systems

To coordinate body homeostatic mechanisms it is necessary to have a communication network that can detect, integrate and adjust the actions of each homeostatic mechanism. The ability of nerves to conduct impulses rapidly along their axons gives them the potential to fulfill the role of rapid conveyers of information in the body. The extensive network of nerves throughout the body enables nerves to realize their potential. The network of nerves is known as the nervous system. Any factor that interferes with their ability to conduct impulses or removes part of the nerve network will result in the potential function of nerves not being realized.

Other forms of communication play an important role in body homeostasis. A mechanism exists whereby information is transferred by regulatory couriers through the blood system from an integrating centre to a region which responds to the courier. This system is called the endocrine system and the couriers are known as hormones. This system requires less energy to function than the high energy demanding nerves. The endocrine system is generally involved in the coordination of the body's metabolic activity.

A third form of communication is the slow flow of substances down the axons of nerves. These substances are referred to as trophic factors and the movement of substances is called axoplasmic flow. This communication path is used to maintain the homeostasis of tissues with which nerves connect. It plays an important role in the development, maintenance and regeneration of the nerve network. EXAMPLE: Individuals who have their vagus nerve severed in an attempt to combat peptic ulcers will have regrowth of the nerve to its original connection site over a period of time. The trophic factors guide the nerve back to the original tissue. Axoplasmic flow also serves the purpose of connecting the nervous and endocrine systems in the region of the brain known as the hypothalamus.

26.2 General organization of the nervous system

This chapter is an overview of the nervous system for the purpose of familiarizing the student with the major pathways and commonly referred to structures. A general knowledge of the nervous system is essential for understanding how the body coordinates homeostatic mechanisms in response to altered internal and external environments.

There are several ways of classifying the various networks in the nervous system. The following is one of the most popular classifications (Figure 26.1):
1. Central nervous system (CNS)—consists of the brain and spinal cord. The CNS integrates incoming information and relays responses to the peripheral nervous system.
2. Peripheral nervous system—consists of the spinal nerves, cranial nerves and the autonomic nervous system (ANS). The spinal nerves convey information to the CNS from receptors and relay the messages from the CNS to the effectors via the somatic nervous system, thus eliciting a response. The autonomic nervous system carries information to and from the CNS that involves smooth muscle, cardiac muscle and glands. The ANS can be subdivided into the sympathetic nervous system and the parasympathetic nervous system. These two systems have opposite actions on smooth muscle, cardiac muscle and glands. The sympathetic nervous system has a stimulatory action whereas the parasympathetic has an inhibitory action.

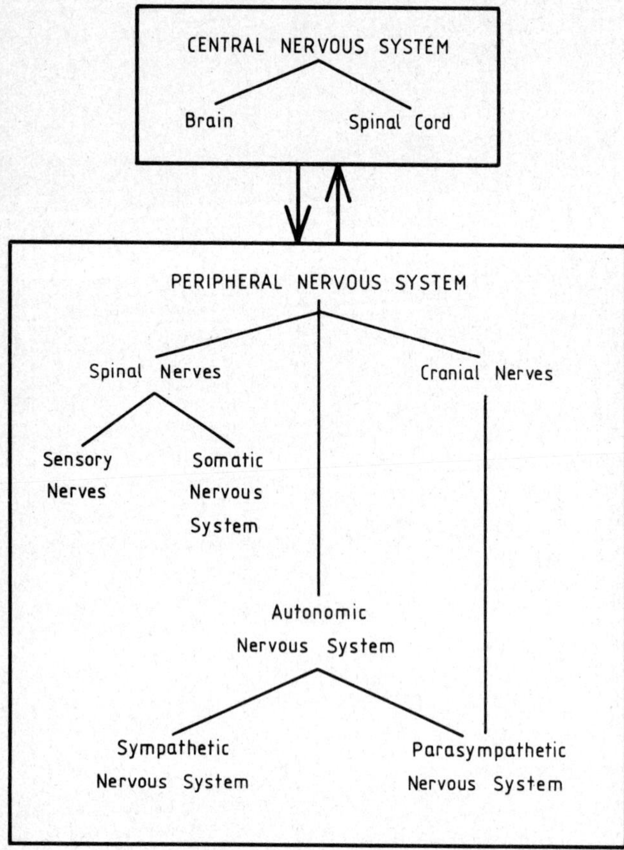

Figure 26.1 General organization of the nervous system. Note that some cranial nerves form part of the parasympathetic nervous system.

26.3 Central nervous system

The brain and spinal cord constitute the nerve network referred to as the CNS. The brain has the ability to regulate and coordinate homeostatic mechanisms. It can perform this action by responding directly to incoming information or by initiating an action according to previous experience. In addition, it can initiate learning, exploratory behaviour and exercise.

The spinal cord is a continuation of the brain. It acts as a distributor of information for the brain to the various regions of the body where it connects with the regional peripheral nervous system.

The CNS can be divided for convenience into the following four divisions according to its anatomical arrangement (Figure 26.2).

1. forebrain,
2. midbrain,
3. hindbrain,
4. spinal cord.

FOREBRAIN

The forebrain or cerebrum is the anterior region of the brain which contains the large lobes referred to as the cerebral hemispheres or telencephalon. It also contains a region known as the diencephalon which consists of the thalamus, hypothalamus and pineal gland.

Cerebral Hemispheres (Telencephalon)

The cerebral hemispheres are two lobes that are identical in shape and size. They are the site of conscious thought and are referred to as the higher centres. The higher centres are located in the outer region consisting of grey matter which is called the cerebral cortex. The cerebral cortex is the principal control centre for the body. It receives information, integrates and compares it with relevant past information and sends out decisions that controls a great proportion of body function.

White matter, that is, myelinated nerve tracts, are located beneath the cerebral cortex. They convey information to and from the various regions of the cerebral cortex. Located within the cerebral hemispheres are aggregations of grey matter known as basal ganglia. The basal ganglia or cerebral nuclei assist the cerebral cortex in maintaining smooth purposeful skeletal muscle contractions. Damage to nerves in this region results in shaking of the hands and loss of discrete movements of the body. EXAMPLE: Parkinson's disease.

Thalamus

The thalamus acts as a giant relay centre that distributes sensory information to the cerebral cortex. A particular piece of sensory information, such as a signal indicating pain, is distributed to the relevant regions of the cerebral cortex that deals with pain.

The thalamus has extensive connections with the cerebral cortex and the hypothalamus. Since the hypothalamus connects the endocrine system to the CNS, the thalamus acts as a relay centre for the transmission of endocrine actions to the cerebral cortex.

Hypothalamus

The hypothalamus links the endocrine and nervous systems. It has connections with the cerebral cortex, thalamus, retina, posterior part of the pituitary gland and the sympathetic nervous system. It also contains aggregations of grey matter or nuclei that are involved in the control of thirst, appetite and temperature. The hypothalamus also coordinates hindbrain homeostatic centres such as the cardiovascular and respiratory centres. Receptors in the hypothalamus monitor the level of individual substances in the blood and the osmotic balance.

The hypothalamus appears to be the central integrating area for initiating homeostatic mechanisms in response to stress.

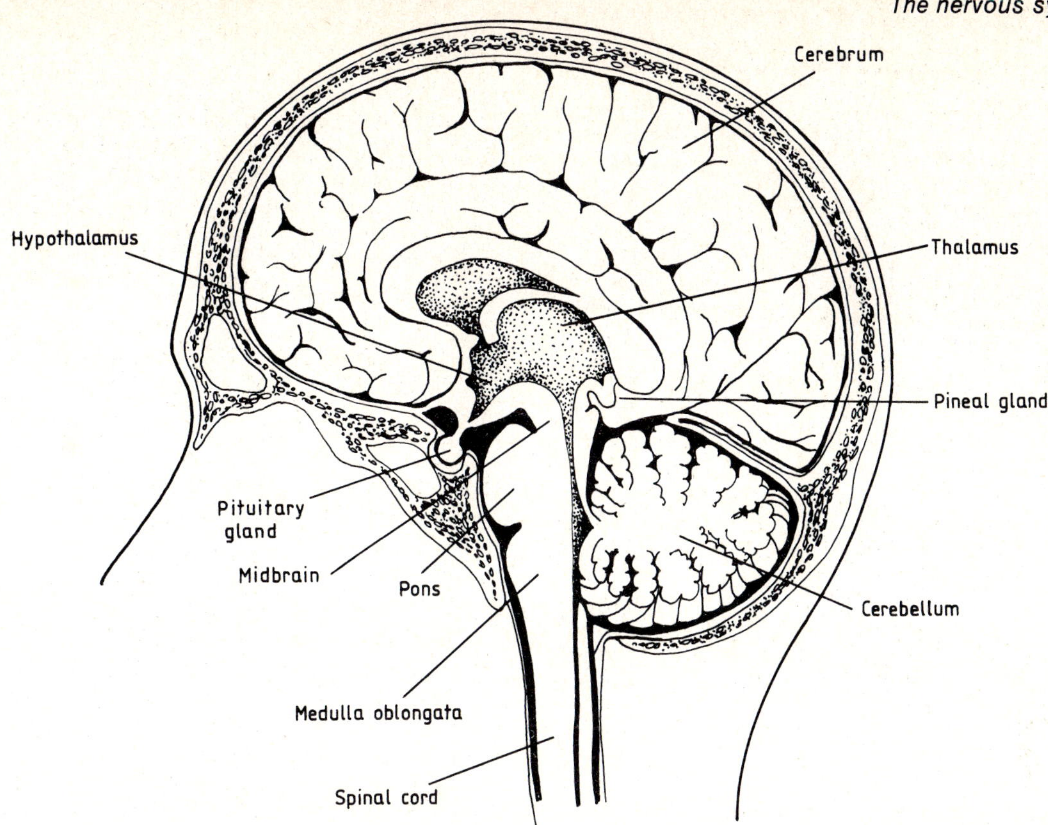

Figure 26.2 General organization of the central nervous system.

The relationship between the hypothalamus and the endocrine system is described in Section 27.4.

Pineal Gland

The pineal gland is a tiny gland about 0.5 cm long that has recently been found to be an important link between the nervous and endocrine systems. This gland is linked to CNS via the sympathetic nervous system. A description of its nervous connection and the actions of this gland occurs in Section 27.4.

MIDBRAIN

The midbrain or mesencephalon portion of the brain connects the forebrain to the hindbrain. Briefly, this segment of the brain contains conduction pathways for both sensory and motor messages between the forebrain and the hindbrain. An additional function of this region is the coordination of visual and auditory reflexes.

HINDBRAIN

The hindbrain or rhombencephalon contains three principal structures known as the cerebellum, pons and medulla oblongata.

Cerebellum

The cerebellum is the second largest region of the brain. It is situated over the medulla and pons. It is a highly specialized part of the CNS, being involved in the subconscious coordination of movements of skeletal muscle such as those required for the maintenance of posture and balance. This function of the cerebellum is shown by the effect of damage to the cerebellum. Damage results in a lack of skeletal muscle coordination.

Pons

The pons links the nerve pathways from the medulla oblongata and cerebellum with the midbrain. It contains a respiratory centre that works in conjunction with a respiratory centre in the medulla oblongata to regulate the rate of breathing.

Medulla Oblongata

The medulla oblongata acts as a link between the spinal cord and the remainder of the brain. It also plays an important role in homeostasis by containing coordinating centres for the following homeostatic mechanisms:

1. cardiac centre—regulates heart rate,
2. vasomotor centre—regulates blood vessel diameter and thus blood pressure,
3. respiratory centre—acts with the pons respiratory centre to regulate the rate of breathing.

Most motor nerve pathways originating from each cerebral hemisphere cross in the medulla, resulting in the right side of brain controlling the left side of the body.

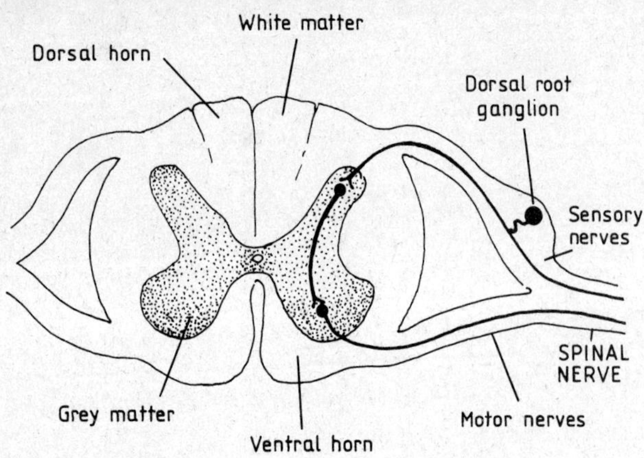

Figure 26.3 Cross-section of the spinal cord showing the point of entry and exit of spinal nerves.

SPINAL CORD

The spinal cord is a long rod of nervous tissue that connects the medulla with the body. It is located in a canal that is found near the centre of the vertebral column. Figure 26.3 shows a typical cross-section through the spinal cord. Note that the relative position of the white and grey matter is reversed when compared with cerebral hemispheres, that is, the white matter now surrounds the grey matter.

The white matter contains groups of long nerve fibres called tracts. These tracts either carry information to the brain (ascending) or transmit messages from the brain to segments of the spinal cord (descending). The ascending and descending tracts originate and terminate respectively at specific points along the cord in the grey matter.

The grey matter is essentially a region consisting of nerve connections or synapses (see 19.2) and short neurons that link other neurons originating or terminating within the grey matter.

At thirty-one points along the cord, spinal nerves enter and leave the spinal cord and pass through natural holes in the vertebral column to the body. These nerves are either sensory (afferent) nerves conveying information from a particular segment of the body or they are motor (efferent) nerves that carry messages to skeletal muscle (effectors). The spinal efferent nerves form the somatic nervous system. The spinal cord is effectively divided into functional segments for different regions of the body.

The point of entry of sensory neurons at a particular segment is located at the dorsal (posterior) horn of the spinal cord (Figure 26.3). These neurons form synapses in the grey matter with either motor neurons, ascending neurons or short interconnecting neurons. Motor nerves form a synapse in the ventral (anterior) horn of the grey matter and leave the spinal cord. The cell bodies of those neurons occur near the cord in a region called the dorsal root ganglion. The sensory and motor nerves form a common group of fibres known as the spinal nerve.

Each spinal cord segment consists of a pair of spinal nerves, one of which communicates with a segment of the left side of the body and the other the corresponding segment on the right side of the body.

Connections exist within the spinal cord between the sensory and motor neurons of the same segment. This results in the formation of a localized communication system. This enables rapid local responses to changes in the environment that are referred to as spinal reflexes.

APPLICATION: STRETCH REFLEXES

Tapping the knee elicits a knee jerk reflex response. Sensory neurons detect the tap and stimulate the motor neurons with which they synapse. The motor neurons stimulate a contraction in the muscle that extends the lower leg (Figure 26.4). This reflex has not required any action by the brain and is thus a rapid acting homeostatic mechanism.

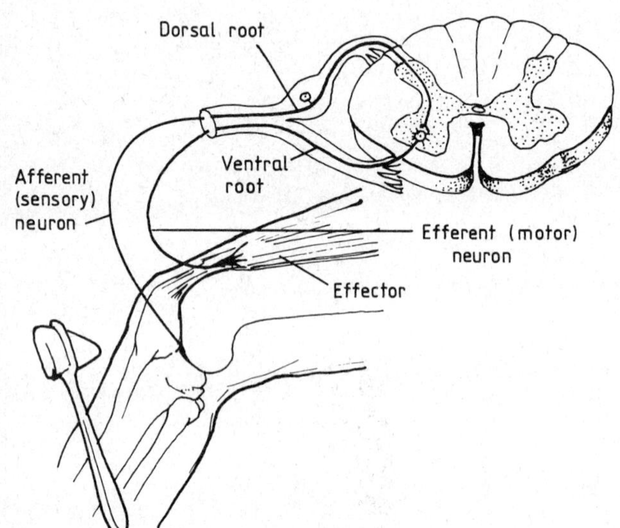

Figure 26.4 Pathway of the knee jerk spinal reflex.

26.4 Peripheral nervous system

The peripheral nervous system (PNS) consists of the spinal nerves, cranial nerves and the autonomic nervous system. The nerves of the PNS conduct information from the body to the CNS where it is analysed and effector messages are delivered to the relevant region of the body. The PNS is an important part of many homeostatic mechanisms since it acts as the afferent and efferent pathways (see 1.2).

SPINAL NERVES

A spinal nerve enters and leaves a particular segment of the spinal cord. Each spinal nerve is composed of a afferent or sensory neuron and a efferent or motor neuron.

The afferent neurons contain sensory nerve endings or receptors. EXAMPLES: Temperature, stretch. Receptors play an essential part in body homeostasis as they detect alterations in the body's internal and external environments. EXAMPLES: Chemical receptors, stretch receptors. When stimulated, these receptors send impulses along the axon entering a spinal cord segment at the dorsal (posterior) horn and are then transmitted via a synapse to either its complementary motor neuron or to nerves of the CNS.

The afferent neurons leave the spinal cord at the ventral (anterior) horn and then connect with muscle or glands. These neurons influence many body homeostatic mechanism responses.

CRANIAL NERVES

There are twelve pairs of nerves that are directly connected with the brain (Figure 26.5). They are associated with functions such as sight (optic nerve), hearing (auditory nerve), smell (olfactory nerve), facial expression (facial nerve) taste (facial, glossopharyngeal nerves) and many other sensory and motor functions related to the head.

The cranial nerve known as the vagus leaves the head and innervates many organs. It plays a major role in the autonomic nervous system.

AUTONOMIC NERVOUS SYSTEM

The autonomic nervous system is a nervous system that connects the CNS to smooth muscle, cardiac muscle and glands. It is composed of two separate systems that exert opposite effects on effector tissue. The sympathetic nervous system is an excitatory or stimulatory system whereas the parasympathetic system is generally considered to be an inhibiting system. The sympathetic system is a highly energy demanding system that enables the

Figure 26.5 The twelve cranial nerves and the structures with which they connect. (Lankford, *Integrated Science for Health Students*, p.221, 1979, reprinted with permission of Reston Publishing Co., a Prentice Hall Co., 11480 Sunset Hills Road, Reston Virginia 22090)

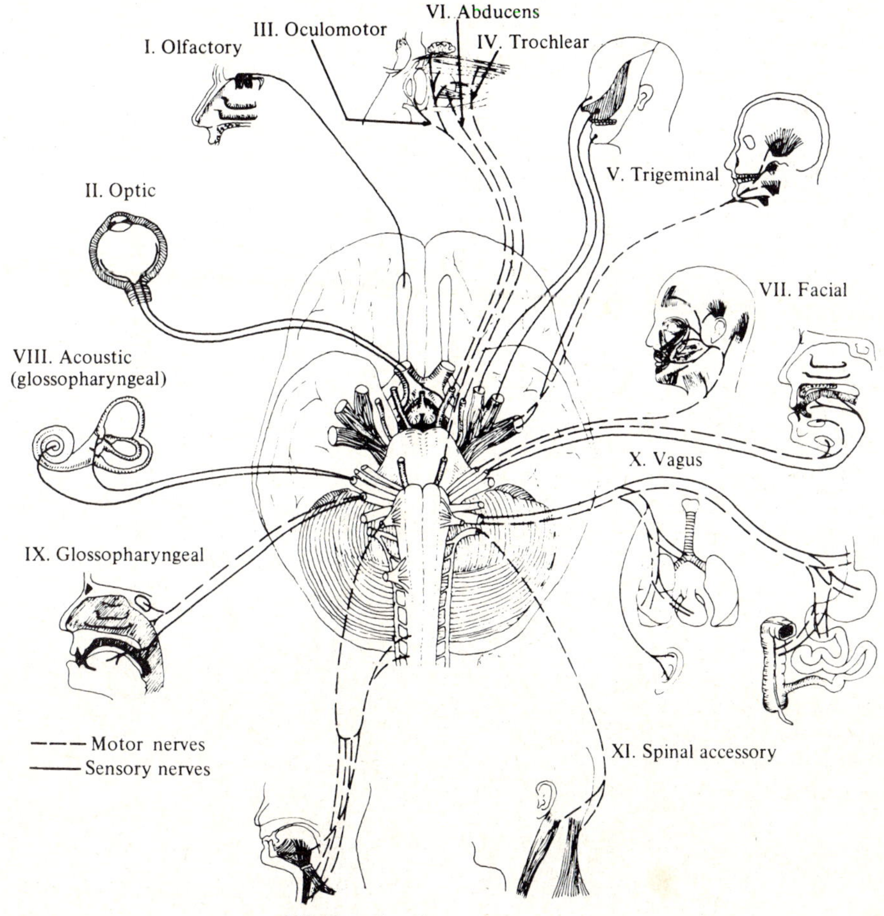

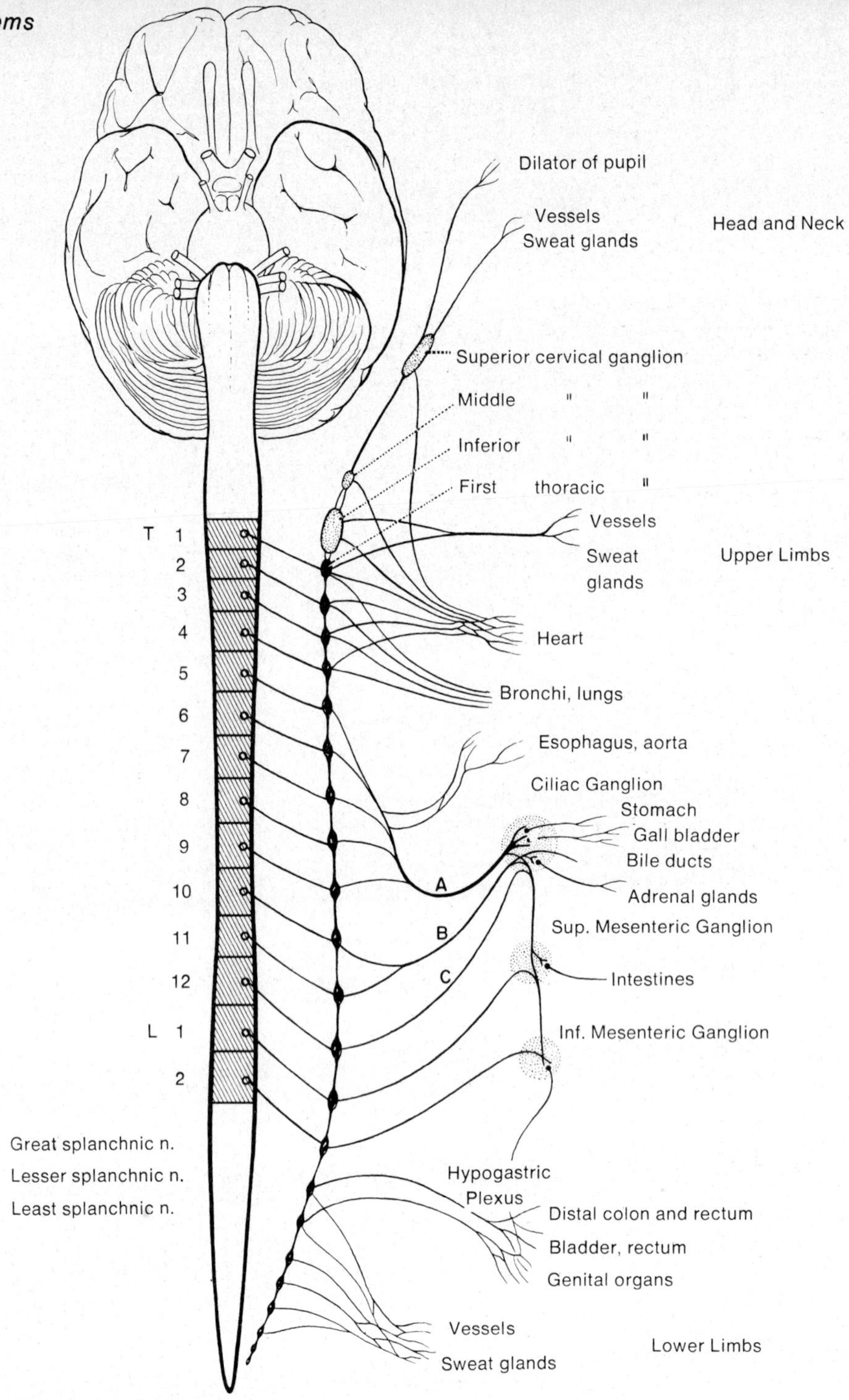

Figure 26.6 Organization of the sympathetic nervous system. (Reprinted with permission of Macmillan Publishing Co., Inc. from *The Human Body: its Structure and Physiology*, 4th edition, copyright © 1978 by Sigmund Grollman)

body to overcome imbalances within the body. The parasympathetic is generally a deactivating system that requires less energy.

The autonomic nervous system was considered to be independent of conscious control. An increased understanding of the autonomic system has led to the ability to train people to consciously influence homeostatic mechanisms that maintain the body's internal environment. This process is referred to as biofeedback and will be referred to later in this chapter.

Sympathetic Nervous System

The sympathetic nervous system originates from the thoracic and lumbar regions of the spinal cord. The neurons enter one of two chains of ganglia that run parallel to the spinal cord (Figure 26.6). The cell bodies of these sympathetic nerves are located in the grey matter of the spinal cord. These neurons form synapses with the cell bodies of post-ganglionic nerves in the ganglia.

Ganglia are merely junction regions between the neurons that synapse with the CNS and those that innervate the internal muscles and glands. Some of the ganglia occur near the organs with which they connect. Others such as the superior cervical ganglia are extensions of the ganglionic chain. This latter ganglia is the junction box for the sympathetic nerves located in the upper regions of the body.

The general effect of the sympathetic nervous system is constriction of smooth muscle tubes, increased rates of muscle contraction, elevated breathing rates and increased secretions from glands. These effects result from the use of the stimulatory neurotransmitter norepinephrine (noradrenaline). The neurons are thus sometimes referred to as adrenergic nerves.

Parasympathetic Nervous System

The parasympathetic nervous system connects with the CNS via some of the cranial nerves and a group of nerves leaving the spinal cord in the sacral region (Figure 26.7). These nerves form synapses in ganglia close to or in the organs that they are innervating. The post-ganglionic neurons are thus short. There appears to be no direct connection between the cranial and sacral parts of the parasympathetic nervous system. Not all organs and internal muscles are connected with the parasympathetic system.

The principal nerve involved in parasympathetic function is the vagus. This nerve innervates a large number of organs including the heart, kidneys and liver, along with the respiratory and digestive systems.

The parasympathetic nerves generally exert an inhibitory action on tissue. The neurons use the neurotransmitter acetylcholine and are thus often referred to as cholinergic nerves.

Biofeedback

Biofeedback is the process whereby a patient is taught to consciously control some actions of the autonomic nervous system such as heart rate or muscle tension. This is achieved by monitoring the action with visual or auditory evidence such as sounding a tone when the blood pressure reaches a certain point.

The steps involved in biofeedback are:

1. Detection of subconscious physiological activity and the connection of a transducer that converts the activity into a visual or auditory signal. This signal must be a representative signal of the physiological activity, that is, variations in the physiological activity must result in a relative alteration in the signal.
2. The patient is then instructed on how to respond to the signal. EXAMPLE: Increased muscle tension

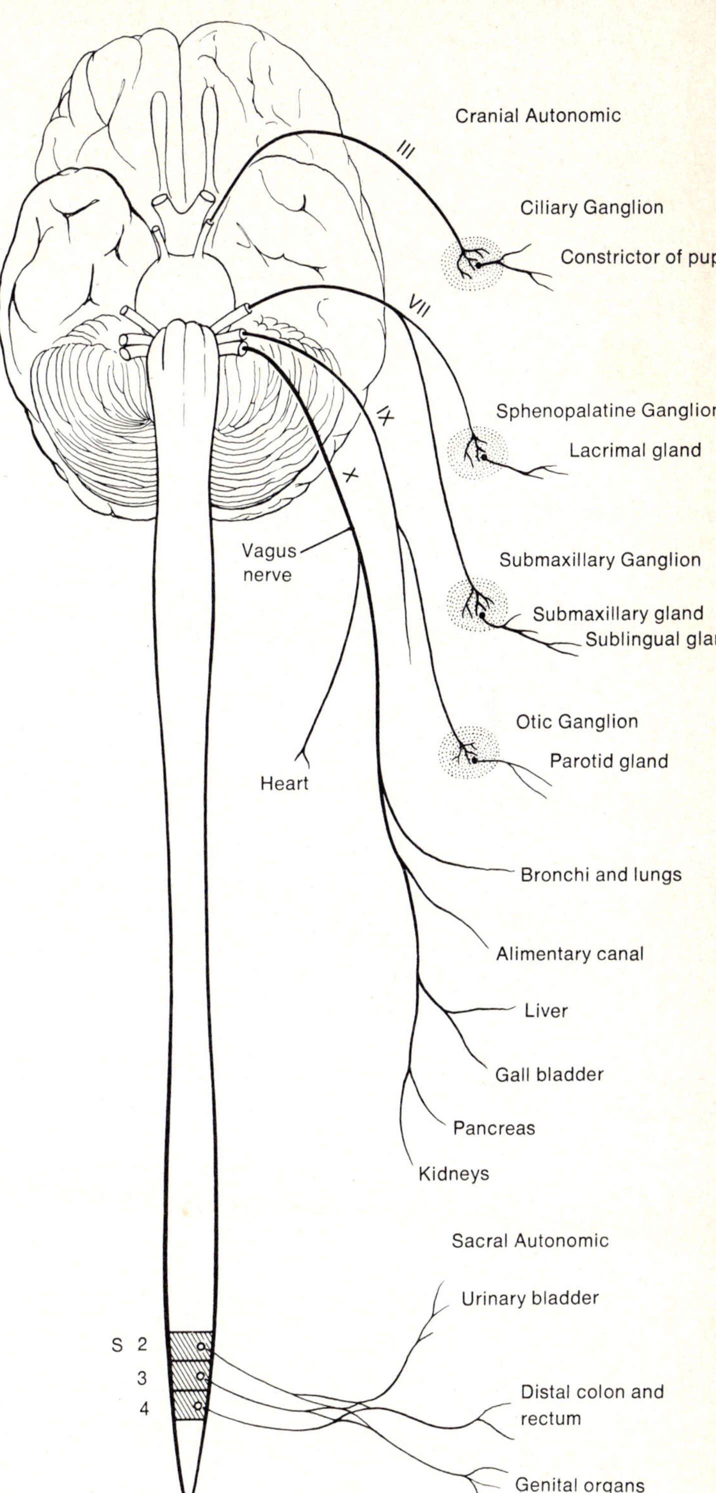

Figure 26.7 Organization of the parasympathetic nervous system. (Reprinted with permission of Macmillan Publishing Co., Inc. from *The Human Body: its Structure and Physiology*, 4th edition, copyright © 1978 by Sigmund Grollman)

produces an elevated volume of sound. The patient is asked to reduce this volume by performing the previously shown methods of relaxation.

Biofeedback is a conscious homeostatic mechanism that relies upon signals to act as the feedback mechanism. It is an attempt to override the subconscious automatic homeostatic mechanism of the autonomic nervous system.

Biofeedback is presently used in an attempt to control many stress-related physiological imbalances. EXAMPLES: Tension headache, anxiety, essential hypertension and gastrointestinal disorders. Its application and effectiveness is certain to increase in the future as an understanding of body homeostatic mechanisms increases.

Summary

The nervous system can be divided into the:
1. Central nervous system which contains the brain and spinal cord.
2. Peripheral nervous system which contains the spinal nerves, cranial nerves and the autonomic nervous system. The efferent spinal nerves represent the somatic nervous system, that is, the nerves going from the CNS to skeletal muscle.

The brain consists of the:
1. Forebrain which contains the cerebral hemispheres, basal ganglia, thalamus, hypothalamus and pineal gland.
2. Midbrain which connects the forebrain to the hindbrain.
3. Hindbrain which contains the cerebellum, pons and medulla oblongata.

Some of the functions of the brain are:
1. Brain generally acts as an integrating centre for many homeostatic mechanisms.
2. Hypothalamus and pineal gland connect the nervous system to the endocrine system.
3. Thalamus and cerebellum play an important part in the coordination of skeletal muscle movements.
4. Medulla oblongata acts as a coordinating centre for cardiovascular and respiratory homeostatic mechanisms.

The spinal cord connects the medulla with the body. It contains ascending and descending tracts which carry information to and from the brain.

Spinal nerves form part of the peripheral nervous system. A pair of spinal nerves enters each of the thirty-one spinal cord segments. These nerves carry sensory information to the spinal cord and instructions from the spinal cord to the effector tissue.

Cranial nerves connect directly with the brain. The twelve pairs are mainly responsible for sensory and motor functions in the head, except for the vagus which has a wide body distribution.

The autonomic nervous system is subdivided into the sympathetic and parasympathetic nervous systems. The sympathetic nervous system generally has a stimulatory action whereas the parasympathetic nervous system is usually an inhibiting system. Sympathetic nerves originate from the thoracic and lumbar spinal cord segments. Parasympathetic nerves consist of several cranial nerves, and a group of nerves leaving the spinal cord in the sacral region. The vagus is the principal parasympathetic nerve.

Biofeedback is the process whereby a patient is taught to consciously control some actions of the autonomic nervous system.

Chapter 27

Neuroendocrine connections

Objectives

At the completion of this chapter the student should be able to:
1. define the terms hormone, target tissue, target gland and tropic hormone,
2. describe the relationship between the nervous system and the endocrine system,
3. describe how the adrenal medulla, pineal gland, hypothalamus and posterior pituitary act as neuroendocrine connections.

27.1 Introduction

The endocrine system is composed of a series of glands that secrete substances called hormones directly into the bloodstream. This system plays a vital role in the maintenance of homeostasis. An endocrine gland performs the function of secreting substances into the bloodstream. Other secretory organs, known as exocrine glands, secrete substances into ducts for transmission to their site of action.

The endocrine system contains a variety of homeostatic mechanisms. These mechanisms rely upon hormones to relay messages from the integrating centres to the effector sites in order to ellicit the desired response. The action of a particular hormone is often supported or inhibited by other hormones. The interaction of hormones in the body results in a coordinated body response to any imbalance in the internal environment.

27.2 Terms associated with the endocrine system

The following terms are frequently used in any discussion of the endocrine system.

HORMONE

A hormone is any substance that is released into the blood in small amounts and which on arrival at a particular tissue elicits a response. Each hormone regulates specific cell reactions by increasing or decreasing particular cellular processes. They never initiate actions and their regulatory action on a cell is based on either altering enzyme activity or membrane transport.

Hormones are either derivatives of lipid or amino acid metabolism.

Lipid Hormones
Lipid hormones occur as steroids secreted from either the adrenal cortex or gonads. EXAMPLES: Cortisol and aldosterone from the adrenal cortex; oestrogen and testosterone from gonads. The prostaglandins are considered to be tissue hormones by some physiologists. An overview of their actions in the body is described in 23.2.

Amino Acid Based Hormones
Hormones derived from amino acids can either exist as peptides or modifications of amino acids. The peptide hormones range in size from small peptide molecules to large polypeptides. Examples of peptide hormones are insulin and antidiuretic hormone. Hormones such as thyroxin and adrenaline are modifications of an amino acid.

TARGET TISSUE

Target tissue refers to the tissue that is acted upon by a hormone. This tissue exhibits a specific physiological response to the hormone. When the tissue is another endocrine gland, the gland is referred to as a target gland. EXAMPLES: The adrenal cortex, thyroid and gonads are target glands that respond to hormones secreted from the anterior pituitary gland.

TROPIC HORMONES

Tropic hormones regulate the release of hormones from a target gland. EXAMPLE adrenocorticotropic hormone (ACTH) is secreted from the anterior pituitary and influences the secretion of hormones from the adrenal cortex.

27.3 Endocrine dependence on the nervous system

The endocrine system plays a major role in body homeostasis, most of its actions being involved in the control of metabolism. For the endocrine system to function in a coordinated manner, it requires an integrating centre and receptors. These receptors inform the integrating centre of how effectively it is functioning (see 1.2). In most endocrine homeostatic mechanisms the brain plays a direct or indirect role as the integrating centre.

The body's metabolism needs to be adjusted according to both the internal and external environment. The nervous system is used to detect changes in the environment. It also forms the afferent part of the communication network.

Connections exist between the nervous system and the endocrine system, that is, neuroendocrine connections occur. It is important to realise that most endocrine actions are influenced by the nervous system. This influence of the nervous system is the basis of how psychological stress can influence normal body metabolism.

27.4 Neuroendocrine connections

The following four pathways show how the nervous and endocrine systems are connected.

HYPOTHALAMUS — ADRENAL MEDULLA

The hypothalamus and the endocrine gland known as the adrenal medulla are connected by a nerve pathway (Figure 27.1). This pathway consists of a nerve tract in the spinal cord which connects with sympathetic nerves that innervate the adrenal medulla. Remember that the hypothalamus is connected to the thalamus which in turn is linked to the cerebrum. Command signals can thus also pass from the cerebrum or forebrain to the adrenal medulla.

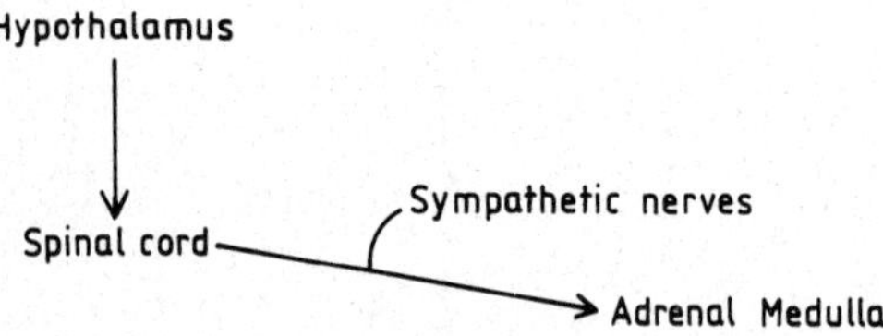

Figure 27.1 The neuronal connection of the hypothalamus and adrenal medulla.

Stimulation of the adrenal medulla by nerve impulses results in the release of the potent stimulatory hormones epinephrine (adrenaline) and norepinephrine (noradrenaline). These hormones are responsible for increasing the ability of the body to cope with stress. This is achieved by increasing the performance of the heart and lungs and increasing blood sugar levels and the breakdown of glucose. The combined effect of these actions is an increased quantity of adenosine triphosphate (ATP).

The connection of the adrenal medulla via a nerve pathway to the cerebral cortex enables the adrenal medulla to respond to stress rapidly and effectively. The cerebrum receives information from the external environment through receptors in the eye, ear, nose, mouth and skin. Through these receptors, the body is able to detect potential stressful situations and prepare itself almost immediately for the stress.

RETINA — HYPOTHALAMUS — PINEAL GLAND

A pathway of nerves connects the pineal gland with the retina via the hypothalamus (Figure 27.2). This pathway originates at the retina where light/dark stimuli are converted into nerve impulses. Some of these nerve impulses travel via a nerve tract to the lateral regions of the hypothalamus. Nerve fibres then pass down the spinal cord and synapse with sympathetic nerves that connect with the pineal gland via the superior cervical ganglion.

The pineal gland appears to be involved with, or may even be the site of, the biological clock, since hormones secreted by the pineal gland are influenced by the presence of light.

Until recently, this gland was thought to only function in children as it calcifies before puberty. Presently, evidence suggests that it functions throughout life and that calcification of the gland does not interfere with its function.

The neuroendocrine connection involving the pineal gland appears to play a major role in endocrinology since the pineal gland is now considered to be the "master" endocrine gland. It is involved in the regulation of body metabolism on a twenty-four hour cycle. This cycle is called the circadian rhythm.

The influence of the nervous system on the pineal gland is shown when the pineal gland secretes the hormones melatonin and arginine-vasotocin (AVT) in response to darkness. Their levels of secretion at midnight are five times that found at noon. This suggests that light stimuli being detected by the retina are inhibiting the production and release of hormones in the pineal gland.

Hormones from the gland such as melatonin and arginine-vasotocin influence the endocrine function of the hypothalamus, pituitary and various other endocrine glands. The relationship between the pineal gland and the endocrine system is discussed in 28.2.

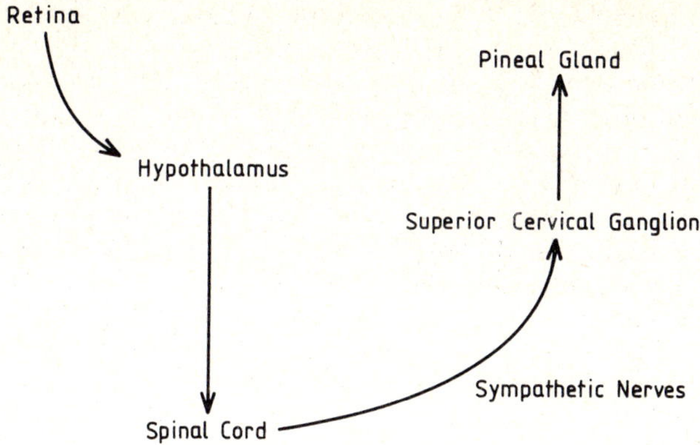

Figure 27.2 The neuronal connection of the retina and hypothalamus with the pineal gland.

HYPOTHALAMUS — PITUITARY GLAND

The hypothalamus forms a neuroendocrine connection with the posterior region of the pituitary gland or neurohypophysis (Figure 27.3). Nerves originating from the hypothalamus terminate in the posterior pituitary. These nerves produce the hormones oxytocin and anti-diuretic hormone (vasopressin) in response to neural and hormonal information. Following production in the cell body of these neurons, the hormones travel via axoplasmic flow to the posterior pituitary gland (see 26.1). This journey takes between one and two hours. The hormones are stored in the posterior pituitary and released in response to stimulation of the hypothalamic nerves.

This neuroendocrine pathway plays an important role in the homeostasis of body fluid concentration.

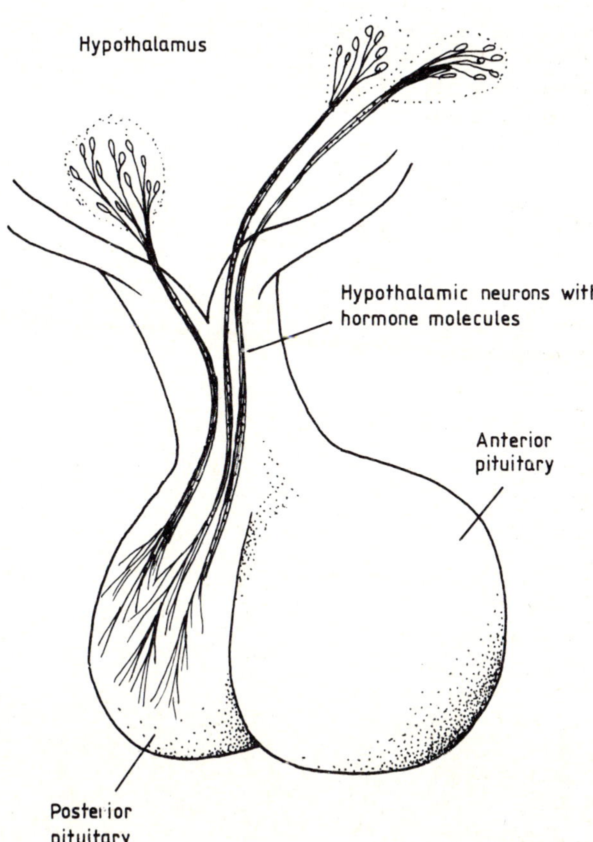

Figure 27.3 Neuronal connection of the hypothalamus with the posterior pituitary gland.

HYPOTHALAMUS

All of the above neuroendocrine connections involve the hypothalamus. The hypothalamus influences the release of hormones from the major endocrine gland known as the anterior pituitary or adenohypophysis.

The hypothalamus secretes a group of regulating hormones that travel via a localized vascular network to the anterior pituitary gland (Figure 27.4). Each particular hormone has the property of stimulating or inhibiting the release of a hormone with a similar structure from the anterior pituitary.

Stimulation of particular neurosecretory cells in the hypothalamus results in the release of a specific regulating hormone. This substance then travels via the blood to the anterior pituitary where it regulates

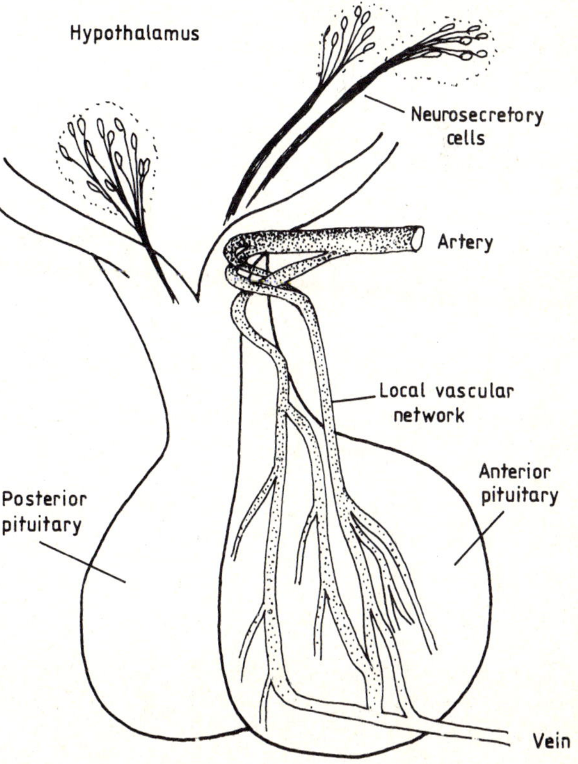

Figure 27.4 Connection of the hypothalamus with the anterior pituitary gland.

the release of its complementary hormone. A description of this mechanism and the hormones released from the anterior pituitary is described in 28.2.

In the case of the anterior pituitary, the hypothalamus contains the neuroendocrine connection within it, since regulating hormones are secreted directly into the blood stream. The anterior pituitary is indirectly connected to the nervous system via the controlled release of regulating hormones from the hypothalamus.

Summary

The endocrine system consists of an inter-related system of glands. Endocrine glands secrete hormones directly into the blood whereas exocrine glands secrete substances such as enzymes into ducts. A hormone is any substance that is released into the blood in small amounts and elicits a response on its target tissue. If the target tissue is another endocrine gland the hormone is referred to as a tropic hormone. Nerve receptors are used to detect alterations in the environment. Information of any alterations is conveyed to the integrating centre (usually the brain). The instruction is then conducted via nerve pathways to neuroendocrine connections where the impulses are converted into the secretion of the relevant hormone.

The important neuroendocrine connections are the adrenal medulla, pineal gland, hypothalamus and posterior pituitary.

Chapter 28

General organization and function of the endocrine system

Objectives

At the completion of this chapter the student should be able to:
1. describe the general functions of the endocrine system,
2. describe the vertical and horizontal hierarchies of endocrine organization,
3. describe the endocrine functions of the hypothalamus and pineal gland,
4. describe how the hypothalamus regulates the secretion of hormones from the anterior pituitary,
5. explain how feedback influences hormonal secretion,
6. describe how stress influences the endocrine system, and in particular the adrenal glands.

28.1 General function of the endocrine system

The endocrine system performs the following functions:
1. regulation of the composition and volume of extracellular fluid by altering the ionic composition and volume of extracellular fluid,
2. regulation of ATP production by promoting metabolic activity,
3. regulation of digestion and absorption,
4. adapting the body to changes in the external environment,
5. adjusting the body's defence mechanisms to cope with the penetration of harmful foreign substances into the body,
6. control of growth and development,
7. development of sperm and the maturation and release of ovum for reproduction,
8. development of the embryo, foetus and newborn baby.

28.2 Organization of the endocrine system

The endocrine system is organized as a highly integrated network where the actions of one gland can greatly influence the function of other glands. The major endocrine glands and their principal actions are listed in Table 28.1. These glands can be arranged into two hierarchies or levels of organization which can be referred to as the vertical and horizontal hierarchies.

Table 28.1 The major endocrine glands, their hormones and the principal actions of the hormones

Gland	Hormone	Actions
Adrenal-cortex	Glucocorticoids	Cortisol and corticosterone inhibit the inflammatory response, maintain blood glucose levels during starvation by increasing protein catabolism and decreasing glucose catabolism
	Mineralocorticoids	Aldosterone stimulates the reabsorption of sodium ions in the kidney and the elimination of potassium ions
	Gonadocorticoids	Male androgens and female oestrogens play a role in secondary sexual characteristics particularly the androgens
Adrenalmedulla	Epinephrine or adrenaline	Reinforces actions of sympathetic nervous system; increases cardiac output, blood sugar level and metabolic rate
	Norepinephrine or noradrenaline	Constricts blood vessels, increases systolic and diastolic pressure
Gonads	Oestrogen	Growth and maintenance of female reproductive organs, breasts and secondary sex characteristics, initiates the preparation of the uterus for embryo implantation
	Progesterone	Secreted by the corpus luteum, essential for completion of the uterus lining in preparation for embryo implantation
	Testosterone	Secreted by the testis, promotes the development of male secondary sex characteristics and development of sex organs
Pancreas (islets of Langerhans)	Insulin	Increases glucose transport into cells and accelerates the conversion of glucose into glycogen, the net effect is a lowered blood glucose level
	Glucagon	Raises blood sugar level by increasing glycogen breakdown secreted by the ovaries
Parathyroid	PTH (parathyroid hormone) or parathormone	Elevates blood calcium concentration by increasing its absorption from the small intestine and its breakdown in bone; it decreases blood phosphate concentration
Pineal gland	Melatonin	Inhibits hypothalamus and the pituitary gland
	AVT (arginine vasotocin)	As above but more potent
Pituitary-anterior	ACTH (adrenocorticotropic hormone)	Stimulates the secretion of hormones from the adrenal cortex
	TSH (thyroid-stimulating hormone)	Stimulates the secretion of hormones from the thyroid gland)
	FSH (follicle-stimulating hormone)	Stimulates the maturation of follicles in the ovaries, the secretion of oestrogen from the follicles and the formation and maturation of sperm
	LH (luteinizing hormone) or ICSH (interstital cell-stimulating hormone)	Stimulates the maturation and release of follicles, the development of the corpus luteum, the secretion of testosterone in males
	HGH (human growth hormone) or somatotropin	Acts on tissues to promote growth by increasing the rate of amino acid transport in cells, stimulates the preferential breakdown of fat instead of glucose
	Prolactin	Acts on the mammary gland to stimulate milk secretion
	MSH (melanocyte-stimulating hormone)	Increases skin pigmentation
Pituitary-posterior	ADH (antidiuretic hormone)	Maintains or expands blood volume by increasing water reabsorption from the kidneys, maintains blood osmolality
	Oxytocin	Stimulates muscular contraction in the pregnant uterus and the ejection of milk
Thyroid	Thyroxin	Regulates catabolism and anabolism, accelerates body growth, glucose absorption from the small intestine
	Triiodothyronine	Same actions as thyroxin but is three to five times more potent
	Calcitonin (thyrocalcitonin)	Inhibits breakdown of calcium from bone and accelerates calcium absorption by bone

Table 28.2 Anterior pituitary tropic hormones and their respective target glands

Tropic hormone	Target gland
ACTH	Adrenal cortex
FSH	Ovaries and testes
LH	Ovaries
ICSH	Testes
TSH	Thyroid

VERTICAL HIERARCHY

The vertical hierarchy consists of a chain of target glands and tropic hormones. Each gland regulates the release of hormones from the next gland in the hierarchy. ANALOGY: In a series of waterfalls, the water from the top waterfall causes water to flow out of the next water fall and so on in a "cascade effect". The hypothalamus and pineal gland appear to be at the top of the hierarchy.

Endocrine Functions of the Hypothalamus

The hypothalamus performs the vital task of being the receptor or detector for many endocrine homeostatic mechanisms. The hypothalamus is able to detect chemical variations in the blood since it is not protected by the blood–barrier (see 12.1).

Information from the higher centres of the brain can influence the response that the hypothalamus may make. EXAMPLE: Anxiety, emotional disturbances can influence the hypothalamus and alter the expected homeostatic response. EXAMPLE: Elevated blood cortisol levels would normally result in the hypothalamus initiating a series of endocrine steps that lead to a decreased blood cortisol concentration. However, a patient under stress stimulates the release of cortisol. Since higher centres have the ultimate authority, this can result in either the cortisol level being maintained or increased. The normal homeostatic response of a decreased blood cortisol concentration has been overriden and has resulted in an abnormal blood cortisol level.

Hypothalamus–Anterior Pituitary

The influence that the hypothalamus exerts over the anterior pituitary is of particular importance since the latter gland acts as an intermediary gland between the hypothalamus and several peripheral glands. The tropic hormones secreted by the anterior pituitary and the glands that they influence are listed in Table 28.2. An additional hormone known as human growth hormone or somatotropin is secreted by the anterior pituitary. This hormone influences a variety of metabolic effects (see Table 28.1).

The hypothalamus secretes specific regulating hormones for each of the anterior pituitary hormones. These regulating hormones are referred to as either releasing hormones or inhibiting hormones according to their actions on the anterior pituitary. These hormones ellicit a specific action by only influencing the release of their complementary anterior pituitary hormone.

The hypothalamus hormones are secreted from the hypothalamus in response to nerve activity. They then flow to the anterior pituitary via a localized vascular network (see 27.4 and Figure 27.4). At the anterior pituitary gland, releasing hormones stimulate the release of their corresponding pituitary hormone. EXAMPLE: Corticotropic-releasing hormone stimulates the secretion of ACTH. A release-inhibiting hormone restrains the release of its corresponding anterior pituitary hormone. EXAMPLE: Prolactin release-inhibiting hormone (PIH) prevents the release of prolactin from the anterior pituitary. Absence of PIH results in the release of prolactin since the restraint on the secretion of prolactin has been removed.

Pineal Gland

In humans, the pineal gland functions throughout life. Calcification of the gland is probably an indicator of the secretory activity of the gland. Calcification appears to have little effect on the endocrine secretory cells. These cells are called pinealocytes.

An indication of the function of the pineal gland is its extensive vascular network. Endocrine glands require a highly vascular network for the efficient release of their hormones into the body's bloodstream. The pineal gland's importance in the endocrine system can be gauged by the observation that it is the second-most vascular organ in the human body (the kidney is first).

The pineal gland has been referred to as the "regulator of regulators", that is, it regulates the endocrine system. It appears that the pineal gland is involved in the regulation of the hypothalamus, pituitary, parathyroid and pancreas. The pineal gland exerts an inhibitory effect on the hypothalamus via the hormones melatonin and arginine vasotocin and thus opposes the actions that the hypothalamus elicits on the endocrine system.

An important aspect of pineal gland function is its neuronal connection with the retina. The pineal gland responds to increases in light intensity by decreasing the secretion of its hormones. Therefore, during the day minimal inhibition of the secretion of hypothalamic regulating hormones exists. At night, elevated levels of melatonin and arginine vasotocin decrease hypothalamic secretion. The pineal gland thus initiates a circadian rhythm, that is, a twenty-four hour cycle system. This action suggests that the pineal gland plays an important role in the functioning of the "biological clock".

The function of the pineal gland in the body homeostasis is only beginning to be unravelled. In the near future it is hoped that a greater knowledge of the gland's actions on the body will enhance the

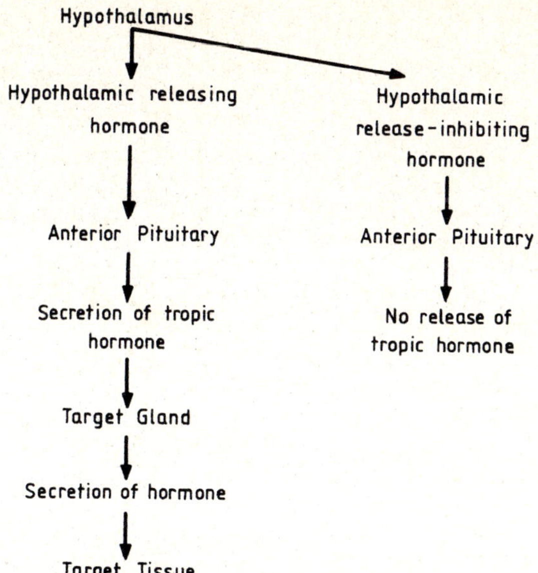

Figure 28.1 Control of peripheral endocrine glands by the hypothalamus and anterior pituitary gland showing a vertical hierarchy.

understanding of body homeostatic mechanisms, and in particular the function of the biological clock in these mechanisms.

Hypothalamus–Anterior Pituitary–Target Glands

The control of peripheral endocrine glands by the hypothalamus and anterior pituitary gland is summarized in Figure 28.1. The following steps are involved in these hormonal pathways:

1. the hypothalamus secretes a regulating hormone that travels via a short localized vascular system to the anterior pituitary gland,
2. the anterior pituitary responds to a releasing hormone by secreting a corresponding tropic hormone; if the regulating hormone is a release-inhibiting hormone the result is a decreased secretion of the corresponding tropic hormone,
3. the tropic hormone travels via the vascular system and stimulates the secretion of a hormone from the target gland,
4. the target gland's hormone then travels via the vascular system to the target tissue where a response occurs.

APPLICATION: CONTROL OF ADRENAL CORTEX

The hypothalamus secretes corticotropic-releasing hormone (CRH) which stimulates the anterior pituitary to secrete adrenocorticotropic hormone (ACTH). The ACTH travels via the vascular system to the adrenal cortex where it elicits the release of a variety of hormones such as cortisol and aldosterone.

Influence of Feedback

In homeostatic mechanisms, a feedback loop is necessary to inform the integrating centre of the effect of its actions on the body. This enables the integrating centre to make further fine adjustments in an attempt to correct any remaining imbalance (see 1.2).

In the hypothalamus–anterior pituitary–target gland endocrine pathway, two negative feedback loops usually exist. These loops are shown by using the example of ACTH and the adrenal cortex.

When the adrenal cortex hormones reach a critical level in the blood, hypothalamic receptors detect this level and suppress the release of CRH. The inhibition of CRH results in no stimulus reaching the anterior pituitary for the release of ACTH. The failure of ACTH to be released results in the adrenal cortex failing to release further quantities of its hormones. As these hormones are utilized, the blood levels decrease until the initial inhibition of CRH is removed. Once this inhibition is removed CRH, ACTH and the adrenal cortex hormones are in turn secreted (Figure 28.2).

A second shorter feedback loop also functions. This loop involves the adrenal cortex hormones directly inhibiting the secretion of ACTH from the pituitary.

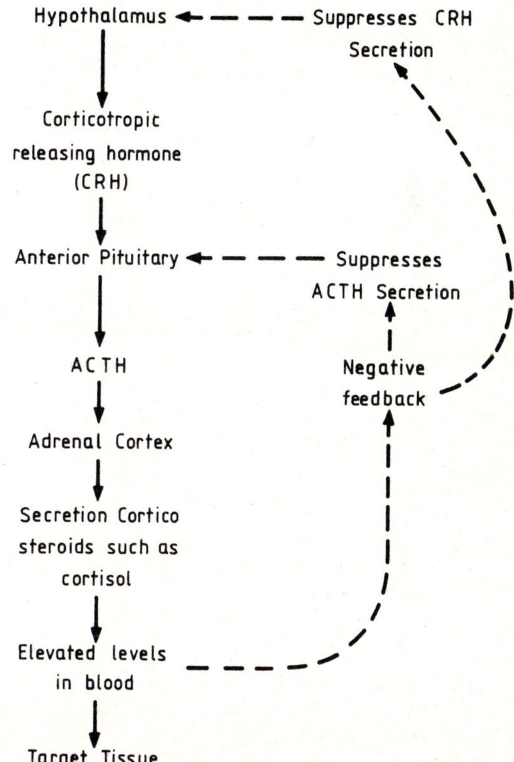

Figure 28.2 Relationship of adrenal cortex secretion to hypothalamic and anterior pituitary gland function.

HORIZONTAL HIERARCHY

The endocrine system also functions in a horizontal hierarchy, that is, a variety of hormones may influence the levels of a particular substance. EXAMPLE: Blood glucose levels can be altered by the actions of insulin, glucagon, adrenaline, human growth hormone, thyroxin and cortisol.

The effect of a particular hormone in altering the level of a substance in the blood will therefore influence the secretion of other hormones involved in the regulation of that substance. EXAMPLE: A decreased blood sugar due to the action of insulin stimulates the release of glucagon.

The action of many hormones is considered to be interrelated with other hormones. It is therefore necessary for the integrating centre to take into account the levels of other hormones when responding to an altered chemical level.

APPLICATION: HORMONAL INFLUENCE ON BLOOD GLUCOSE LEVELS

Insulin promotes increased movement of glucose across cell membranes along with increased utilization of glucose within the cell. The net effect of these actions is a decreased blood glucose level.

Glucagon acts in the liver to increase the quantities of glucose in the blood by increasing glycogen breakdown in the liver. Glucagon and insulin act together to regulate the storage of incoming glucose and the release of glucose between meals. Other hormones can also influence blood glucose levels in certain circumstances.

Epinephrine (adrenaline) is released in response to stress. This hormone elevates blood glucose levels by increasing the breakdown of muscle and liver glycogen. The increased quantities of glucose enhance energy production in time of need.

Thyroxine increases blood glucose levels by facilitating glucose absorption in the small intestine.

Cortisol and ACTH maintain blood glucose levels for the brain during starvation by decreasing glycogen and glucose catabolism, and increasing protein catabolism for energy production.

Somatotropin (human growth hormone) elevates blood glucose levels by accelerating fat catabolism and inhibiting non-essential use of glucose.

28.3 Influence of stress on the endocrine system

Stress is any stimulus that causes an imbalance in the body's environment. The stress may result from such things as injury, starvation, emotional disturbance and excessive noise. Any factor that causes stress will initiate a homeostatic response. Most of these homeostatic responses will exert either a direct or indirect stimulation of hormones in the endocrine system. The hypothalamus appears to be involved in the integration of the nervous and endocrine responses to stress. The effect of this integration is shown in the "fight or flight" response. This response involves the action of the hypothalamus on sympathetic nerves. The effect of this action is increased heart rate, constriction of blood vessels in the skin and abdomen, dilation of blood vessels in skeletal muscle and dilation of bronchioles. The rapid action of these nerves prepares the body for fight or flight.

A slower sustained and more generalized effect is achieved by releasing epinephrine (adrenaline) and norepinephrine (noradrenaline) from the adrenal medulla. The release of these hormones is due to the action of the hypothalamus on sympathetic nerves that connect with the adrenal medulla. The release of hormones from the adrenal medulla as a consequence of stress, is thus due to a direct stimulation of this peripheral gland by the hypothalamus.

Stress can also influence the secretion of hormones from the peripheral endocrine glands by exerting an influence on the secretion of regulating hormones from the hypothalamus. The result of stress on the hypothalamus is the altered secretion of tropic hormones and subsequently altered target gland function. Note that stress can override the feedback loop which can result in a resetting of the homeostatic mechanism. If the stress remains, the resetting will lead to an altered level of circulating hormone.

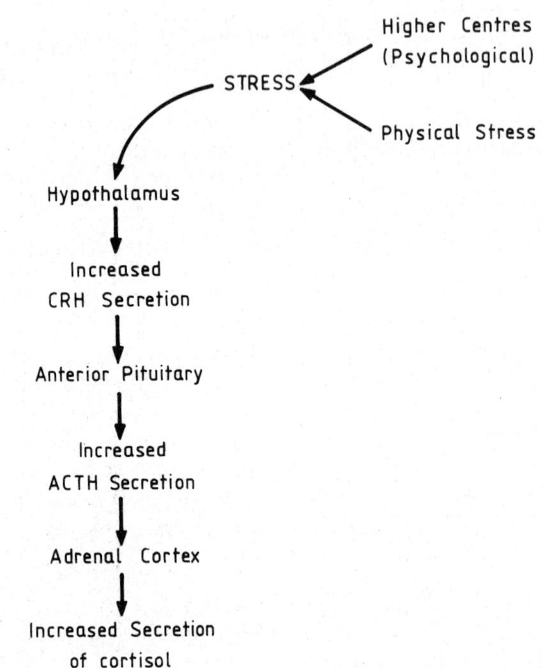

Figure 28.3 Effect of stress on cortisol secretion.

APPLICATION: STRESS AND THE ADRENAL CORTEX

Stress exerts an indirect effect on the adrenal cortex via the vertical hierarchy of the hypothalamus–anterior pituitary (Figure 28.3). Stress stimulates the release of corticotropin-releasing hormone (CRH) from the hypothalamus. The CRH stimulates the release of ACTH from the anterior pituitary. The ACTH then travels via the blood stream to the adrenal cortex where it stimulates the release of cortisol. The stimulation of ACTH release may even occur when an elevated blood ACTH level exists, that is, stress overrides the normal homeostatic mechanism.

28.4 Diabetes mellitus: an example of endocrine malfunction

Diabetes mellitus is the most prevalent endocrine malfunction. It can be described as a chronic disorder due to a relative or absolute deficiency of insulin. Symptoms include the presence of glucose in the urine (glycosuria) and elevated blood glucose levels. These symptoms usually result from a malfunction in the secretion of insulin from the beta cells of the pancreas. The relative lack of insulin results in an inability of glucose to enter most cells and, as a consequence, blood glucose levels remain elevated for an abnormally long time after a meal. When the blood glucose levels exceed the T_m for kidney reabsorption, glucose appears in the urine (see 17.3).

The high blood glucose levels stimulate the hypothalamic blood glucose receptors to initiate the normal homeostatic mechanism for decreasing blood glucose, that is, pancreatic secretion of insulin. In patients where the pancreatic beta cells are destroyed, a failure of the homeostatic mechanism occurs.

The relative intracellular deficiency of glucose can cause an acid/base imbalance as a result of an elevated breakdown of fats and protein. This catabolism occurs in an attempt to supply sufficient energy for cell function.

Excessive catabolism of fats and protein can lead to metabolic acidosis. The acidosis results from the abnormally high levels of ketone bodies produced as a consequence of fat and protein catabolism.

Summary

The endocrine system is involved in the homeostatic control of many important body functions including:

1. regulation of the composition and volume of extracellular fluid,
2. regulation of ATP production,
3. regulation of digestion and absorption,
4. adapting the body to changes in the external environment,
5. adjusting the body's defence mechanisms,
6. control of growth and development,
7. development of mature sperm and ovum,
8. development of the embryo and foetus.

The endocrine system is arranged into two hierarchies known as the vertical and horizontal hierarchies.

The vertical hierarchy usually involves a neuroendocrine connection that controls the release of a series of tropic hormones. The end result of a vertical hierarchy is the secretion of hormones from a peripheral endocrine gland. Feedback and higher centres can influence the secretion of tropic hormones.

The horizontal hierarchy involves the secretion of several hormones for a particular substance. The effect of one hormone usually influences the secretion of other hormones. The interrelationship of hormone actions necessitates an integrating centre to take into account the levels of other hormones when responding to an altered chemical level.

The pineal gland appears to play an important role in the control of the endocrine system. It secretes hormones in response to changes in light intensity. These hormones are involved in the regulation of hypothalamic function.

Stress is any stimulus that causes an imbalance in the body's environment. The hypothalamus appears to play a major part in the integration of the nervous and endocrine responses to stress. The feedback loop in homeostatic mechanisms can be overriden by stress originating from the higher centres, for example psychological stress.

Alterations in the function of part of a homeostatic mechanism results in the inability of the mechanism to perform its normal regulatory role. In diabetes mellitus the relative or absolute deficiency of insulin results in glycosuria and elevated blood sugar levels.

APPENDIX
Normal values of frequently used blood and plasma pathology tests.

	S.I. unit	Old unit
BLOOD		
Haemoglobin		
Males	7.4–11.2 mmol/L or 12.0–18.0 g/dL	12–18 mg/100mL
Females	7.14–10.00 mmol/L or 11.5–16.0 g/dL	11.5–16.0 mg/100mL
Red blood cells		
Females	3.9-5.6 × 10^{12}/L	3.9-5.6 × $10^6\mu$ L/mm^3
Males	4.5-6.5 × 10^{12}/L	4.5-6.5 × $10^6\mu$ L/mm^3
White blood cells	4.0–11.0 × 10^9/L	4000–11,000/mm^3
Platelets	150–400 × 10^9/L	150,000–400,000/mm^3
PLASMA		
Bicarbonate (HCO_3^-)	21–32 mmol/L	21–32 meq/L
Calcium (Ca^{2+})	2.25–2.60 mmol/L	4.5–5.2 meq/L or 9.0–10.4 mg/100mL
Chloride (Cl^-)	95–105 mmol/L	95–105 meq/L
Cholesterol	2.6–6.5 mmol/L	100–250 mg/100mL
Creatinine		
Females	0.04–0.10 mmol/L	0.45–1.10 mg/100mL
Males	0.06–0.12 mmol/L	0.68–1.40 mg/100mL
Glucose	5.0–6.6 mmol/L	80–120 mg/100mL
Magnesium (Mg^{2+})	0.80–1.05 mmol/L	1.6–2.1 meq/L or 1.95–2.55 mg/100mL
Phosphate (inorganic)	0.8–1.5 mmol/L	2.5–4.6 mg/100mL
Potassium	3.8–5.0 mmol/L	3.8–5.0 meq/L
Sodium	137–149 mmol/L	137–149 meq/L
Urea	3.2–7.5 mmol/L	19–45 mg/100mL
Uric acid	0.18–0.47mmol/L	3–8 mg/100mL

Bibliography

Biochemistry and Cell Physiology

BROWN, W. H. and ROGERS, E. P. *General, Organic, and Biochemistry.* William Grant Press, Boston, 1980.

CONN, E. E. and STUMPF, P. K. *Outlines of Biochemistry.* John Wiley, New York, 1978.

GIESE, A. C. *Cell Physiology,* Fifth edition. W. B. Saunders, Philadelphia, 1979.

GOLDSBY, R. A. *Cells and Energy,* Second edition. Macmillan, New York, 1977.

MCELROY, W. D. *Cell Physiology and Biochemistry,* Third edition. Prentice-Hall, Englewood Cliffs, New Jersey, 1971.

MCGILVERY, R. W. *Biochemistry: A Functional Approach,* Second edition. W. B. Saunders, Philadelphia, 1979.

ROBINSON, C. H. *Fundamentals of Normal Nutrition,* Second edition. Macmillan, New York, 1973.

ROUTH, J. I. *Introduction to Biochemistry,* Second edition. W. B. Saunders, Philadelphia, 1978.

STANBURY, J. B., WYNGAARDEN, J. B. and FREDERICKSON, O. S. (eds). *The Metabolic Basis of Inherited Disease,* Third edition. McGraw-Hill, New York, 1972.

VARLEY, H., GOWENLOCK, A. H. and BELL, M. *Practical Clinical Biochemistry, Vol. 2, Hormones, Vitamins, Drugs and Poisons,* Fifth edition. W. Heinemann Medical Books, London, 1976.

WHITE, A., HANDLER, P., SMITH, E. L. HILL, R. L. and LEHMAN, I. R. *Principles of Biochemistry,* Sixth edition. McGraw-Hill, New York, 1978.

YOST, H. T. *Cellular Physiology.* Prentice-Hall, Englewood Cliffs, New Jersey, 1972.

Chemistry

BAUM, S. J. and SCAIFE, C. W. *Chemistry. A Life Science Approach,* Second edition. Macmillan, New York, 1980.

BRADY, J. E. and HUMISTON, G. E. *General Chemistry. Principles and Structure,* Second edition. John Wiley, New York, 1978.

CHANG, R. *Physical Chemistry with Applications to Biological Systems.* Macmillan, New York, 1977.

GRILLOT, G. F. *A Chemical Background for the Paramedical Sciences,* Second Edition. Harper and Row, New York, 1974.

HUHEEY, J. E. *Inorganic Chemistry. Principles of Structure and Reactivity,* S. I. units edition. Harper and Row, London, 1975.

KILGOUR, O. F. G. *An Introduction to the Chemical Aspects of Nursing Science.* W. Heinemann Medical Books, London, 1972.

KOROLKOVAS, A. and BURCKHALTER, J. H. *Essentials of Medicinal Chemistry.* John Wiley, New York, 1976.

LIPPINCOTT, W. T., GARRETT, A. B. and VERHOEK, F. H. *Chemistry. A Study of Matter,* Third edition. John Wiley, New York, 1977.

SACKHEIM, G. I. and SCHULTZ, R. M. *Chemistry for the Health Sciences,* Third edition. Macmillan, New York, 1977.

SEARS, C. T. and STANITSKI, C. L. *Chemistry for Health-related Sciences. Concepts and Correlations.* Prentice-Hall, Englewood Cliffs, New Jersey, 1976.

SEARS, C. T. and STANITSKI, C. L. *Aspects of Chemistry for Health-related Sciences.* Prentice-Hall, Englewood Cliffs, New Jersey, 1979.

ZILVA, J. F. and PANNALL, P. R. *Clinical Chemistry in Diagnosis and Treatment,* Second edition. Lloyd-Luke, London, 1978.

Pathology

BLOOM, W. and FAWCETT, D. W. *A Textbook of Histology,* Tenth edition. W. B. Saunders, Philadelphia, 1975.

HAM, A. W. and CORMACK, D. H. *Histology,* Eighth edition. J. B. Lippincott, Philadelphia, 1979.

ROBBINS, S. L. and COTRAN, R. S. *Pathologic Basis of Disease,* Second edition. W. B. Saunders, Philadelphia, 1979.

TIGHE, J. R. *Pathology,* Third edition. Baillière Tindall, London, 1972.

Pharmacology

BAILEY, R. E. *Pharmocology for Nurses,* Fourth edition. Baillère Tindall, London, 1975.

GRINGAUZ, A. *Drugs. How They Act and Why.* C. V. Mosby, St. Louis, 1978.

JOHNS, M. P. *Drug Therapy and Nursing Care.* Macmillan, New York, 1979.

Remington's Pharmaceutical Sciences, Fifteenth edition. Mack Publishing Co., Easton, Pennsylvania, 1975.

Physics

BRUER, H. *Physics for Life Science Students.* Prentice-Hall, Englewood Cliffs, New Jersey, 1975.

CROMER, A. H. *Physics for the Life Sciences*, Second edition. McGraw-Hill, New York, 1977.

Duncan, G. *Physics for Biologists.* Blackwell, Oxford, 1975.
Gray, H. J. and Isaacs, A. (eds) *A New Dictionary of Physics,* Second edition, Longman, London, 1975.
Greenberg, L. H. *Physics for Biology and Pre-med Students.* W. B. Saunders, Toronto, 1978.
Greenberg, L. H. *Physics with Modern Applications.* W. B. Saunders, Toronto, 1978.
Gustafson, D. R. *Physics: Health and the Human Body.* Wadsworth Publishing Co., Belmont, California, 1980.
Halliday, D. and Resnick, R. *Physics Parts 1 and 2,* Third edition. John Wiley, New York, 1978.
Horsfield, R. S. *Biomechanics.* Science Press, Sydney, 1978.
Horsfield, R., Solomons, S. and Ward, A. *Physics and Chemistry for the Health Sciences.* Science Press, Sydney, 1978.
Kilgour, O. F. G. *An Introduction to the Physical Aspects of Nursing Science,* Third edition. W. Heinemann Medical Books, London, 1978.
Kane, J. W. and Sternheim, M. M. *Physics.* John Wiley, New York, 1978.
MacDonald, S. G. and Burns, D. M. *Physics for the Life and Health Sciences.* Addison—Wesley, London, 1975.
Shortley, G. and Williams, D. *Elements of Physics,* Fifth edition. Prentice-Hall, Englewood Cliffs, New Jersey, 1971.

Physiology

Anthony, C. P. and Thibodeau, G. A. *Textbook of Anatomy and Physiology,* Tenth edition. C. V. Mosby, St. Louis, 1979.
Bell, G. H. Emslie-Smith, D. and Paterson, C. R. *Textbook of Physiology and Biochemistry,* Ninth edition. Churchill Livingston, Edinburgh, 1976.
Bennett, A. and Sanger, G. J. Prostanoids and their relationship to the non-pregnant uterus. *Research and Clinical Forums* **1** (2), 21–25, 1979.
Campbell, E. J. M., Dickinson, C. J. and Slater, J. D. H. (eds) *Clinical Physiology,* Fourth edition. Blackwell Scientific Publications, Oxford, 1974.
Cioffi, D. The fundamentals of blood gases. *The Lamp* August, 5–12, 1979.
Grollman, S. *The Human Body: Its Structure and Physiology,* Fourth edition. Macmillan, New York, 1978.
Guyton, A. C. *Physiology of the Human Body,* Fifth edition. W. B. Saunders, Philadelphia, 1979.
Guyton, A. C. *Textbook of Medical Physiology,* Fifth edition. W. B. Saunders, Philadelphia, 1976.
Hardisty, R. M. and Weatherall, D. J. (eds) *Blood and Its Disorders.* Blackwell Scientific Publications, Oxford, 1974.
Iveson-Iveson, J. The nervous system 1. *Nursing Mirror* January 25th, 20–23, 1979.
Iveson-Iveson, J. The nervous system 2. *Nursing Mirror* February 1st, 26–28, 1979.
Iveson-Iveson, J. The nervous system 3. *Nursing Mirror* February 8th, 32–33, 1979.
Kappers, J. and Ariens Pevet, P. (eds) *The Pineal Gland of Vertebrates Including Man,* Progress in Brain Research, Vol.52. Elsevier/North Holland Biomedical Press, Amsterdam, 1979.
Luft, R. Diabetes: the outlook is bright. *World Health* May, 3–7, 1979.
Miller, M. A., Drakontides, A. B. and Leavell, L. G. *Kimber-Gray-Stackpole's Anatomy and Physiology,* Seventeenth edition. Macmillan, New York, 1977.
Mountcastle, V. B. *Medical Physiology,* Volumes 1 and 2, Fourteenth edition. C. V. Mosby, St. Louis, 1980.
Pearce, E. *Anatomy and Physiology for Nurses.* Sixteenth edition. Faber and Faber, London, 1975.
Robinson, J. R. *Fundamentals of Acid–Base Regulation,* Fifth edition. Blackwell Scientific Publications, Oxford, 1975.
Sodeman, W. A. and Sodeman, T. M. *Pathologic Physiology. Mechanisms of Disease,* Sixth edition. W. B. Saunders, Philadelphia, 1979.
Strand, F. L. *Physiology: A Regulatory Systems Approach.* Macmillan, New York, 1978.
Tortora, G. J. and Anagnostakos, N. P. *Principles of Anatomy and Physiology,* Second edition, Canfield Press, San Francisco, 1978.
Tortora, G. J. and Becker, J. F. *Life Science,* Second edition. Macmillan, New York, 1978.
Vander, A. J. Sherman, J. H. and Luciano, D. S. *Human Physiology: The Mechanisms of Body Function,* Second edition. McGraw-Hill, New York, 1975.
Watson, J. E. *Medical Surgical Nursing and Related Physiology.* W. B. Saunders, Toronto, 1979.
Williams, R. H. (ed) *Textbook of Endocrinology,* Fifth edition. W. B. Saunders, Philadelphia, 1974.
Williams, S. Physiological aspects of stress, *Australian Nurses Journal* **9**(1), 45–48, 1979.

General

Brooks, S. M. *Integrated Basic Science,* Fourth edition. C. V. Mosby, St. Louis, 1979.
Carter, C. O. Recent advances in genetic counselling. *Nursing Times* October 18th, 1795–1798, 1979.
Champney, B. and Smiddy, F. G. *Symptoms, Signs and Syndromes. A Medical Glossary.* Ballière Tindall, London, 1979.
Clarke, M. and Montague, S. (eds) Stress. *Nursing,* Feb., 1980.
Cochaud, G. A. M. and Cochaud, J. E. *Basic Science for the Health Team.* Pitman Medical, Melbourne, 1978.
Dorland's Illustrated Medical Dictionary, Twenty-fifth edition. W. B. Saunders, Philadelphia, 1974.

Health Commission of N.S.W. *S. I. Units in Clinical Pathology.* Sydney, 1977.

HOUGHTON, M. and JEE, L. A. *Baillière's Pocket Book of Ward Information,* Twelfth edition. Baillière Tindall, London, 1971.

JEFFERIES, P. M. (ed) *Baillière's Nurses' Dictionary,* Nineteenth edition. Baillière Tindall, London, 1979.

KILGOUR, O. F. G. *An Introduction to the Biological Aspects of Nursing Science.* W. Heinemann Medical Books, London, 1978.

LANKFORD, T. RANDALL *Integrated Science for Health Students,* Second edition. Reston Publishing Company, Reston, Virginia, 1979.

Mims Annual, 1979 Incorporating the Australian Drug Compendium, Australian edition. Intercontinental Medical Statistics, Sydney.

RYAN, B. W. and KEEN, J. *Basic Science for Nurses.* McGraw-Hill, Sydney, 1970.

WILSON, M. E. and MIZER, H. E. *Microbiology in Patient Care,* Second edition. Macmillan, New York, 1974.

UVAROV, E. B., CHAPMAN, D. R. and ISAACS, A. *The Penguin Dictionary of Science,* Fourth edition. Penguin Books, Harmondsworth, Middlesex, 1977.

Glossary

Absolute zero Lowest temperature that is theoretically possible and is equal to 0 K or -273°C
Acceleration Change of velocity with time
Acellular Non-cellular
Acetyl CoA A substance that is composed of an acetyl group and coenzyme A and is involved in the production of energy
Acid Any substance that releases hydrogen ions into water
Acidity The relative number of free hydrogen ions in water; the greater the number of hydrogen ions the more acidic is that solution
Acidosis Condition resulting from the accumulation of acid or the loss of base from the body
  Metabolic Acidosis that results from either an increase in acidic metabolic products other than carbon dioxide or an excessive loss of base
  Respiratory Acidosis resulting from excess retention of carbon dioxide in the body
Action potential A wave of current that is conducted along a nerve fibre
Active process Energy-dependent movement of substances across membranes
Active transport An active process that moves substances across membranes by means of carriers
Adipose tissue A specialized form of loose connective tissue which is involved in the production and storage of fat
Adrenergic nerve Neuron that has norepinephrine (noradreneline) as its transmitter
Adsorption The holding of a substance to a surface
Aerobic A process that occurs in the presence of oxygen
Aerosol A colloid composed of liquids or solids dispersed in a gas
Alcohol A molecule with an —OH functional group attached to a carbon atom
Aldehyde A molecule that contains the functional group

$$-C-C\begin{matrix}\nearrow O \\ \searrow H\end{matrix}$$

Alkali *See* Base
Alkalosis Condition resulting from the accumulation of base or the excessive loss of acid from the body
  Metabolic—Alkalosis that results from either a decreased hydrogen ion concentration or an elevated base concentration
  Respiratory—Alkalosis resulting from the excess loss of carbon dioxide from the body
Alkanes Hydrocarbons that have carbon atoms linked to each other by single bonds
Alkenes Hydrocarbons that contain one or more carbon-carbon double bonds
Alkynes Hydrocarbons that contain one or more carbon-carbon triple bonds.
Alleles Different forms of a particular gene
Allergen An antigen that causes an allergy
Allergy Oversensitivity of a person to a particular antigen
Alpha particle A combination of two protons and two neutrons which is emitted from the nucleus of some radioactive isotopes
Amines A group of organic compounds derived from ammonia (NH_3); the one, two or three hydrocarbon chains are attached to it according to the number of hydrogens that have been displaced
Amino acids Organic acids that contain an amine group
  Essential Amino acids that are unable to be produced by the body in sufficient quantities for normal metabolism
  Non-essential Amino acids readily manufactured by the body
Amniocentesis Collection of amniotic fluid using a syringe and needle
Ampere S.I. unit for electric current
Anabolism Synthesis of a biochemical
Anaemia A relative deficiency of haemoglobin
Anaerobic A process that occurs in the absence of oxygen
Anaphylactic shock A rapidly developed allergic reaction
Anion An atom that has a relative excess of electrons and has a negative charge of at least -1
Anode Positively charged electrode
Anti-codon Nucleotide triplet on tRNA
Antidiuretic hormone A hormone secreted from the posterior pituitary that increases the reabsorption of water by the kidneys
Antigen Any substance that induces the formation of antibodies or reacts with them
Aqueous solution Contains water as the solvent
Arteries Vessels that carry blood from the heart to capillaries
Arteriole Small artery, connects arterial system to a capillary
Atheroma A mass of material, usually fat, which accumulates on an arterial wall
Atom Smallest structure of matter that represents an element; it consists of protons, neutrons and electrons
Atomic number The number of protons in an atom
Atomic weight The relative weight of an atom compared with an atom of carbon; it is approximately equal to the number of protons and neutrons in an atom
ATP Adenosine triphosphate is a high-energy

molecule that provides energy for cell functions
Atria Filling chambers of the heart
Autoimmunity The formation of an antigen within the body and the subsequent formation and reaction of antibodies to that antigen
Autonomic nervous system The nervous system that connects the central nervous system to smooth muscle, cardiac muscle and glands
Axon A long projection of the cell body that transmits action potentials away from the cell body
Axoplasmic flow Slow movement of substances down an axon

Bacilli Rod-shaped bacteria
Barr body A small dark staining area that occurs when more than one X chromosome is present in a cell
Basal ganglia Aggregations of grey matter located within the cerebal cortex
Base . A substance that donates hydroxyl ions (OH^-) into water or accepts hydrogen ions (H^+) from water
Base of support The region of support for holding an object in a balanced position
Basophil A type of white blood cell
B-Cells Lymphocytes that are involved in the immune system and which originate from recognition lymphocytes; they are later transformed into plasma cells.
Beta oxidation The process of forming acetyl CoA by removing acetyl groups from a fatty acid
Beta particle An electron emitted from the nucleus of an atom as a result of the breakdown of a neutron
Bile A secretion of the gall bladder which has the ability to emulsify fat
Biofeedback The process whereby a patient is taught to consciously control some actions of the autonomic nervous system
Biological value Term relating to the content of essential amino acids in a protein
Blood pressure Hydrostatic pressure that forces blood through the vascular system
  Diastolic Pressure that occurs during filling of the ventricle
  Systolic Pressure that results from contraction of heart ventricles; this pressure is greater than diastolic pressure
Bone tissue A form of connective tissue where the intercellular matrix is calcified
Boyle's Law The volume of a gas varies inversely with pressure when the temperature is constant
Brownian movement The constant motion of particles in random directions
Buffer Chemical that resists changes in pH on addition of acids or bases

---

Cancer A cellular disease that is characterized by uncontrollable cellular multiplication
Capillary Smallest blood vessel, site of nutrient and waste exchange between the blood and tissues
Capsule Slimy mucus-like coat surrounding the bacterial cell wall
Carbohydrate An organic compound containing carbon, hydrogen and oxygen arranged into sugar molecules
Carbonic anhydrase An enzyme that rapidly forms carbonic acid from water and carbon dioxide
Carcinogen An agent that initiates the uncontrollable multiplication of cells, i.e. it initiates cancer
Cardiac arrhythmia Abnormal rhythmic contraction of the heart
Cartilage A form of connective tissue that consists of a gel-like matrix enclosed in a layer of dense connective tissue
Catabolism The breakdown of biochemicals into simpler structures
Cathode Negatively charged electrode
Cation An atom with a positive charge of at least +1; the charge results from the loss of electrons
Cell The smallest and simplest unit of matter that has the potential to duplicate itself
  Eucaryotic Contains a highly organized internal structure
  Procaryotic Contains a simple internal structure
Cell membrane *See* Plasma membrane
Central nervous system Comprises the brain and spinal cord
Centre of gravity The point where the resultant gravitational force acts on an object
Centrioles Small cylindrical cytoplasmic structures that play an important role in cell division
Cerebellum Second largest region of the brain located over the medulla and pons; it is involved in subconscious coordination of skeletal muscle movements
Cerebral cortex Region of grey matter in the cerebral hemispheres
Cerebral hemispheres Two large lobes of the brain that are the site of conscious thought
Cerebrospinal fluid (CSF) Fluid that surrounds the brain and spinal cord
Cerebrum *See* Forebrain
Charles' Law At constant pressure the volume of gas is directly proportional to its absolute temperature
Chemical energy Energy required to bind atoms together
Cholesterol A steroid that is an important component of cell membranes
Cholinergic nerve Neuron that uses the neutrotransmitter acetylcholine
Chromosome Thick, thread-like structure found in the nucleus of a cell that contain the cells hereditary information
Chylomicrons Fat droplets that form an emulsion in lymph and blood
Cilia Short, hair-like, motile projections that

occur in numerous quantities on the plasma membrane of some cells
Circadian rhythm A twenty-four hour cycle
Coagulation Formation of suspended particles due to the combination of positively and negatively-charged colloidal particles
Cocci Spherical-shaped bacteria
Codon Nucleotide triplet in mRNA
Coenzyme A Coenzyme that acts as a carrier and assists enzyme function in the catabolism of carbohydrates and fats
Colloid A fluid mixture in which the particles dispersed in the fluid are too large to be solutes and too small to form suspensions
Compensation Return of pH towards normal due to body homeostatic mechanisms
Compound A combination of atoms in a fixed proportion which confers definite physical and chemical properties to that compound
  Inorganic Chemical compound that lacks carbon, with the exception of substances such as carbon dioxide
  Organic Chemical compound that contains carbon, with the exception of substances such as carbon dioxide
Concentration The quantity of substance present in a given volume
Concentration gradient A difference in concentration between two regions
Connective tissue Tissue that binds, supports and protects other tissue
Connective tissue proper A form of connective tissue that consists of cells and extracellular fibres embedded in a semi-fluid matrix
Covalent bond A linkage of atoms joined by sharing electrons
Covalent compound A compound formed of atoms linked by covalent bonds
Cranial nerves A set of twelve pairs of nerves that connect directly with the brain
Cystinuria An inborn error which results in an inability to transport cysteine and similar structured amino acids across membranes
Cytochrome chain *See* Electron transport system
Cytology The study of cells
Cytoplasm General name for all the contents of a cell apart from the nucleus; it consists of organelles, inclusions and a cytoplasmic matrix
Cytoplasmic matrix Cellular fluid in which organelles and inclusions lie

---

Dalton's Law Each gas in a mixture of gases exerts its own pressure as if all other gases were not present
Deamination Removal of amine groups from the amino acid pool
Defibrillation Process whereby defibrillators apply strong electric shocks to the heart
Dehydration Excessive loss of water from the body resulting in a hypertonic extracellular fluid
Depolarization A decrease in potential difference
Diabetes mellitus A chronic disorder due to a relative or absolute deficiency of insulin
Dialysate Bathing solution used in dialysis
Dialysis The separation of suspended particles and colloids from solutes by a membrane
  Haemodialysis An artificial membrane is used to remove soluble toxic substances from the blood
  Peritoneal The capillary network of the peritoneal cavity is used to remove toxic waste substances from the blood.
Differentiation The process whereby cells acquire specialized functions
Diffusion The net movement or flow of particles from a region of high concentration to a region of low concentration
Dipole-Dipole interaction Linking of a polar molecule with other polar molecules as a result of the attraction of the slightly charged regions in each molecule
Disaccharide A compound consisting of two sugar molecules
Dissociation Breakdown of a substance into ions when that substance is placed in a solvent
Dissolving The loose linking of solute molecules to solvent molecules
DNA Deoxyribonucleic acid; contains the genetic code
Dorsal horn Point of entry of sensory neurons into the spinal cord

---

Earthing Process whereby the metal casing and other areas of a piece of equipment are connected to the earth
ECG *See* Electrocardiogram
Elasticity The property of matter that returns it to its original shape after a force has altered the shape
Electric current The flow of positively charged ions in a liquid or the flow of electrons through a metal
Electrical conductor Substance that enables charged particles to move to their oppositely charged regions
Electrical energy Results from a flow of charged particles
Electrical gradient Difference in charge between two regions
Electrocardiogram A tracing made of various phases of the heart's electrical activity by a machine known as an electrocardiograph
Electrochemical gradient The combined differences of chemical concentrations and electrical charges across a membrane
Electrolysis The movement of ions to oppositely charged electrodes in fluids
Electrolytes Substances that move to oppositely charged electrodes when placed in a liquid
Electromagnetic radiation Waves or rays of energy that require no medium for propagation
Electron Subatomic particle with a small mass and a electrical charge of -1; located in orbits around the nucleus

Electron transport system A system where the transfer of electrons along a chain of substances results in the release of energy that can be used to form ATP
Element Matter that is composed of atoms with the same atomic number; it may be changed by physical means but usually not by chemical methods
Emulsion A colloid where one liquid in the shape of small droplets is dispersed throughout a second liquid
Permanent The two liquids exist as a homogenous mixture
Temporary The two liquids form separate layers upon standing
Endocrine glands Secrete substances referred to as hormones directly into the bloodstream
Endoplasmic reticulum Membraneous network of canals that runs through the entire cytoplasm
Rough Surface contains ribosomes
Smooth Surface lacks ribosomes
Endoscopes Instruments that use optic fibres to examine hollow organs or internal cavities
Endospores Dormant cells within the cytoplasm of some bacteria
Energy Creates a potential to do work or enables matter to do work
Enzymes Proteins that affect the speed of chemical reactions
Eosinophil A type of white blood cell
Epithelial tissue Covers the surface of the body and organs, lines body cavities and organs along with forming glands
Erythrocytes Red blood cells
Erythropoiesis The production of erythrocytes (red blood cells)
ESR The erythrocyte sedimentation rate is a measure of the rate at which erythrocytes settle; used in the diagnosis of various pathological states
Ester Molecule that contains the functional group

$$-\mathrm{C}-\overset{\overset{\displaystyle \mathrm{O}}{\|}}{\mathrm{C}}-\mathrm{O}-\mathrm{C}-$$

Exocrine glands Secrete substances via ducts into the body
Extracellular fluid Body fluid located outside cells

Facilitated diffusion A process by which a lipid-insoluble substance is passively transported across a membrane by a carrier
Fats A group of lipids that consist of three fatty acids bonded to a glycerol molecule
Feedback, Negative Mechanism that results in a reversal of a physiological response back to the normal range following an imbalance
Fibrillation Asynchronous contraction and quivering of the atria or ventricles
Fibrinogen Plasma protein involved in the formation of blood clots
Flagella Long motile projections of the plasma membrane
Fluid A substance that takes the shape of its container and can be a liquid or a gas
Force Any influence that either changes the position, the state of rest or the motion of an object. The S.I. unit of force is the newton
Forebrain Anterior region of the brain which contains the cerebral hemispheres, thalamus, hypothalamus and pineal gland
Formulae:
Molecular States the actual number of atoms of each element present
Structural Shows the actual number of atoms of each element present and their structural arrangement
Friction Force that resists the movement of one object over another
Functional group An arrangement of atoms that has very similar chemical and physical properties wherever it is found in an organic molecule

---

Gamma rays Electromagnetic radiation emitted from radioactive isotopes with a shorter wavelength than X-rays
Ganglion Junction region where a mass of nerve cell bodies exist
Gas The state of matter where molecules are able to move independently of each other; the molecules have a greater kinetic energy than those of liquids
Gel A colloid composed of solid particles set in a semi-solid that has jelly-like properties
Gene The smallest unit of heredity which represents a single physical characteristic
Genetics The study of genes and their relationships to hereditary characteristics
Genetic code The language that is used by a cell to transfer genetic information into chemical manufacture
Genetic predisposition An individual that has a genetic potential for a specific disease but this potential is only realized when certain environmental conditions exist
Genotype The genetic composition of an individual
Glycerol A three-carbon alcohol that combines with fatty acids to form fats
Glycogen A polysaccharide that acts as a temporary storage of body carbohydrate; it is composed of chains of glucose molecules
Glycogen storage disease A disease where glycogen is unable to be reconverted into glucose
Glycolysis The first step in glucose catabolism whereby glucose is broken down to form pyruvic acid
Glycosuria The appearance of glucose in urine
Golgi apparatus Network of flattened tubes that act as the cell's storage and distribution centre
Gravity The force of attraction between objects; it is often used to describe the attraction of objects to the earth

Grey matter Regions of the brain and spinal cord that consist mainly of cell bodies
Guthrie Test Used to detect the presence of phenylalanine in the blood and urine

---

Haeme group A chemical structure that has iron located in its centre
Haemoglobin A pigment that consists of the protein globin wrapped around four haeme groups
Oxygenated The loss of hydrogen ions from haemoglobin or the addition of oxygen to haemoglobin
Reduced The addition of hydrogen ions to haemoglobin or the loss of oxygen
Haemophilia A sex-linked genetic disorder that results in either the failure of blood to clot or the very slow clotting of blood following bleeding
Haemostasis The arrest of bleeding
Half-life Time taken for half a given number of atoms of an isotope to decay into a new element
Heat Energy possesed by a substance as a result of vibration of atoms and molecules
Henry's Law The amount of gas dissolved in a liquid varies proportionately with the partial pressure of the gas, when the temperature remains constant
Herpes Simplex Virus responsible for the development of cold sores
Hertz The S.I. unit of frequency which represents the number of waves per second
Heterocyclic compound Cyclic organic compounds that contain another element within the carbon ring
Heterozygous The presence of two different alleles of the same gene
Hindbrain Contains three principal structures known as the cerebellum, pons and medulla oblongata
Homeostasis A relative state of equilibrium between the internal environment of an organism and the changing external environment
Homozygous The presence of a pair of chromosomes with the same allele
Hormone Any substance that is released from a gland into the blood in small amounts and which on arrival at a particular tissue elicits a response
Hydrated diameter The diameter of the combination of a solute and its attached water molecules
Hydrocarbon Compounds composed of carbon and hydrogen
Aromatic A compound that contains a ring of six carbon atoms joined together in an alternating sequence of single and double bonds
Hydrogen bond A dipole-dipole interaction involving hydrogen
Hydrophilic Water attracting
Hydrophobic Water repelling
Hydrostatic pressure The prssure that results from the particles of a liquid colliding with the walls of a container
Hydroxyl ion The radical OH^-
Hypertonic solution A solution that has a greater osmotic pressure than intercellular fluid; these solutions cause cells to shrink
Hyperventilation A rate of breathing that is faster than normal
Hypothalamus A major neuroendocrine connection located in the forebrain
Hypoventilation A rate of breathing that is slower than normal
Hypoxaemia An insufficient concentration of oxygen in the blood
Hypoxia Diminished amount of oxygen in the tissues

Immiscible A liquid which is insoluble in another liquid
Immunity
Cell mediated Specific defence mechanism that involves the production of specialized lymphocytes
Humoral Specific defence mechanism that involves the production of immunoglobulins
Inclusions A diverse group of structures found in cells
Inertia The reluctance of a body to start moving after a force has been applied to it and the reluctance of the body to stop once it has begun to move
Inflammation The response of tissues to infection or injury which is characterized by localized dilation of blood vessels, the accumulation of fluid, pus formation and the formation of a clot
Inflammatory response A body response to harmful foreign material that involves the combined action of non-specific defence mechanisms and the repair of any damaged region
Infra-red rays Electromagnetic radiation represented by waves that have a wavelength that ranges between microwaves and light rays
Insulator Substance or medium incapable of carrying charged particles to oppositely charged regions, or a substance that is incapable of transmitting heat
Intercellular fluid *See* Interstitial fluid
Interphase Period between cell divisions
Interstitial fluid Extracellular fluid that surrounds tissue cells
Intracellular fluid Fluid found within cells
Intravascular fluid Extracellular fluid located in the vascular system; commonly referred to as blood
Ion An atom or group of atoms that contain an electrical charge of at least +1 or −1
Ionic bond A linkage that binds ions together as a result of the attraction of opposite charges
Ionic compound A compound formed from the combination of ions
Isomer Substances that have the same molecular formula but different structural formulae

Isosmotic solution *See* Isotonic solution
Isotonic solution Has the same osmotic pressure as the solution with which it is compared; cells neither shrink nor swell when placed in such a solution
Isotopes Atoms of the same element that contain different mass numbers

---

Joule The S.I. unit for energy

Karyotype A formal arrangement of chromosomes used to identify chromosomal abnormalities and identify the sex of an individual
Kelvin The S.I. unit for temperature
Ketone Organic substances that contain the functional group

```
  |     |
–C–C–C–
  |  ||  |
     O
```

Ketone body A group of ketones that are formed during fatty acid and amino acid catabolism; most ketone bodies are acids
Ketosis A form of metabolic acidosis due to the excessive production of ketone bodies
Kilogram The S.I. unit for weight
Kinetic energy Energy resulting from action or performing work
Krebs Cycle Aerobic catabolism of acetyl group to carbon dioxide and hydrogen

---

Lactose intolerance A deficiency in the enzyme lactase which results in an inability to digest and absorb milk
Leucocyte White blood cell
Leukaemia A blood disease in which an uncontrolled production of white blood cells occur
Lever A bar which connects an object and a force
Light Electromagnetic radiation that contains a smaller wavelength than infra-red rays but a greater wavelength than ultraviolet rays; it is able to stimulate the retina of the human eye to produce images in the brain
Lipids A diverse group of organic substances which share the property of being relatively insoluble in water and readily soluble in organic solvents
Liquid A state of matter in which molecules have limited movement; the molecules have a greater kinetic energy than those in solids
Litre The S.I. unit for volume
Loose connective tissue Consists of a loose network of collogen and elastic fibres
Lumbar puncture The insertion of a needle into the C.S.F. in the lumbar region of the spinal cord
Lymph Extracellular fluid located in the lymph vessels
Lymphatic system Composed of the spleen, tonsils and thymus, tissue called lymph nodes, and a network of lymphatic vessels
Lymphocytes A type of white blood cell
Lysosomes Small organelles that contain powerful digestive substances for the breaking down of unwanted material

---

Maltose A disaccharide formed in the mouth and small intestine
Mass The quantity of matter of an object
Mass number The number of protons and neutrons present in an atom
Matter Anything that occupies space and has some mass
Mechanical energy Energy associated with moving parts
Medulla oblongata Hindbrain structure that links the spinal cord and the remainder of the brain
Meiosis A process that results in the formation of cells with half the number of chromosomes of the parent cell
Memory cells Cells involved in immune process; recognition lymphocytes that remain unchanged following an infection
Mesencephalon *See* Midbrain
Metabolism The manufacture and breakdown of body chemicals
Metastasis Transfer of disease from one organ or part of an organ to another not directly connected to it
Metre The S.I. unit for length
Microvilli Small cytoplasmic extensions that increase the surface area of cells
Microwaves Electromagnetic radiation with wavelengths ranging from short radio waves to rays near the infra-red region
Micturition The process of expelling urine from the bladder; also known as urination or voiding
Midbrain Connects the forebrain to the hindbrain and is involved in the coordination of visual and auditory reflexes
Milliequivalent The amount of substance required to combine with, or displace, one milligram of hydrogen
Minerals Inorganic substances that function in the body in combination with other minerals or with organic substances
Miscible A liquid which is soluble in another liquid
Mitochondria Double-membraned organelles in which the inner membrane is folded to form cristae; site of aerobic respiration
Mitosis A series of stages in which a parent cell divides to form two cells that are identical with the parent cell
Mixture Contains two or more elements that are not bound to each other and which may exist in variable proportions
Mole The atomic or molecular weight of a substance in grams; one mole contains 6.023×10^{23} particles
Molecular weight The total weight of all the constituent atoms

Molecule Smallest combination of atoms that can combine together and still exhibit the properties of a compound
Moment The measure of the tendency of a force to rotate an object about a point
Momentum Motion of a body as a result of its mass and velocity
Monocyte A type of white blood cell
Monosaccharide A single sugar molecule
Mutliple alleles When more than two alleles exist for the one gene
Muscle atrophy Muscle wasting
Muscular tissue Tissue that has the ability to contract
Myelinated axon Axon covered with sheaths of myelin
Myopathy Disorder of skeletal muscle

NAD A coenzyme that plays an important role in the transfer of hydrogen to the cytochrome chain
Nebulizer A device that uses either compressed air or ultrasonic waves to form an aerosol by dispersing liquids or solids into a gas
Nerve Combination of many bundles of nerve fibres
Nerve fibre Nerve axon and its surrounding sheaths
Nervous tissue Composed of neurons and neuroglia
Neuroglia A group of cells that support, protect and assist the function of neurones
Neuromuscular junction Region where a nerve and muscle make functional contact
Neuron Cell capable of conducting and transmitting impulses; it consists of a cell body with short protrusions called dendrites and a long projection called an axon
Neurotransmitter Chemicals released from presynaptic nerves that indirectly transfer a nerve impulse to postsynaptic nerves or muscle cells
Neutralization Process whereby acids and bases are combined until neither is in excess and the solution has a neutral pH
Neutron Uncharged subatomic particle of similar mass to a proton; located in the nucleus
Neutrophil A type of white blood cell
Nitrogen balance The amount of nitrogen in the urine compared with nitrogen intake over a period of time
  Negative More nitrogen is being excreted than is being taken into the body
  Positive Intake of nitrogen exceeds excretion
Non-disjunction The failure of chromatids to separate during cell division
Non-polar molecule A molecule that is composed of atoms that share their electrons approximately fifty per cent of the time with other atoms
Nucleolus Spherical body found in the nucleus of cells
Nucleotide A combination of a nitrogen base, a sugar and a phosphate group which acts as the building block of DNA and RNA
Nucleus
  Atom Central part of an atom containing protons and neutrons
  Cell A relatively large membrane-bound structure that controls and regulates cell multiplication and maintains cell function

---

Obesity A condition in which more kilojoules than the body requires are taken in over a period of time
Optic fibres Flexible glass rods that are capable of conducting light from one end to the other
Organ Group of tissues that act in a unique, coordinated and functional manner
Organelles Small membrane-bound functional structures found in cells
Osmol Unit of measurement that measures a solute's ability to induce osmosis and osmotic pressure; it is a measure of the total number of particles present in a solution
Osmolality The osmols of particles per kilogram of water
Osmolarity The osmoles of particles per litre of solution
Osmosis The net movement of water through a semi-permeable membrane from an area of high water concentration to an area of low water concentration
Osmotic pressure The pressure required to exactly oppose the net movement of water by osmosis, that is, the resistance that is required to just stop the net flow of water
Oxaloacetic acid An important compound in the Krebs cycle involved in protein, carbohydrate and fat metabolism
Oxidation The process of either combining oxygen with a substance or removing hydrogen from a substance
Oxidative phosphorylation A process where energy from the electron transport system is used to form ATP in the presence of oxygen

---

Parasympathetic nervous system A division of the autonomic nervous system that generally acts as an inhibiting system
Parathyroid glands Tiny glands located on the thyroid gland that are involved in the regulation of calcium and phosphate concentrations in the blood
Partial pressure The pressure exerted by a particular gas in a mixture of gases
Pascal The S.I. unit for pressure
Pascal's Law Pressure exerted on a confined liquid is transmitted equally in all directions
Peptide bond A bond that links the amine group of one amino acid with the carboxyl group of an adjacent amino acid
Periodic group The atoms in a periodic group contain the same number of outer shell electrons

and are found in the same vertical rows of the periodic table
Periodic period The elements in the same horizontal row of the periodic table
Peripheral nervous system Consists of the spinal nerves, cranial nerves and the autonomic nervous system
Peripheral resistance The resistance of small diameter blood vessels to blood flow
Peristalsis Rhythmic waves of smooth muscle contractions
Permeability The ease with which a substance passes across a membrane
pH A numerical scale that represents the powers of ten of hydrogen ions; it is a relative indicator of acidity and alkalinity
Phagocytosis The active ingestion of large particles by specialized cells
Phenol A molecule with a -OH functional group attached to a benzene ring
Phenotype Physical expression of the genotype
Phenylketonuria An inborn error of metabolism that results in an inability to convert phenylalanine into tyrosine; it can prevent normal development of the brain
Phospholipids A group of lipids that contain glycerol combined with two fatty acids and one phosphate group; a major constituent of membranes
Pineal gland An important neuroendocrine connection located in the forebrain; its function is linked with light and darkness and believed to play an important role in the biological clock and the control of the endocrine system
Pinocytosis The active movement of substances across a membrane by vesicles
Pitch Conscious perception of a particular frequency of sound wave
Pituitary gland A major endocrine gland that is linked to the hypothalamus
Pivot Point around which rotation occurs
Plasma The non-cellular component of blood
Plasma cells Involved in the immune response; transformed B-cells that have the ability to secrete immunoglobulins
Plasma membrane Membrane that separates the cell's internal contents from the extracellular environment
Platelet Tiny disc-shaped cells that lack a nucleus
Polar molecule A molecule that is composed of slightly-charged atoms due to unequal sharing of electrons between atoms
Polypeptide A chain of many amino acids
Polysaccharide A carbohydrate that consists of many sugar molecules
Pons Hindbrain structure that links the medulla oblongata and cerebellum to the midbrain
Pore A small hole passing through a membrane
Postural hypotension A decreased blood pressure that results from the inability of the heart to force enough blood to the brain on sudden rising from a horizontal position
Potential difference The potential for electricity to flow due to a difference in charge between two regions; it is a form of potential energy
Potential energy The energy stored in an object as a result of its position or chemical composition
Power The rate of doing work; the S.I. units are joules per second or watts
Precipitate Undissolved solute that is found in saturated solutions
Pressure The force exerted by particles on a given surface area of matter
Negative Pressure lower than standard atmospheric pressure
Positive Pressure greater than standard atmospheric pressure
Prostaglandins A group of lipids that consist of twenty carbon fatty acids of which five of the carbon atoms are arranged in a ring; they are sometimes referred to as tissue hormones
Proteins An essential group of biochemicals composed of polypeptide chains
Fibrous Long helical chains of amino acids entwined with each other
Globular Considerably folded polypeptide chains
Proton Subatomic particle which has an electrical charge of +1 located in the nucleus
Pulley A grooved wheel on an axle that is used to change the direction of a force
Purgative agent An agent which assists evacuation of the bowels

---

Qualitative measurement Descriptive signs used to record observations
Quantitative measurement Numbers used to express measurements recorded by an instrument

---

Radical A group of atoms which acts as a unit within a molecule, and when that molecule undergoes a chemical change the radical remains as an intact unit
Radioisotope An unstable isotope that emits radiation
Recognition lymphocyte Special lymphocytes that identify antigens
Regulating hormones Hormones secreted by the hypothalamus that either stimulate the release or inhibit the release of specific hormones from the anterior pituitary
Renal calculi Precipitate found in the kidney that is composed of uric acid or calcium salts, commonly referred to as kidney stones
Repolarization A return of a potential difference towards the resting potential
Resistance, Electrical Opposition to the flow of charges, measured in ohms
Respirator Machine that assists breathing
Respiratory distress syndrome Results from a lack of pulmonary surfactant in newborn babies

Resting potential Potential difference between the inside and outside of a nerve or muscle cell prior to an electrical event
Reticuloendothelial system A functional system that consists of highly phagocytic cells broadly distributed throughout the body
Retina Layer of photoreceptive cells in the eye that convert light images into nerve impulses
Rhombencephalon *See* Hindbrain
Ribosomes Small organelles that are either located on the endoplasmic reticulum or are dispersed throughout the cytoplasm
RNA Ribonucleic acid provides the means by which genetic information can be transposed into actions within the cell.
 mRNA Carries genetic information from the nucleus to the ribosomes
 tRNA Carries amino acids to the ribosomes

Salt An ionic compound formed as a result of combining an acid with a base
Saturated
 Compound Carbon atoms are linked by single bonds
 Solution Contains the maximum quantity of dissolved solute
Scalar Any physical quantity that is described only in terms of magnitude
Semi-permeable membrane A membrane which allows the movement of water through it, but not other substances
Semi-qualitative measurement The use of a relative scale, usually in the form of 'plus' or relative words.
Sensitized Possession of a quantity of immunoglobulin along with a reserve of memory cells for a particular antigen
Serum Fluid component of blood that differs from plasma in that it lacks clotting factors
Signs Evidence of a disease to an observer
Sol A colloid composed of solid particles dispersed in a liquid
Solid A state of matter where atoms, ions and molecules are held in a relatively fixed position
Solubility A measure of the amount of solute that will dissolve in a specified volume of solvent
Solutes Substances dissolved in a solvent
Solution A fluid that consists of a homogenous mixture of two or more substances
Solvent The greater quantity of substance in a solution
Sonic energy Vibration of particles in the form of waves through matter; sound waves require a medium for propagation
Specific gravity The weight of a certain volume of substance compared with the weight of the same volume of water
Sphygmomanometer An instrument that is used to measure arterial blood pressure
Spinal cord A long rod of nervous tissue that connects the brain with the body
Spinal nerve Each nerve contains afferent nerve fibres that conduct information to the spinal cord and efferent nerve fibres that carry information to muscles or glands
Spinal reflex A localized reflex between the sensory and motor neurons of the same spinal cord segment
Spirillia Coiled-shaped bacteria
Stability The ability of an object to resist falling
Steatorrhoea The occurrence of large amounts of fat in faeces
Sterilization Measure which destroys or removes all living organisms
Steroids A group of lipids that contain four carbon rings arranged in a specific configuration; includes cholesterol, bile salts and adrenocortico hormones
Stress A stimulus that causes an imbalance in the homeostasis of an individual and initiates a homeostatic response
Surface tension The pull of the surface particles of a liquid into the body of the liquid; it exists in any boundary surface of a liquid
Surfactant A substance that decreases the surface tension of a liquid
Suspension Non-homogenous fluid mixture that contains large particles that settle when the mixture is left to stand
Sympathetic nervous system A division of the autonomic nervous system that prepares the body to respond to stress
Synapse A region where two neurons make close contact with each other
Synovial fluid A low friction substance found in cavities between bones
System A group of interacting organs that cooperate as a functional unit

---

Target gland A gland that is acted upon by a hormone
Target tissue Tissue acted upon by a hormone
T-Cell Specialized lymphocytes that combine directly with an antigen in a response referred to as cell mediated immunity
Telencephalon Cerebral hemispheres and basal ganglia located within them
Temperature A measure of heat intensity
Thalamus A giant relay system located in the forebrain that distributes sensory information to the cerebral cortex
Thrombocyte *See* Platelet
Tincture A solution where alcohol is used as the solvent
Tissues A combination of many cells that have a similar structure and function
Tissue rejection Immune response to the introduction of transplanted cells
Traction Process where the fractured ends of a bone are kept in position by a system of pulleys and weights
Transamination Process where an amine group

is transferred from one amino acid to an organic compound to form another amino acid
Transcription Formation of RNA from DNA
Translation Mechanism of forming protein from the conversion genetic information
Transport maximum (T_m) The flow rate for a substance that is carried across a membrane; it represents the concentration gradient where all carriers are occupied
Triglyceride Fat
Trisomy The presence of an additional chromosome due to an error in chromosome division
Trophic factors Substances that travel down nerves by axoplasmic flow to assist in the development, maintenance and regeneration of the nerve network
Tropic hormone A hormone that regulates the release of hormones from a target gland
Tumour A swelling due to the abnormal growth of tissue

Ultrasound Extremely high frequency sound waves used for diagnosis and therapy
Ultraviolet rays Electromagnetic radiation with wavelengths ranging in size between X-rays and light rays
Unmyelinated Axon that lacks sheaths of myelin
Unsaturated
Compound A compound where some of the carbon atoms are linked to other carbon atoms by either double or triple bonds
Solution A solution that has the ability to dissolve additional solute
Urea A relatively non-toxic end product of protein catabolism
Urine Fluid that is secreted by the kidneys

Vaccination Artificial introduction of a small quantity of antigen which results in the recipient being immunized for that antigen
Vagus nerve Principal parasympathetic nerve
Valence The combining power of an atom
Vector Any physical quantity that has both magnitude and direction
Veins Vessels that return blood from capillaries to the heart
Velocity A vector describing the distance that something has travelled divided by the time taken
Ventral horn Point of exit of motor nerves from the spinal cord
Ventricles Large chambers where blood is forced out of the heart
Venule Small vein, connects a capillary to the venous system
Virus Small acellular structure that consists of hereditary material surrounded by protein
Viscosity The tendency of a liquid to resist flow; the S.I. unit of viscosity is the poise
Vitamins Organic substances that act in trace amounts to promote enzyme function
Volt S.I. unit for potential difference

Wavelength The distance between two identical points of adjacent waves. EXAMPLE: The distance between the crests of two adjacent waves
Waxes A group of lipids that are less greasy than fats
White matter Regions of the brain and spinal cord that contain a predominance of axons
Wire
Active Carries current from the power source to a instrument or appliance; the wire is covered by brown or red insulating material
Earth A green/yellow or green covered wire that connects equipment directly to the earth
Neutral Blue or black covered wire that returns current from a piece of equipment to the power supply
Work The transfer of energy from one object to another, it is the distance that an object is moved in the direction of a force multiplied by the magnitude of that force; the S.I. unit of work is the joule

---

X-rays Electromagnetic radiation with a shorter wavelength than ultraviolet rays and a longer wavelength than gamma rays; capable of penetrating most matter including human tissue

---

Zwitterion An ion that contains a region with a positive charge and another area with a negative charge

Index

Italic numbers indicate illustrations
glos. indicates that the term is defined in the glossary
ap. indicates that the subject is discussed in an application on that page
passim indicates that the subject occurs throughout

Periodic T

Electron distribution · Calcium (Name) · 20 Ca (Symbol) · 40.08 (Atomic weight) · 2 8 8 2 (Electron distribution) · **Atomic number**

Transition elements

Period	I	II A	III B	IV B	V B	VI B	VII B	VI
1	Hydrogen 1 H 1.00 [1]							
2	Lithium 3 Li 6.94 [2 1]	Beryllium 4 Be 9.01 [2 2]						
3	Sodium 11 Na 22.99 [2 8 1]	Magnesium 12 Mg 24.31 [2 8 2]						
4	Potassium 19 K 39.10 [2 8 8 1]	Calcium 20 Ca 40.08 [2 8 8 2]	Scandium 21 Sc 44.96 [2 8 9 2]	Titanium 22 Ti 47.90 [2 8 10 2]	Vanadium 23 V 50.94 [2 8 11 2]	Chromium 24 Cr 51.99 [2 8 13 1]	Manganese 25 Mn 54.94 [2 8 13 2]	Iron 26 Fe 55.85 [2 8 14 2] · Cobalt 27 Co 58.93 [2 8 15 2]
5	Rubidium 37 Rb 85.47 [2 8 18 8 1]	Strontium 38 Sr 87.62 [2 8 18 8 2]	Yttrium 39 Y 88.91 [2 8 18 9 2]	Zirconium 40 Zr 91.22 [2 8 18 10 2]	Niobium 41 Nb 92.90 [2 8 18 12 1]	Molybdenum 42 Mo 95.94 [2 8 18 13 1]	Technetium 43 Tc 98.91 [2 8 18 14 1]	Ruthenium 44 Ru 101.07 [2 8 18 15 1] · Rhodium 45 Rh 102.91 [2 8 18 16 1]
6	Cesium 55 Cs 132.91 [2 8 18 18 8 1]	Barium 56 Ba 137.33 [2 8 18 18 8 2]	Lanthanum 57 La 138.91 [2 8 18 18 9 2]	Hafnium 72 Hf 178.49 [2 8 18 32 10 2]	Tantalum 73 Ta 180.95 [2 8 18 32 11 2]	Tungsten 74 W 183.85 [2 8 18 32 12 2]	Rhenium 75 Re 186.2 [2 8 18 32 13 2]	Osmium 76 Os 190.2 [2 8 18 32 14 2] · Iridium 77 Ir 192.22 [2 8 18 32 15 2]
7	Francium 87 Fr (223)[a] [2 8 18 32 18 8 1]	Radium 88 Ra (226.03)[b] [2 8 18 32 18 8 2]	Actinium 89 Ac (227)[a] [2 8 18 32 18 9 2]	Kurchatav-ium 104 Ku (257)[a] [2 8 18 32 32 10 2]	Hahnium 105 Ha (260)[a] [2 8 18 32 32 11 2]			

Cerium 58 Ce 140.12 [2 8 18 20 8 2]	Praseodym-ium 59 Pr 140.91 [2 8 18 21 8 2]	Neodymium 60 Nd 144.24 [2 8 18 22 8 2]	Promethium 61 Pm (147)[a] [2 8 18 23 8 2]	Samarium 62 Sm 150.4 [2 8 18 24 8 2]	Europium 63 Eu 151.96 [2 8 18 25 8 2]
Thorium 90 Th 232.04 [2 8 18 32 18 10 2]	Protactinium 91 Pa 231.04 [2 8 18 32 20 9 2]	Uranium 92 U 238.03 [2 8 18 32 21 9 2]	Neptunium 93 Np (237.05)[b] [2 8 18 32 22 9 2]	Plutonium 94 Pu (242)[a] [2 8 18 32 24 8 2]	Americium 95 Am (243)[a] [2 8 18 32 25 8 2]

Key

a. Mass number of most stable or best known isotope

b. Weight of most commonly available, long-lived isotope

Metals (unshaded)

Non-metals (shaded)

# of Elements

	I B	IIB	III A	IV A	V A	VI A	VII A	VII A
								Helium 2 He 4.00 2
			Boron 5 B 10.81 2 3	Carbon 6 C 12.01 2 4	Nitrogen 7 N 14.01 2 5	Oxygen 8 O 15.99 2 6	Fluorine 9 F 18.99 2 7	Neon 10 Ne 20.18 2 8
			Aluminium 13 Al 26.98 2 8 3	Silicon 14 Si 28.08 2 8 4	Phosphorus 15 P 30.97 2 8 5	Sulfur 16 S 32.06 2 8 6	Chlorine 17 Cl 35.45 2 8 7	Argon 18 Ar 39.95 2 8 8
Nickel 28 Ni 58.71 2 8 16 2	Copper 29 Cu 63.55 2 8 18 1	Zinc 30 Zn 65.38 2 8 18 2	Gallium 31 Ga 69.72 2 8 18 3	Germanium 32 Ge 72.59 2 8 18 4	Arsenic 33 As 74.92 2 8 18 5	Selenium 34 Se 78.96 2 8 18 6	Bromine 35 Br 79.90 2 8 18 7	Krypton 36 Kr 83.80 2 8 18 8
Palladium 46 Pd 106.4 2 8 18 18	Silver 47 Ag 107.87 2 8 18 18 1	Cadmium 48 Cd 112.41 2 8 18 18 2	Indium 49 In 114.82 2 8 18 18 3	Tin 50 Sn 118.69 2 8 18 18 4	Antimony 51 Sb 121.75 2 8 18 18 5	Tellurium 52 Te 127.60 2 8 18 18 6	Iodine 53 I 126.91 2 8 18 18 7	Xenon 54 Xe 131.30 2 8 18 18 8
Platinum 78 Pt 195.09 2 8 18 32 17 1	Gold 79 Au 196.97 2 8 18 32 18 1	Mercury 80 Hg 200.59 2 8 18 32 18 2	Thallium 81 Tl 204.37 2 8 18 32 18 3	Lead 82 Pb 207.2 2 8 18 32 18 4	Bismuth 83 Bi 208.98 2 8 18 32 18 5	Polonium 84 Po (210)[a] 2 8 18 32 18 6	Astatine 85 At (210)[a] 2 8 18 32 18 7	Radon 86 Rn (222)[a] 2 8 18 32 18 8

Gadolinium 64 Gd 157.25 2 8 18 25 9 2	Terbium 65 Tb 158.93 2 8 18 27 8 2	Dysprosium 66 Dy 162.50 2 8 18 28 8 2	Holmium 67 Ho 164.93 2 8 18 29 8 2	Erbium 68 Er 167.26 2 8 18 30 8 2	Thulium 69 Tm 168.93 2 8 18 31 8 2	Ytterbium 70 Yb 173.04 2 8 18 32 8 2	Lutetium 71 Lu 174.97 2 8 18 32 9 2
Curium 96 Cm (247)[a] 2 8 18 32 25 9 2	Berkelium 97 Bk (247)[a] 2 8 18 32 26 9 2	Californium 98 Cf (251)[a] 2 8 18 32 28 8 2	Einsteinium 99 Es (254)[a] 2 8 18 32 29 8 2	Fermium 100 Fm (253)[a] 2 8 18 32 30 8 2	Mendelevium 101 Md (256)[a] 2 8 18 32 31 8 2	Nobelium 102 No (254)[a] 2 8 18 32 32 8 2	Lawrencium 103 Lr (257)[a] 2 8 18 32 32 9 2